Progress in Nonhistone Protein Research

Volume III

Editor

Isaac Bekhor, Ph.D.

Professor
Department of Basic Sciences
School of Dentistry
University of Southern California
Los Angeles, California

Associate Editor

C. C. Liew, Ph.D.

Professor
Department of Clinical Biochemistry
The Banting Institute
University of Toronto
Toronto, Ontario, Canada

CRC Press, Inc.
Boca Raton, Florida

Library of Congress Cataloging-in-Publication Data
(Revised for volume 3)

Progress in nonhistone protein research.

Includes bibliographies and indexes.
1. Nonhistone chromosomal proteins.   I. Bekhor,
Isaac.   II. Mirell, Carol J.   III. Liew, C. C.   [DNLM:
1. Chromosomal Proteins, Non-Histone.   QU 58 P964]
QP552.N62P76   1985   574.19'245   84-9408
ISBN 0-8493-5528-1 (v. 1)
ISBN 0-8493-5529-X (v. 2)
ISBN 0-8493-5530-3 (v. 3)

© 1989 by CRC Press, Inc.

International Standard Book Number 0-8493-5528-1 (Volume I)
International Standard Book Number 0-8493-5529-X (Volume II)
International Standard Book Number 0-8493-5530-3 (Volume III)

Library of Congress Card Number 84-9408
Printed in the United States

FOREWORD

Nuclear proteins (excluding the five major histones) are generally defined as nonhistone nuclear proteins. In the past, investigators have prepared chromatin and fractionated its protein components into basic (histones) and acidic proteins (nonhistones) based on their mobility in an electrical field; these approaches have defined the highly heterogeneous nonhistone nuclear proteins as nonhistone chromatin proteins (NHCP). In the 1970s, several groups of investigators attempted to characterize the NHCPs into loosely bound and tightly bound nonhistone chromatin proteins, based on their solubility in solutions of low and high salt molarity. Several NHCPs which were soluble in mineral acids (similar to the histones) were designated as high-mobility group (HMG) proteins based on their electrophoretic mobility in acid-urea gels. Others were designated tightly bound based on their resistance to dissociate from DNA in solutions of 2 M NaCl. Other NHCPs were suspected to be covalently linked to DNA based on their lack of dissociation from DNA under extreme conditions. The loosely bound nonhistones were simply those proteins which were dissociated from chromatin under variable salt concentrations which ranged up to 0.6 M. All of the above procedures did not yield, in present-day molecular biology, data relevant to specific gene regulation or expression.

However, NHCPs still became both interesting and challenging to many investigators due to their continued suspected roles in regulating specific gene expression. Many efforts have been centered upon the search for the structure and function of several individual proteins. For the past two decades the nonhistone chromosomal proteins were categorized into three major classes:

1.   Those possessing enzymatic activity (e.g., kinases, polymerases, methylases, topo-isomerases, etc.)
2.   Those involved in selective structural roles (e.g., lamina scaffold proteins, etc.)
3.   Those which interact directly with DNA and RNA (e.g., hormone receptor-acceptor proteins, sequence-specific DNA-binding proteins, RNA-binding proteins, etc.)

However, due to the rapid advance in recombinant DNA technology, many investigators have abandoned the study of NHCP. In fact, those investigators who are currently studying these groups of proteins have further characterized them as DNA-binding proteins, sRNA processing proteins, trans-acting factors, initiating factors, promoter and enhancer factors, etc.

As a follow-up to Volumes I and II in this series, in this volume we are attempting to provide an update into the current research in NHCP with the hope that these articles may foster a more analytical elucidation of the role of these proteins in gene expression. This volume was initiated at an international meeting sponsored by the University of Camerino, Camerino, Italy, in May of 1985. Many of the contributors to this volume participated at that meeting. At that time it was evident that work on proteins associated with chromatin-DNA was being carried out mostly in Europe and not in the U.S. An attempt was made by every contributor to present his or her contributions in clear English. Translation from one language to another by the authors themselves provided an additional personal touch to what is being presented. The editor insisted on maintaining originality in the various contributions. Some sentences may have been modified slightly for purpose of clarity; however, that was kept to a minimum. We believe that this volume should be dedicated to our non-American colleagues who are working deligently on these nonhistones, or nuclear, or "factors", or DNA-binding proteins, or call them what you wish, and have clearly devoted their lives to this extremely challenging problem, the problem of gene regulatory factors (= proteins = nonhistones).

C. C. Liew
I. Bekhor

THE EDITOR

Isaac Bekhor, Ph.D., is Chief, Laboratory for Molecular Genetics, University of Southern California, and Professor of Biochemistry and Molecular Biology, School of Dentistry, University of Southern California, Los Angeles, California.

Dr. Bekhor graduated from the University of California, Los Angeles, with a B.S. in Chemistry, and obtained his Ph.D. in 1966 in Biochemistry, University of Southern California, Los Angeles, where the Graduate School presented him with the Harry J. Deuel, Jr. Award in Biochemistry in recognition of outstanding research. Between 1966 and 1969 Dr. Bekhor was a postdoctoral Fellow at the California Institute of Technology, Pasadena. In 1970 he joined the University of Southern California School of Dentistry Faculty. In 1971 he won a 5-year Career Development Award from the National Institutes of Health.

Dr. Bekhor is a member of the American Association for The Advancement of Science, American Chemical Society, American Society for Cell Biology, International Biochemical Society, International Society for Developmental Biologists, Southern California Craniofacial Group and West Coast Chromatin Group. He has been a consultant to *Industry on Enzyme Technology* for over 15 years.

His major research interest is the participation of nonhistone chromosomal proteins in gene regulation in eukaryotes. He has published more than 50 papers in *Molecular Biology*, and is currently cloning specific genes to facilitate his studies on the possible regulatory roles of specific nonhistones.

CONTRIBUTORS

Melissa B. Aldrich, M.B.A.
Research Assitant
Department of Pharmacology
Baylor College of Medicine
Houston, Texas

Domenico Amici, Ph.D.
Professor
Department of Cell Biology
University of Camerino
Camerino, Italy

Zoya V. Avramova, Ph.D.
Research Associate
Institute of Molecular Biology
Bulgarian Academy of Science
Sofia, Bulgaria

Isaac Bekhor, Ph.D.
Professor
Department of Basic Sciences
School of Dentistry
University of Southern California
Los Angeles, California

Massimo Bramucci, Ph.D.
Researcher
Department of Cell Biology
University of Camerino
Camerino, Italy

Paola Caiafa, Ph.D.
Associate Professor
Department of Biological Sciences
Institute of Biological Chemistry
Faculty of Pharmacy
University of Rome
Rome, Italy

Pui-Kwong Chan, Ph.D.
Department of Pharmacology
Baylor College of Medicine
Houston, Texas

E.C. Chew, Ph.D.
Department of Anatomy
School of Medicine
The Chinese University of Hong Kong
Shatin, New Territories, Hong Kong

Sandra Coderoni, Ph.D.
Researcher
Department of Cell Biology
University of Camerino
Camerino, Italy

Michel Crepin, Ph.D.
Professor
Department of Molecular Biology
Pasteur Institute
Paris, France

George Dessev, Ph.D.
Visiting Professor
Department of Cell Biology and Anatomy
Northwestern University Medical School
Chicago, Illinois

Franco Felici, Ph.D.
Researcher
Department of Cell Biology
University of Camerino
Camerino, Italy

Gian Luigi Gianfranceschi, Ph.D.
Professor
Department of Cell Biology
University of Camerino
Camerino, Italy

M. J. Halikowski, Ph.D.
Department of Clinical Biochemistry
Banting Institute
University of Toronto
Toronto, Ontario, Canada

Ronald Hancock, Ph.D.
Cancer Research Centre
Laval University
Laval, Quebec, Canada

H. H. Jamal, M.Sc.
Department of Clinical Biochemistry and
 Medicine
Banting Institute
University of Toronto
Toronto, Ontario, Canada

Jean Claude Lelong, Ph.D.
Research Director
Department of Molecular Biology
Pasteur Institute
Paris, France

Choong-Chin Liew, Ph.D.
Professor
Departments of Clinical Biochemistry and
  Medicine
University of Toronto
Toronto, Ontario, Canada

Antonino Miano, Ph.D.
Researcher
Department of Cell Biology
University of Camerino
Camerino, Italy

Beatrice Neuer-Nitsche, Dr. rer. nat.
Institute for Cell and Tumor Biology
German Cancer Research Center
Heidelburg, West Germany

Mario Paparelli, Ph.D.
Associate Professor
Department of Cell Biology
University of Camerino
Camerino, Italy

G. Prevost, M.A.
Department of Biology
University of Tours
Tours, France

Bistra T. Tasheva
Research Associate
Institute of Molecular Biology
Bulgarian Academy of Science
Sofia, Bulgaria

Roumen G. Tsanev, M.D.
Professor
Institute of Molecular Biology
Bulgarian Academy of Science
Sofia, Bulgaria

Dieter Werner, Dr. rer. nat.
Professor
Department of Cell and Tumor Biology
German Cancer Research Center
Heidelberg, West Germany

Benjamin Y.-M. Yung
Department of Pharmacology
Baylor College of Medicine
Houston, Texas

PROGRESS IN NONHISTONE PROTEIN RESEARCH

Volume I

Volume II

TABLE OF CONTENTS

Chapter 1

THE POTENTIAL ROLE OF A PHOSPHOPROTEIN IN NUCLEAR ORGANIZATION

C. C. Liew, M. J. Halikowski, H. H. Jamal, E. C. Chew

TABLE OF CONTENTS

I. SYNOPSIS

The availability of specific groups of nonhistone chromatin proteins and their covalent modifications are critical to the conformational changes of chromatin in gene expression. We provide evidence through a biochemical and immunological approach that a nuclear phosphoprotein has two domains to interact with the nuclear matrix on one end and association with nucleosomal DNA on the other end. This phosphoprotein also exhibits some c-myc oncogene protein characteristics. A potential role of this nuclear phosphoprotein in acting as a regulatory factor for either actively transcribed genes or for DNA replication is discussed.

II. INTRODUCTION

Protein phosphorylation has long been established as a widely occurring post-translational modification involved in the modulation of the structure and function of protein molecules. This phenomenon also includes those proteins of the cell nucleus where phosphorylation has been shown to be involved in selective aspects of enzyme regulation and cellular growth control (i.e., within the scope of replication and transcription).[1,2] This review will summarize and present evidence suggesting that a particular nuclear phosphoprotein (designated B_2, M_r: 68 kDa; pI: 6.5 to 8.2) plays a central role in selective features of nuclear structure and function.

In 1974 we developed a two-dimensional polyacrylamide gel electrophoretic system to analyze nonhistone chromosomal proteins (NHCP)[3,4] and proposed to utilize this technique to systematically identify NHCP with other known structures and functions.[5-7] In 1976 we were first to identify the B_2 phosphoprotein in chromatin subunits (i.e., nucleosomes)[8] which were released from isolated rat liver nuclei by mild micrococcal nuclease digestion and subsequent fractionation by 5 to 30% sucrose density gradient centrifugation. This B_2 phosphoprotein was further characterized by two-dimensional polyacrylamide gel electrophoresis, which revealed some heterogeneity (expected from its inherent pI range from 6.5 to 8.2) and a relative mass of 68 kDa.[9]

The content of the B_2 phosphoprotein in the isolated nuclei was found to be approximately 2 to 4 μg/g of rat liver starting tissue or approximately one molecule of B_2 phosphoprotein for every 200 nucleosomes. In contrast, other known nuclear proteins, i.e., high mobility group (HMG) proteins, were present at levels of up to 10^6 copies per nucleus.[10] This low content of B_2 phosphoprotein would seem to confer upon it more of a regulatory role than a purely structural one. In addition, this phosphoprotein was found to be associated with two classes of nucleosome monomers: those containing (MN_2) and those devoid of histone H_1 (MN_1).[11,12] The acidic to basic amino acid ratio of this nuclear protein was found to be less than one, with a high content of glutamic acid, aspartic acid (perhaps more involved in conveying nucleosome binding properties), lysine, arginine (perhaps more involved in a DNA binding capability), glycine, and alanine.[9]

III. EVIDENCE FOR THE STRUCTURE AND FUNCTION OF A NUCLEAR PHOSPHOPROTEIN

From *in vivo* labeling experiments we have found that this B_2 nuclear protein is highly phosphorylated as revealed by two-dimensional polyacrylamide gel electrophoresis.[13] The phosphate content was estimated to be approximately 0.3% with most of this phosphate associated with serine.[9] In regenerating liver, we found that phosphorylation of this B_2 protein was sigificantly increased (by 1.6 times) compared to sham-operated controls.[13] Moreover, when incorporation of ^{32}P into nucleosome fractions was compared, the 2.5-min micrococcal nuclease digested fraction was found to contain higher levels of ^{32}P than the

5-, 10-, or 15-min fractions.[14] These findings indicate that the B_2 phosphoprotein is preferentially associated with actively transcribed nucleosomes. An increase in phosphorylation may also be involved in activation of gene transcription. There is circumstantial evidence to suggest that covalent modification of nuclear proteins is a prerequisite step for gene activation.[1,2]

To resolve quantitative and qualitative aspects of this B_2 phosphoprotein within the nucleus, we began an immunological investigation by producing and utilizing both polyclonal and monoclonal antibodies raised against this phosphoprotein.[12,15] The specificity of these antibodies made it possible to verify the nuclear location of B_2 phosphoprotein. We have carried out immunofluorescence on both isolated nuclei and various cell cultures to confirm this nuclear localization (Plate 1*). We have also found that B_2 phosphoprotein still remains with *in situ* matrices.[16]

In order to examine the neighboring constituents of the B_2 phosphoprotein, we initiated a chemical cross-linking study in (1) micrococcal nuclease released nucleosomes and (2) in whole nuclei (the membrane permeable cross-linking reagents utilized are shown in Figure 1). The structure of purified monomers was found to be unperturbed by cross-linking (Figure 2B). Moreover, when proteins from cross-linked monomers were immunoblotted with B_2 protein-monoclonal antibody IgG_{2a}, only uncross-linked B_2 protein was detected (Figure 2C). However, when enriched protein complexes from cross-linker exposed nuclei were immunoblotted, a multi-band pattern could be visualized. In addition, when β-mercaptoethanol was used to cleave the cross-links prior to polyacrylamide gel electrophoresis (PAGE), only a single (uncross-linked) B_2 phosphoprotein band was detected (Figure 2D). Based on these results, we postulate a twofold role for this phosphoprotein. Since it is probably located in proximity to the periphery of the nucleus, it may (dynamically) interact with the nuclear envelope at one end while the other end interacts with DNA. This proposed DNA interaction is consistent with our earlier observations in which we demonstrated B_2 phosphoprotein binding to rat liver DNA-cellulose matrix and subsequent elution by 0.4 to 2.0 M NaCl gradient.[17]

Along this line, nonhistone phosphoproteins have been postulated to play some role in selective transcription (likely by means of their association with the DNA itself). This has been particularly well documented in prokaryotes (i.e., lac repressor inhibits lac operon transcription).[18] Earlier studies demonstrated that the greatest degree of binding of nonhistone phosphoproteins was with DNA of the same species, and in general this heterogeneous population of phosphoproteins accounted for approximately 1% of the total NHCP in chromatin.[19] The characterization of individual DNA binding phosphorylated NHCPs was first analyzed using rat liver nuclear proteins which were soluble at low ionic strength. Several of these nuclear proteins (designated B33, C18, CN') were in fact characterized.[20] Bluthman has characterized a DNA binding phosphoprotein (M_r: 30 kDa) from bovine lymphocytes which showed preferential binding to lymphocyte DNA (as opposed to *Escherichia coli* DNA) and contained 1% phosphate by weight.[21] However, this protein has since been shown to recognize DNA with a low degree of selectivity and binds to as much as 30% of lymphocyte DNA (one binding site: 1700 base pairs). This binding characteristic, along with the high content of acidic amino acids, led to further speculation that this protein may be a member of the HMG protein family. However, N-terminal analysis has identified an arginine residue, ruling out this possibility.[1]

Though DNA binding proteins have long been suspected of being involved in the regulation of gene transcription, they may also play a role in DNA replication and recombination. Such proteins have recently been characterized from mouse ascites cells, rat spermatocytes, and animal virus-infected cells.[22,23] Phosphorylation of a DNA-binding protein from mouse

---

* Plate 1 appears following page 120.

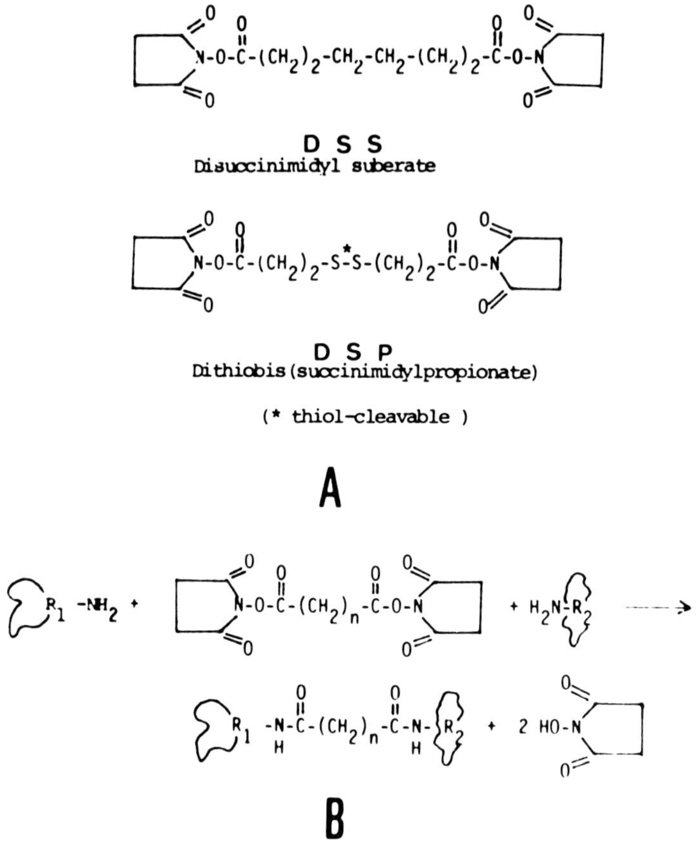

FIGURE 1.   Chemical cross-linking agents. (A) The chemical reagents used for the cross-linking studies on nucleosomes and isolated rat liver nuclei. (B) A schematic presentation to indicate how the polypeptides (R-NH$_2$) react with DSS.

ascites cells reduces its binding to single-stranded DNA but not duplex DNA and abolishes the ability of this protein to stimulate DNA polymerase activity. This is highly suggestive of some kind of "helix-destabilizing" role. Similarly, a DNA-binding protein from adenovirus and the T-antigen of SV 40 (both necessary in the initiation of DNA synthesis) have been found to be phosphoproteins.[24] In fact, if a protein kinase is added to cultured rat liver cells, DNA synthesis is stimulated (and subsequently abolished if kinase inhibitor is introduced) suggesting that specific phosphorylation of DNA binding proteins may play a key role in initiation of DNA synthesis.[25,26] In addition, rat spermatocyte meiotic cells have also been shown to contain a phosphorylatable helix-destabilizing DNA binding protein (which is present mainly between the S-phase of the meiotic cell cycle and the termination of chromosome pairing[27]). Its high level of phosphorylation during this period suggests a possible function in genetic recombination.

Specific phosphorylation may also be involved in DNA replication and mitosis due to the close association of this phosphoprotein with the lamina complex. Although strong correlation between phosphorylation of histone H$_1$ and H$_3$ at the onset of chromosome condensation and mitotic initiation has been established,[28,29] it has been suggested that H$_1$ superphosphorylation is by itself not sufficient to bring about chromosome condensation since phosphorylation can occur without any condensation.[30] Attention has now shifted to the role of NHCP phosphorylation in this scheme of mitotic-related events. For example, phosphorylation of HMG proteins has been suggested to be responsible for "turning off" gene transcription during mitosis.[31] Also, increased intermediate filament protein (i.e., vimentin) phosphorylation[32] as well as phosphorylation/dephosphorylation of lamina proteins [33] have

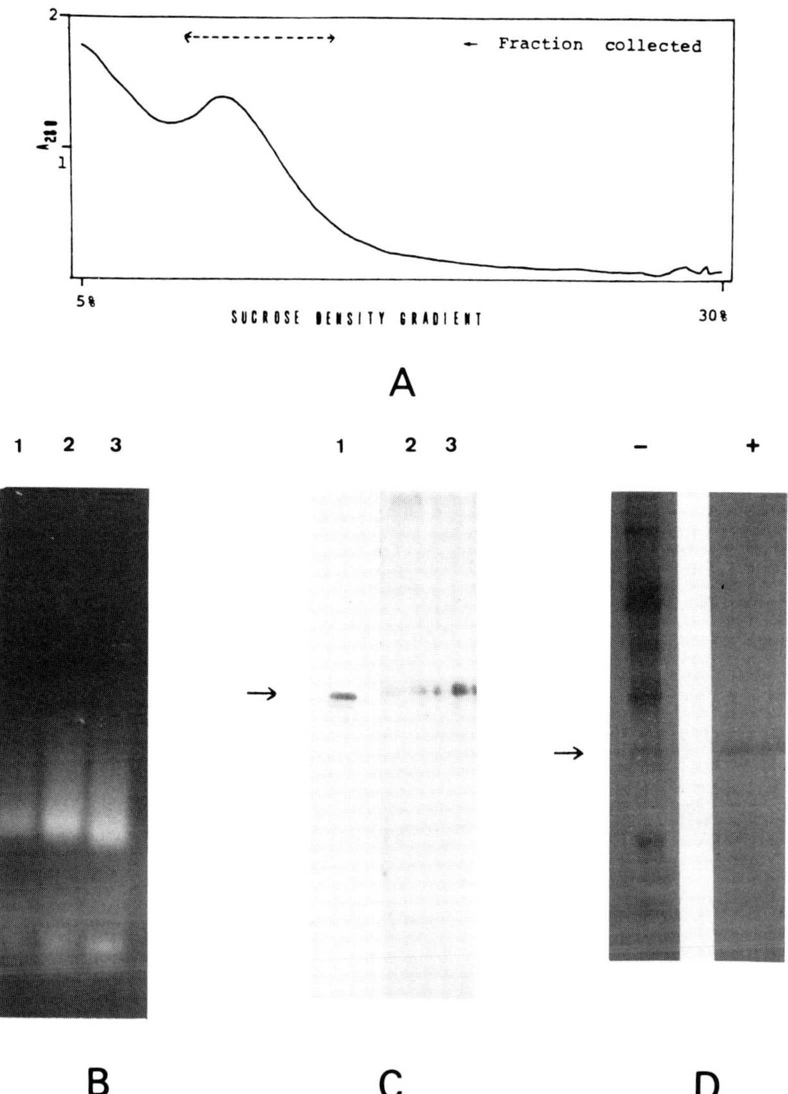

FIGURE 2.   Crosslinking. (A) Rat liver nuclei were digested with micrococcal nuclease for 2.5 min and the released nucleosomes then fractionated on a linear 5 to 30% sucrose density gradient. (B) The monomer fraction was collected from the gradient, cross-linked with disuccinimidyl suberate (DSS) or dithiobis(succinimidyl propionate) (DSP) (for 1 h at 20° using 1.3 μmol of cross-linker per mg protein), and subsequently fractionated on a 1% agarose gel. Lane 1: control-untreated monomers, Lane 2: DSP cross-linked monomers, Lane 3: DSS cross-linked monomers. (C) Proteins extracted from the monomers were immunoblotted after fractionation by 7.5 to 20% polyacrylamide gel electrophoresis. Lane 1: purified B_2 phosphoprotein, Lane 2: control of untreated monomers containing B_2 phosphoprotein, Lane 3: DSS cross-linked (immunoblots were probed with B_2 protein-monoclonal antibody IgG_{2a}). (D) 10 A_{260} U of whole nuclei were treated with 0.5 μmol of the (thiol-cleavable) DSP for 5 min at 0°C. The B_2 phosphoprotein-containing species were enriched by immunoaffinity chromatography and immunoblotted following fractionation by 5 to 20% polyacrylamide gel electrophoresis (immunoblots were probed with B_2 protein monoclonal antibody IgG_{2a}); (+) and (−) represent presence or absence of β-mercaptoethanol, respectively. Arrow indicates the position of uncross-linked B_2 phosphoprotein.

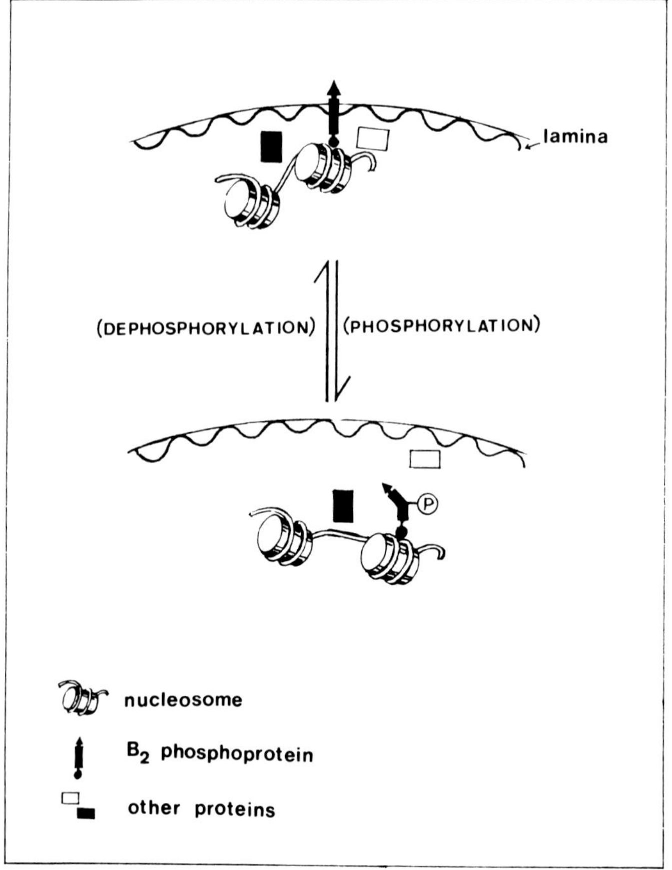

FIGURE 3.   Schematic representation of the B$_2$ phosphoprotein involvement with nuclear organization.

been implicated in the dissolution/reformation of the nuclear envelope. Moreover, phosphorylation of ribosomal protein S6, along with increased phosphorylation of other NHCP (i.e., maturation promoting factor) have been shown to be associated with meiotic maturation of *Xenopus laevis* oocytes.[34] Recent work has demonstrated that phosphorylation/dephosphorylation is intimately involved during oocyte maturation brought about by HeLa cell mitotic factors.[35] Those phosphorylated NHCP that are extractable with 0.2 M NaCl are in fact causally related to the entry of cells into mitosis while their selected dephosphorylation occurs during exit from mitosis. Specific monoclonal antibodies have been shown to react with these phosoproteins only during mitosis, while when they are dephosphorylated during G$_1$ and S their antigenicity is somehow altered.

IV. HYPOTHESIS AND CONCLUSION

These findings together with our earlier observations allowed us to infer that the availability of some chromosomal proteins (e.g., heterogeneous RNA particles, B$_2$ phosphoprotein) and their covalent modifications were critical for the conformational changes of chromatin which allow for specific gene expression.[12,13,36] This has allowed us to formulate a possible role for the B$_2$ phosphoprotein in nuclear organization (Figure 3). Our schematic model suggests that this phosphoprotein may play an important role in nuclear organization by anchorage of actively transcribed nucleosomes in the nuclear matrix. Once being phosphorylated, this nuclear protein dissociates from the nuclear matrix (together with the nucleosomes), thus allowing for more efficient transcription.

Our hypothesis on the role of the B_2 phosphoprotein in nuclear organization was further substantiated by the use of a scanning electron microscope coupled with an immuno-gold technique. We have demonstrated that the B_2 protein monoclonal antibody IgM is distinctively localized peripherally as shown in Figures 4A and B. The interaction of IgM monoclonal antibody and the phosphoprotein at the periphery of the nuclear matrix suggests that the site specific epitope of this antibody is exposed on the outer edges of the nuclear matrix.

Finally, we have also demonstrated that this phosphoprotein is present in the nucleus of many other cell types including human tumor cell lines (e.g., HeLa). In addition we have shown that this phosphoprotein cross-reacts with a human c-myc polyclonal antibody (anti myc-12C) using both "dot" blotting and "Western" blotting.[37] Further immunoprecipitation studies using c-myc expressing cell lines (e.g., Daudi) have also demonstrated that a protein in the region of 66 to 68 kDa is selectively immunoprecipitated by two of the IgG monoclonal antibodies raised against the B_2 protein. However, a more definitive relationship between the c-myc protein and the B_2 phosphoprotein awaits the results of further peptide analysis. It is thus conceivable that this B_2 phosphoprotein first identified in association with the nucleosome [8] and later with the nuclear matrix [16] may also be related to the myc protein oncogene family.

In conclusion, in the context of nucleosome structure and overall nuclear organization, we have had a unique opportunity to identify, isolate, and characterize one of the many nuclear phosphoproteins. Evidence has been presented showing that this B_2 phosphoprotein is closely associated with both actively transcribed chromatin and the nuclear matrix. It is also tightly coupled to cell proliferation and possesses a tentative myc protein immunological cross reactivity. In the future, our studies will be directed towards the (1) elucidation of some degree of the primary structure of this phosphoprotein (as an aid in the eventual cloning and sequencing of its gene) and (2) continuing the delineation of the specific role it may play within the context of growth and differentiation.

ACKNOWLEDGMENT

The generous support by grants from the Medical Research Council, Ontario Heart and Stroke Foundation, Canada and the National Institutes of Health are greatly appreciated.

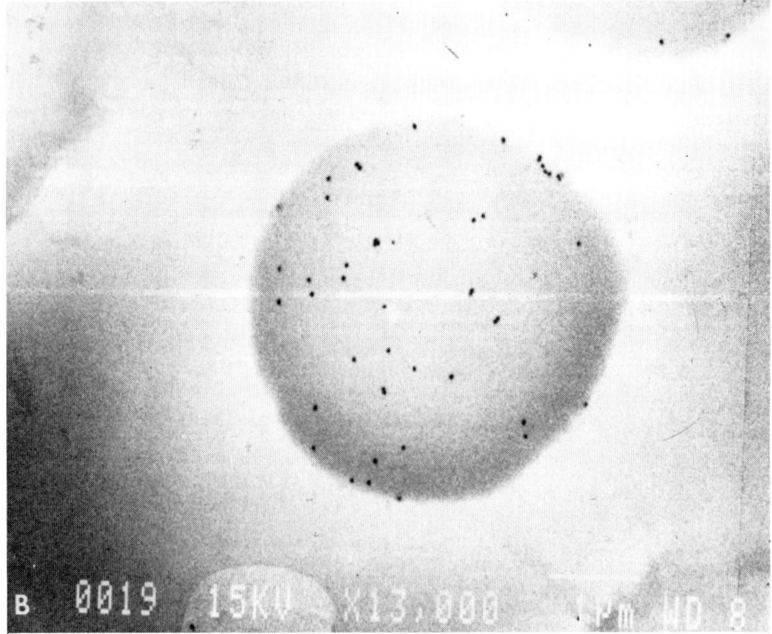

FIGURE 4.   Immunoelectron microscopy demonstrating the localization of the B_2 phosphoprotein in the nuclear periphery. Adult rat liver nuclei were prefixed with 4% paraformaldehyde and then saturated with 1% ovalbumin in PBS. The specimens were then incubated with B_2 phosphoprotein monoclonal antibody IgM (diluted 1:250). After extensive washing the nuclei were then incubated with goat anti-mouse IgM conjugated with colloidal gold (1:10). After post-fixation with 2.5% glutaraldehyde, the nuclei were dehydrated, critical-point dried, coated in a vacuum evaporator by carbon evaporation, and examined with a JEOL JSM 840 scanning electron microscope. (A) Adult rat liver nuclei labeled with 40 nm colloidal gold particles after incubation with IgM monoclonal antibody (secondary electron image). (B) The same immunostained nuclei viewed with backscattered electron image with reversed polarity (note the lack of staining of cellular debris).

REFERENCES

1. **Olson, M. O. J.,** Nonhistone nuclear phosphoproteins, in *Chromosomal Nonhistones Proteins,* Vol. III, Hnilica, L. S., Ed., CRC Press, Boca Raton, FL, 1984, chap. 3.
2. **Stein, G., Stein, J.,and Kleinsmith, L. J.,** Nonhistone proteins and gene regulation, in *Eukaryotic Genes* MacLean, N., Gregory, S. P., and Flavell, R. A., Eds., Butterworths, London, 1983, chap. 3.
3. **Suria, D. and Liew, C. C.,** Isolation and analysis of nonhistone chromatin proteins from rat-liver nuclei by three different methods, *Can. J. Biochem.,* 52, 1143, 1974.
4. **Jackowski, G., Suria, D., and Liew, C. C.,** Fractionation of nucleolar proteins by two-dimensional gel electrophoresis, *Can. J. Biochem.,* 54, 9, 1976.
5. **Suria, D. and Liew, C. C.,** Characterization of proteins associated with nuclear ribonucleoprotein particles by two-dimensional polyacrylamide gel electrophoresis, *Can. J. Biochem.,* 57, 32, 1979.
6. **Liew, C. C.,** Chromosomal nonhistone proteins, in *Concepts of the Structure and Function of DNA, Chromatin and Chromosomes,* Dion, A. S., Ed., Year Book Medical Publishers, Chicago, 1979, 153.
7. **Liew, C. C., Sole, M. J., Silver, M. D., and Wigle, E. D.,** Electrophoretic profiles of nonhistone nuclear proteins of human hearts with muscular subaortic stenosis, *Circ. Res.,* 46, 513, 1980.
8. **Liew, C. C. and Chan, P.,** Identification of nonhistone chromatin proteins in chromatin subunits, *Proc. Natl. Acad. Sci. U.S.A.,* 73, 3458, 1976.
9. **Chan, P. K. and Liew, C. C.,** Purification of a phosphoprotein from chromatin of rat liver, *Biochem. J.,* 183, 143, 1979.
10. **Johns, E. W.,** HMG: history, definitions and problems, in *The HMG Chromosomal Proteins,* Johns, E. W., Ed., Academic Press, New York, 1982, chap. 1.
11. **Chan, P. K. and Liew, C. C.,** Identification of nonhistone chromatin proteins in chromatin subunits (or mononucleosomes) devoid of histone H_1, *Can. J. Biochem.,* 57, 666, 1979.
12. **Zhao, M. S. and Liew, C. C.,** Production of antibody to phosphoprotein associated with nucleosome structure, *Can. J. Biochem.,* 60, 356, 1982.
13. **Liew, C. C., Halikowski, M. J., and Zhao, M. S.,** Two specific groups of NHC proteins involved in gene expression, in *Progress in Nonhistone Protein Research,* Vol. I, Bekhor, I., Ed., CRC Press, Inc., Boca Raton, FL, 1984, chap. 2.
14. **Liew, C. C., Halikowski, M. J., and Zhao, M. S.,** A chromosomal phosphoprotein is preferentially released by mild micrococcal nuclease digestion, *Biochem. J.,* 220, 539, 1984.
15. **Halikowski, M. J. and Liew, C. C.,** Monoclonal antibodies to a phosphoprotein from chromatin of rat liver, *Biochem. J.,* 225, 357, 1985.
16. **Halikowski, M. J. and Liew, C. C.,** Identification of a phosphoprotein in the nuclear matrix by monoclonal antibodies, *Biochem. J.,* 241, 693, 1986.
17. **Chan, P. K.,** A Biochemical Study of Nonhistone Chromatin Proteins in Chromatin Subunits of Rat Liver, Ph.D. thesis, University of Toronto, Toronto, 1978.
18. **Adler, K., Beyreuther, K., Fanning, E., Geisler, N., Gronenborn, B., Mueller-Hill, B., Phahl, M., and Schmitz, A.,** How lac repressor binds to DNA, *Nature (London),* 237, 322, 1972.
19. **Kleinsmith, L. J.,** Specific binding of phosphorylated nonhistone chromatin proteins to deoxyribonucleic acid, *J. Biol. Chem.,* 248, 5648, 1973.
20. **Prestayko, A. W., Crane, P. M., and Busch, H.,** Phosphorylation and DNA Binding of Nuclear Rat Liver Proteins Soluble at Low Ionic Strength, *Biochemistry,* 15, 414, 1976.
21. **Bluthmann, H.,** Specific binding of a nonhistone chromosomal protein from lymphocyte to DNA, *Eur. J. Biochem.,* 70, 233, 1976.
22. **Henning, R. and Montenarh, M.,** Simian virus 40 T-antigen phosphorylation is variable, *FEBS. Lett.,* 114, 107, 1980.
23. **Knippers, P., Otto, B., and Baynes, M.,** A single-strand specific DNA-binding protein from mouse cells that stimulates DNA polymerase. Its modification by phosphorylation, *Eur. J. Biochem.,* 73, 17, 1977.
24. **Collins, J. K., Tegtmeyer, P., and Rundell, K.,** Modification of simian virus 40 protein A, *J. Virol.,* 21, 647, 1977.
25. **Weisbrod, S., Groudine, M., and Weintraub, H.,** Interaction of HMG 14 and 17 with actively transcribed genes, *Cell,* 19, 289, 1980.
26. **Whitfield, J. F. and Boynton, A. L.,** A possible involvement of type H cAMP-dependent protein kinase in the initiation of DNA synthesis by rat liver cells, *Exp. Cell Res.,* 126, 477, 1980.
27. **Hotta, Y. and Stern, H.,** The effect of dephosphorylation on the properties of a helix-destabilizing protein from meiotic cells and its partial reversal in physaarum polysephalum, *Eur. J. Biochem.,* 90, 29, 1978.
28. **Gurley, L. R., D'Anna, D. A., Halleck, M. S., Barham, S. S., Walters, R. A., Jett, J. J., and Tobey, R. A.,** Relationships between histone phosphorylation and cell proliferation, in *Protein Phosphorylation,* Book B, Rosen, O. M., and Krebs, E. G., Eds., Cold Spring Harbor Laboratory, New York, 1981, 1073.
29. **Hohmann, P., Tobey, R. A., and Gurley, L. R.,** Phosphorylation on distinct regions of f_1 histone. Relationship to the cell cycle, *J. Biol. Chem.,* 251, 3685, 1976.

30. **Jungmann, R. A., Laks, M. S., Harrison, J. J., Suter, P., and Jones, C. E.,** Modulation of nuclear protein kinases at times of gene activity, in *Protein Phosphorylation,* Book B, Rosen, O. M., and Krebs, E. G., Eds., Cold Spring Harbor Laboratory, New York, 1981, 1109,

31. **Bhorjee, J. S.,** Phosphorylation of the high mobility group nonhistone proteins, in *Progress in Nonhistone Protein Research,* Vol. II, Bekhor, I., Mirell, C. J., and Liew, C. C., Eds., CRC Press, Boca Raton, FL, 1985, chap. 4.

32. **Evans, R. M. and Fink, L. M.,** An alteration in the phosphorylation of vimentin-type intermediate filaments is associated with mitosis in cultured mammalian cells, *Cell,* 29, 43, 1982.

33. **Gerace. L. and Blobel, G.,** The nuclear envelope lamina is reversibly depolymerized during mitosis, *Cell,* 19, 277, 1980.

34. **Nielson, R. J., Thomas, G., and Maller, J. L.,** Increased phosphorylation of ribosomal protein S6 during meiotic maturation of *Xenopus* Oocytes, *Proc. Natl. Acad. Sci. U.S.A.,* 79, 2937, 1982.

35. **Rao, P. N., Adlakha, R. C., Sahasrabuddhe, C. G., Wright, D. A.,and Bigo, H.,** Role of nonhistone protein phosphorylation in the regulation of mitosis in mammalian cells, in *Experimental Biology and Medicine,* Skehan, P. and Friedman, S. J., Eds., Humana Press, Clifton, N.J., 1985, 59.

36. **Liew, C. C., Jackowski, G., Ma, T., Jung, Y. C., and Sole, M. J.,** Possible role of nonhistone chromatin proteins associated with heterogeneous nuclear RNA in myocardial differentiation and in the genesis of cardiomyopathy, in *Perspectives in Cardiovascular Research,* Vol. 7, Raven Press, New York, 1983, 497.

37. **Halilowski, M. J.,** An Immunological Study of the B_2 Phosphoprotein from Rat Liver Chromatin, Ph.D. thesis, University of Toronto, Toronto, 1987.

Chapter 2

STRUCTURAL AND FUNCTIONAL STUDIES OF PROTEIN B23

Pui K. Chan, Benjamin Y-M. Yung, and Melissa B. Aldrich

TABLE OF CONTENTS

I. SYNOPSIS

A major phosphopeptide labeled *in vivo* was identified in nucleolar protein B23 (M_r, pI = 37 kDa, 5.1) after tryptic digestion. This peptide was purified on high pressure liquid chromatography (HPLC) using two reverse-phase (C8 and C18) columns. This phosphopeptide contains 20 amino acids, including 1 phosphoserine, 7 glutamic acids, and 4 aspartic acids. The amino acid sequence is: HLVAVEEDAES(p)EDEDEEDVK. This phosphopeptide amino acid sequence is similar to phosphopeptide sequences of protein C23 and the R_{II} subunit of cAMP-dependent protein kinase. There are eight consecutive identical amino acids in these protein sequences. This homologous sequence is as follows: Ser(P)-Glu-Asp-Glu-Asp-Glu-Glu-Asp.

An antigenic peptide obtained after partial V8 protease digestion of protein B23 was identified and purified. The amino acid sequence was determined. The antigenic peptide contains 68 amino acids and is located at the carboxyl terminal of protein B23. The NH$_2$-terminal region of this peptide is lysine-rich, and it may interact with RNA. This region also contains two repeated tripeptide sequences, Lys-Thr-Pro.

Three cloned cDNAs (hpB1, hpB2, and hpB7) which code for the 82 amino acids at the C-terminal of protein B23 were identified and characterized. All three clones have almost identical nucleotide sequences. Mutation of six nucleotide bases of one clone (hpB2) has caused changes in four amino acids in the sequence just preceding the immunoreactive region. These results suggest the presence of at least two immunologically similar, but structurally distinct genes.

Under native conditions, a hexameric form of protein B23 was purified by affinity chromatography. The hexamer is composed of four α and two β monomeric units. The Stoke's radius and the sedimentation coefficient of this hexamer are 51 Å and 10S, respectively. The molecular weight is estimated to be 230 kDa. The hexameric form of protein B23 is a structural element which associates with pre-RNP particles. The hexameric protein B23 is dissociated into monomers by 7 M urea treatment.

When HeLa cells were cultured in serum-free medium for 48 h, protein B23 translocated from nucleoli to the nucleoplasm, as detected by immunofluorescence. Relocation of protein B23 to the nucleoli was observed after refeeding the cells with serum-containing medium. Quantitation of protein B23 by the enzyme-linked immunosorbent (ELISA) assay confirmed the results of the immunofluorescence studies. Translocation of protein B23 was observed using certain antitumor agents, such as actinomycin D, toyocamycin, luzopeptins, and doxorubicin. "B23 translocation" correlates very well with the antitumor activity of drugs and can be used to study drug efficacy and detect drug-resistant cells.

II. INTRODUCTION

One striking feature of cancer cells is their enlarged, pleomorphic nucleoli, which are the primary sites of ribosome synthesis.[5] Various agents such as chemical carcinogens, UV radiation, and oncogene products, in one way or another induce a cascade mechanism which eventually transforms normal cells into tumor cells. A virtually universal feature of cancer cells is the activation of the rDNA and synthesis of ribosomes to meet their high protein demand. The cascade of reactions initiated by growth factors or hormone stimuli that eventually turns on the synthetic machinery of nucleoli is not well defined.

Protein phosphorylation and dephosphorylation have been implicated in the regulation of cellular processes.[20,21,27,31,42,50,63] Phosphorylation of nuclear proteins has been correlated with increased chromatin template activity.[2,17,18,40,41,46,67,76] Some of these nonhistone proteins are tightly bound to chromatin,[33,47] and are involved in gene activation.

Protein B23 is a nucleolar nonhistone phosphoprotein which is more abundant in tumor and growing cells than in normal resting cells.[6] This protein is associated with preribosomal particles,[61,65] and is localized in the granular region of the nucleolus,[74] where ribosomes are assembled. This article summarizes the recent progress of the structural and functional studies of protein B23 and its translocation property, which is induced by certain antitumor agents.

III. EXPERIMENTAL PROCEDURES

A. Cells

HeLa S3 cells were grown in Eagle's minimum essential medium (MEM) supplemented with 10% fetal calf serum, glutamine, and antibiotics (100 μg/ml penicillin, 100 μg/ml streptomycin) in a 5% CO_2-humidified incubator at 37°C. For immunofluorescence studies, cells were grown on slides in a Petri dish. Novikoff hepatoma ascites cells were implanted intraperitoneally in adult male albino Holtzman rats (200 g), and were left to grow for 6 d.

B. Preparation of Nucleoli

Nucleoli were isolated using the NP-40 method previously described.[10] Cells were suspended in 20 volumes of RSB buffer (0.01 M Tris HCl, 0.01 M NaCl, 1.5 mM $MgCl_2$ (pH 7.2)) for 30 min and then centrifuged at 630 × g for 8 min. Swollen cells were resuspended in 20 volumes of RSB buffer containing 0.5% NP-40. The cells were homogenized with a

Dounce homogenizer (10 up-and-down strokes), and the crude nuclei were collected by centrifugation at 630 $\times$ g for 8 min. The nuclei were resuspended in 10 volumes of 0.25 M sucrose, 10 mM MgCl$_2$, and underlayered with an equal volume of 0.88 M sucrose, 0.05 mM MgCl$_2$. After centrifugation at 1700 $\times$ g for 10 min, the nuclei were resuspended in 10 volumes of 0.34 M sucrose, 0.05 mM MgCl$_2$. The suspension was sonicated for 1 min, underlayered with an equal volume of 0.88 M sucrose, 0.05 mM MgCl$_2$, and centrifuged at 3000 $\times$ g for 18 min. The pellet (nucleoli) was then collected.

C. Purification of Protein B23

Protein B23 was purified from nucleoli following the method described by Michalik et al.[53] with modifications. Nucleoli were extracted with 4 M urea, 3 M LiCl, 1 mM PMSF (phenylmethyl sulfonyl fluoride), 1 mM leupeptin, 1 mM pCMPS(p-chloromercuriphenyl sulfonic acid) for 16 h and centrifuged at 27,000 $\times$ g for 20 min. The supernatant was dialyzed against Tris-Maleate buffer (20 mM Tris-maleate, 1 mM DTT (dithiothreitol), 1 mM EDTA, 4 M urea, 1 mM PMSF, 1 mM leupeptin, 1 mM PCMPS, pH 6.5). The precipitate, which contained less than 5% of the protein B23 as determined by ELISA, was removed after centrifugation. The supernatant, which contained protein B23, was further purified by DEAE cellulose chromatography. Protein B23 was eluted from the column with 0.14 M NaCl in the Tris-maleate buffer.

D. *In Vivo* Labeling of Protein B23 with ^{32}Pi

Novikoff hepatoma ascites cells were implanted intraperitoneally into adult male albino Holtzman rats (200 g) 6 d before collection. Cells were washed twice with 0.13 M NaCl, 5 mM KCl, 8 mM MgCl$_2$ (NKM) buffer to remove most of the red blood cells. Cells were then incubated in MEM phosphate-free medium containing ^{32}Pi (0.1 mCi/ml) at 37°C for 3 h.

E. Isolation of the Phosphopeptide

Protein B23, which was previously labeled *in vivo* with ^{32}Pi, was digested with trypsin (50 μg/ml/mg B23) in 50 mM NH$_4$HCO$_3$, pH 8, at 37° for 18 h. The ^{32}P-tryptic peptides were applied onto a C8 reverse-phase column (LiChrosorb, EM) that was equilibrated with 0.05% trifluoroacetic acid (TFA). Peptides were eluted from the column by a linear 0 to 40% gradient of acetonitrile in 80 min with a flow rate of 1 ml/min. One major ^{32}P-radioactive peak was eluted at about 20% acetonitrile (arrow, Figure 3A). Further purification of the phosphopeptide fraction was achieved by rechromatography on a C18 reverse-phase column (Bondapak, Waters Associates) equilibrated with the same solvent systems. A much shallower gradient (13 to 17% acetonitrile in 30 min with flow rate 1 ml/min) was used. The fraction was separated into two major components, of which the second component was the purified phosphopeptide.

F. Identification and Purification of Antigenic Peptide

Purified protein B23 (1 mg/ml) was incubated with staphylococcal V8 protease (10 μg/ml) in 0.1 M NH$_4$HCO$_3$, pH 7.8 at 37°C for 30 min. The mixture was applied onto a DEAE cellulose column (1.5 $\times$ 10 cm) equilibrated with 25 mM pyridine/acetic acid, pH 6.0. The flow-through fractions which contained the antigenic peptide were collected and concentrated by ultrafiltration (YCO5 membrane; 500 molecular weight cutoff; Amicon Corporation).

The peptides in the flow-through fraction of the DEAE column were separated by HPLC on a reverse-phase C18 column, which was equilibrated with 0.05% trifluoroacetic acid (TFA). Peptides were eluted from the column by the following acetonitrile gradients: 0 to 35% in 35 min with a flow rate of 2 ml/min, and 35 to 50% in 20 min. The antigenic peptide was eluted at 42% acetonitrile (P42).

G. Complete V8 Protease or Tryptic Digestion

Antigenic peptide P42 was lyophilized and redissolved in 50 mM NH$_4$HCO$_3$, pH 8. V8 protease or trypsin (TPCK-treated) was added (final concentration of 10 μg/ml), and the mixtures were incubated at 37°C for 6 h. The digests were then applied onto a C18 reverse-phase column equilibrated with 0.05% TFA. The acetonitrile gradient for the V8 peptides was 0% for 10 min with a flow rate from 2 to 1 ml/min, then 0 to 25% in 25 min with a flow rate of 1 ml/min, and finally, 25 to 35% in 35 min with a flow rate of 0.5 ml/min. Acetonitrile gradients for tryptic peptides were the same as for isolation of the antigenic peptide.

H. Amino Acid Sequence

The amino acid sequence of the phosphopeptide was determined by the method of Edman and Begg[26] in a Beckman 890C liquid phase or Applied Biosystem 470A gas phase sequenator as described in Reference 22. Polybrene was used as a nonprotein carrier. About 10 to 20 nmol or 0.5 to 1 nmol of peptide were applied into the liquid or gas phase sequenator, respectively. The resulting PTC-amino acids from the liquid phase sequenator were hydrolyzed in 5.7 N HCl, 0.1% SnCl$_2$ at 150°C, for 4 h. The ^{32}P-radioactivity after each step of sequencing was determined by liquid scintillation counting. PTH-derivatives from the gas phase sequencer were identified by HPLC on a Waters Nova Pak C18 column.[22] Norleucine was added to each sample and served as an internal standard. All PTH-derivatives were monitored at 265 nm and 313 nm (for serine and threonine).

I. Peptide Mapping

The Coomassie blue-stained protein B23 was excised from a 2-D gel. The gel pieces were washed with 10% methanol overnight before lyophilization. One milliliter of trypsin solution (50 μg trypsin/ml) in 50 mM NH$_4$HCO$_3$, pH 8, was added to the dried gel pieces and incubated for 18 h at 37°C. The peptide solution (excluding the gel pieces) was lyophilized and the mapping was performed according to the method described by Zweig.[86]

J. Screening of the λgtll cDNA Library

A λgtll cDNA library that was prepared from human placental mRNA was kindly provided by Dr. Brian Knoll. About 2 × 10^6 phages from the library were plated onto 40 large (150 mm) L-agar plates and screened with ^{125}I-labeled antibody using the method of Young and Davis with slight modifications.[29,81] Positive clones were isolated and plaque-purified, and the insert size was estimated by polyacrylamide gel electrophoresis after digestion by EcoR1 endonuclease.

K. DNA Sequence Analysis

Phage DNA was prepared from positive clones using a plate-lysis method,[49] followed by banding on a cesium chloride step gradient as described by Degen et al.[24] The cDNA inserts were released by digestion with EcoR1 and subcloned into M13mp18 and M13mp19 vectors for sequencing by Sanger's dideoxy chain termination method. All sequences were determined three or more times as well as from different M13mp18 subclones. DNA sequences were stored and analyzed by the computer programs of Larson and Messing.[45]

L. Purification of Protein B23 with Native Conditions

HeLa nucleoli were extracted (stirred at 10°C) with 10 mM Tris HCl (pH 7.5), 0.5 mM MgCl$_2$, 1 mM PMSF, 1 mM leupeptin, 1 mM pCMPS for 15 min and centrifuged at 27,000 × g for 20 min. The supernatant (Tris extract of HeLa nucleoli), containing the partially purified protein B23, was applied to an immunoaffinity column for further purification.

M. Immunoaffinity Chromatography

The monoclonal antibody to protein B23 was coupled to CNBr-activated Sepharose 4B. The Tris extract of HeLa nucleoli was mixed with B23 antibody-Sepharose overnight at 4°C by tumbling end-over-end. The unbound proteins were removed from the column by washing with 20 bed volume of 10 mM Tris, 0.5 mM MgCl$_2$, 1 mM PMSF, 1 mM leupeptin, 1 mM pCMPS, and the bound protein B23 was eluted with 4 M MgCl$_2$. All fractions, including the unbound, the washed, and the eluted samples, were analyzed for protein B23 by ELISA and immunoblot assays. A control column, made by coupling pre-immune serum to Sepharose 4B, was also incubated with the Tris extract of HeLa nucleoli, washed, and eluted similarly.

N. Molecular Sieve Chromatography

An Ultrogel AcA 22 column (0.8 × 45 cm) was equilibrated with 10 mM Tris HCl (pH 7.5), 0.5 mM MgCl$_2$, 100 μg/ml, ml bovine serum albumin. Affinity-purified protein B23 in the above buffer was applied to the column and eluted in 0.25 ml fractions at a flow rate of 10 ml/h. The fractions containing protein B23 were identified by ELISA analysis.

O. Determination of the Sedimentation Coefficient of Protein B23

The immunoaffinity-purified native protein B23 was layered on a sucrose density gradient (5 to 20% in 10 mM Tris HCl (pH 7.5), 0.5 mM MgCl$_2$, 1 mM PMSF, 1 mM pCMPS, 1 mM leupeptin) and centrifuged at 80,000 × g for 24 h at 4°C in an SW28 rotor. Fractions of 0.6 ml were collected and assayed for ELISA activity.

P. ELISA Procedures

The proteins were bound to microtiter plates overnight at 4°C. Blocking of the unbound sites was done with 3.0% bovine serum albumin, 0.05% Tween 20, phosphate-buffered saline, pH 7.5, and 10% chicken serum to reduce background. The primary antibody was antiprotein B23 monoclonal antibody (diluted 1:25,000). The second antibody was a peroxidase-labeled goat antimouse antibody. Color was developed after the addition of 0.02% of H$_2$O$_2$ and 300 μM of ABTS (2, 2′-azino-di-3-ethyl-benzthiazoline sulfonic acid diammonium salt). The reaction was terminated in 30 min with the addition of 1 M NaF. The absorbance was read at 415 nm on an ELISA reader.

Q. Polyacrylamide Gel Electrophoresis (PAGE)

One-dimensional SDS-PAGE was performed according to the method of Laemmli.[44] Samples were mixed with an equal volume of SDS sample buffer before application to an 8% SDS-polyacrylamide gel. For two-dimensional gel analysis, the samples were incorporated into isoelectrofocusing gels, and electrophoresis was performed as described by Chan et al.[10]

R. Immunoblot (Western) Procedures

Proteins were fractionated by SDS polyacrylamide gel electrophoresis and subsequently transferred electrophoretically to nitrocellulose papers (Schleicher & Schuell type BA-85, 0.45 μm) at 10 V for 16 h or at 50 V for 1 h (high molecular weight proteins could not be transferred efficiently using the latter conditions). The nitrocellulose paper was soaked in blocking solution (3% BSA, 0.9% NaCl, 10 mM Tri-HCl, pH 7.5, 10% chicken serum) overnight. The paper was then incubated with protein B23 antibody in NP-40 buffer (0.015 M NaCl, 5 mM EDTA, 5 mM Tris-HCl, pH 7.5, 0.05% NP-40, 0.25% gelatin) for 2 h at 25°C. The antibody was removed by two washes with NP-40 buffer (25 ml, 10 min each wash). The paper was subsequently incubated with the second antibody (rabbit antimouse IgG), washed (2×), incubated with ^{125}I-protein A solution, and washed (2×). The paper was finally washed 12 times (25 ml, 20 min each) with sarkosyl buffer (1 *M* NaCl, 5 m*M* EDTA, 5 m*M* Tris-HCl, pH 7.5, 0.4% sarkosyl, 0.25% gelatin). The dried nitrocellulose paper was then exposed to Kodak® X-ray film (XRP-5).

S. Immunofluorescence

HeLa cells (grown on slides) were fixed immediately after incubation with the drugs. Novikoff cells and mouse leukemia cells were cytocentrifuged onto clean slides. The slides were fixed first in 2% *p*-formaldehyde in PBS (8.45 m*M* Na_2HPO_4, 1.6 m*M* NaH_2PO_4, 145 m*M* NaCl) for 20 min at room temperature, and then were washed three times in PBS with gentle shaking for 5 min per wash. The slides were fixed in acetone at $-20°C$ for 3 min, rinsed briefly 4 times in PBS, and blown dry with cool air. The first antibody (a monoclonal IgG that recognizes only protein B23 in total HeLa cell extracts) was diluted 1/16 in PBS (to a concentration of 3.4 mg/ml) and applied to the slides. For an experimental control, NS-1 supernatant or PBS alone was added to some slides to determine if there would be any background fluorescence. The slides were incubated in a moist chamber overnight at 4°C. The slides were washed 4 times, for 15 min per wash, in PBS, with gentle shaking. The slides were blown dry with cool air. The second antibody (1 mg/ml fluorescien-conjugated goat antimouse IgG, heavy and light chains specific, affinity purified) was diluted 1/20 in PBS and applied to the slides. The slides were incubated in a moist chamber for 1 h at room temperature. The slides were washed 4 times with PBS, blown dry, and mounted in 50% glycerol in PBS (pH = 9). The slides were viewed under a fluorescence microscope.

T. B23 Translocation Assay

The procedure for immunofluorescence is described in References 10 and 13. Control cells without drug treatment have bright nucleolar fluorescence(Figure 14A), which is defined as pattern A. Total B23 translocation shows a uniform nucleoplasmic fluorescence (Figure 14c), which is defined as pattern C. Pattern B represents intermediate translocation, in which both nuclear and nucleolar fluorescence are observed. The number of cells with these different fluorescence patterns was tallied. More than 500 cells per slide were counted.

IV. RESULTS AND DISCUSSION

A. Phosphorylation Sites of Protein B23[11]

1. Identification of Phosphopeptides of Protein B23

Purified protein B23 was analyzed by two-dimensional gel electrophoresis. Figure 1A shows the Coomassie blue-stained protein B23. Two major protein spots (M_r = 37,000 and 35,000, pI = 5.1), designated as the α and β forms of protein B23, were observed. Both α and β were ^{32}P-labeled as shown in the autoradiograph of the 2-D gel (Figure 1B). To analyze the phosphorylated peptide of protein B23, the protein spots (α and β) were excised separately from the 2-D gel and digested with trypsin. The tryptic phosphopeptides were separated on a cellulose sheet by electrophoresis and chromatography. As shown in Figure 2A, one major·(#3) and four minor (#1, 2, 4, and 5) ^{32}P-labeled peptides were identified. The phosphopeptide maps for α and β were the same. Figure 2B shows the purified phosphopeptide (#3) after HPLC chromatography.

2. Isolation of the Major Phosphopeptide

Figure 3A shows the elution profile of the tryptic peptides of protein B23 fractionated by HPLC on a C8 reverse phase column. Of the 40 peptide peaks, the major ^{32}P-labeled peptide peak was eluted at 20% acetonitrile (arrow). The first ^{32}P peak that was eluted at the solvent front contained very small amounts of peptide, and was not investigated further. The peptides that were eluted with 20% acetonitrile were further separated on another C18 reverse phase

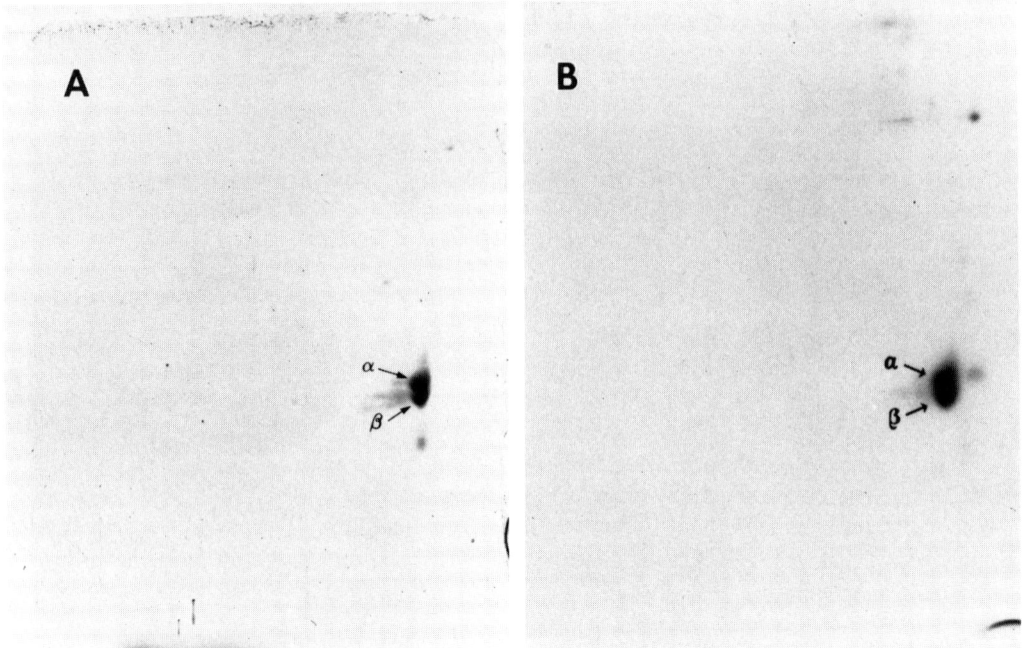

FIGURE 1.   Two-dimensional gel electrophoresis analysis of protein B23. Purified protein B23 was analyzed on two-dimensional gel electrophoresis. (A) Coomassie blue stain, and (B) ^{32}P-autoradiography. Two Coomassie blue stained spots of protein B23 (α and β) which differ slightly in molecular weight were observed. Both α and β were phosphorylated. (From Chan, P. K., Aldrich, M., Cook, R. G., and Busch, H., *J. Biol. Chem.*, 261, 1868, 1986. With permission.)

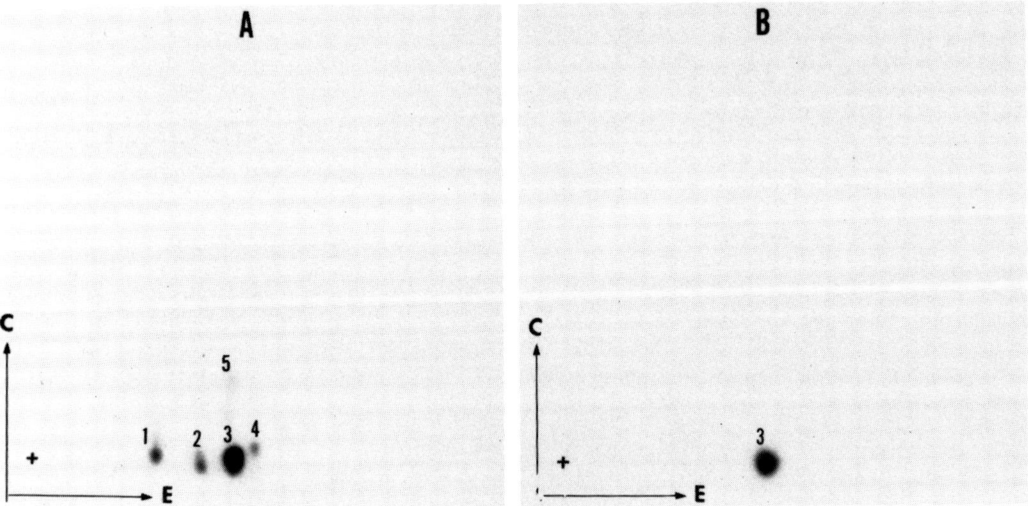

FIGURE 2.   Tryptic phosphopeptide map of protein B23. (A) Protein B23 spots (both α and β) were excised from a 2-D gel (Figure 1) and digested with trypsin (TPCK-treated) for 18 h in 50 mM NH$_4$HCO$_3$, pH 8. The tryptic peptides were fractionated by electrophoresis (E) and chromatography (C). ^{32}P-radioactivity was detected by autoradiography. One major (#3) and four minor ^{32}P-labeled peptides (#1, 2, 4, and 5) were identified. (Both α and β spots of protein B23 have identical peptide maps.) (B) ^{32}P-Peptide map of major phosphopeptide (#3) after purification by HPLC (Figure 3B). (From Chan, P. K., Aldrich, M., Cook, R. G., and Busch, H., *J. Biol. Chem.*, 261, 1868, 1986. With permission.)

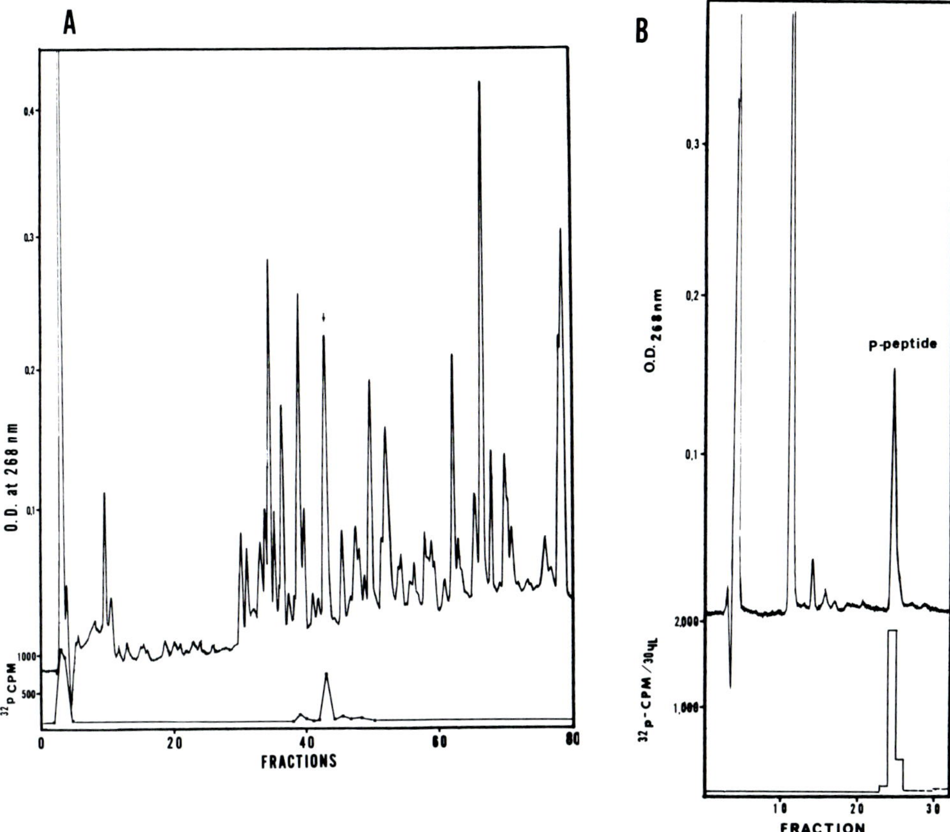

FIGURE 3.   Purification of the phosphopeptide by HPLC in reverse phase columns. (A) The ^{32}P-tryptic peptides of protein B23 were applied onto a C8 reverse phase column (EM) that was equilibrated with 0.05% trifluoroacetic acid (TFA). Peptides were eluted from the column by a linear 0 to 40% gradient of acetonitrile. The phosphopeptide (arrow, fraction A) was eluted at about 20% acetonitrile. (B) Further purification of the phosphopeptide (fraction A) was achieved by rechromatography on a C18 reverse phase column equilibrated with the same solvent systems, using a much shallower gradient (13 to 17% acetonitrile in 30 min). Fraction A was separated into two components, of which the second component was the purified phosphopeptide. (From Chan, P. K., Aldrich, M., Cook, R. G., and Busch, H., *J. Biol. Chem.*, 261, 1868, 1986. With permission.)

column (Figure 3B) with an extended acetonitrile gradient. The ^{32}P-labeled peptide was the second major peptide eluted from the column (Figure 3B). The purified phosphopeptide was identified as peptide #3 (Figures 2A and 2B) on cellulose sheets (Figure 2B). Phosphoserine was the only phosphorylated amino acid identified in the peptide. Table 1 shows the amino acid composition of the peptide, which has 19.5 and 34.1 mol percentages of aspartic and glutamic acids, respectively.

3. Amino Acid Sequence

The amino acid sequence of the phosphopeptide was determined by both liquid and gas phase protein sequenators. The amino acid sequence is His-Leu-Val-Ala-Val-Glu-Glu-Asp-Ala-Glu-Ser(P)-Glu-Asp-Glu-Asp-Glu-Glu-Asp-Val-Lys (Figure 4). This phosphopeptide contains 20 amino acids, including 1 phosphoserine, 7 glutamic acids, and 4 aspartic acids. Figure 4 also shows the amino acid sequences of phosphopeptides of protein C23,[48] the R_{II} subunit of cAMP-dependent protein kinase,[9] and a nonphosphorylated site in DNA-dependent RNA polymerase.[7] There are eight consecutive amino acids identical in all four protein sequences. This homologous sequence is as follows: Ser(P)-Glu-Asp-Glu-Asp-Glu-Glu-Asp.

Table 1
AMINO ACID COMPOSITION OF THE
PHOSPHOPEPTIDE OF PROTEIN B23

	Mol (%)	Number of residues	
		Expected	Experimental
Asp	19.5 ± 0.5	3.8	4
Ser	3.4 ± 0.4	0.7	1
Glu	34.1 ± 1.1	6.7	7
Gly	2.0 ± 0.3	0.4	0
Ala	9.7 ± 0.5	1.9	2
Val	14.3 ± 0.5	2.8	3
Leu	5.1 ± 0.1	1.0 (assigned)	1
Lys	6.3 ± 0.6	1.2	1
His	5.3 ± 0.2	1.0	1
Total	99.7	19.5	20

From Chan, P. K., Aldrich, M., Cook, R. G., and Busch, H., *J. Biol. Chem.*, 261, 1868, 1986. With permission.

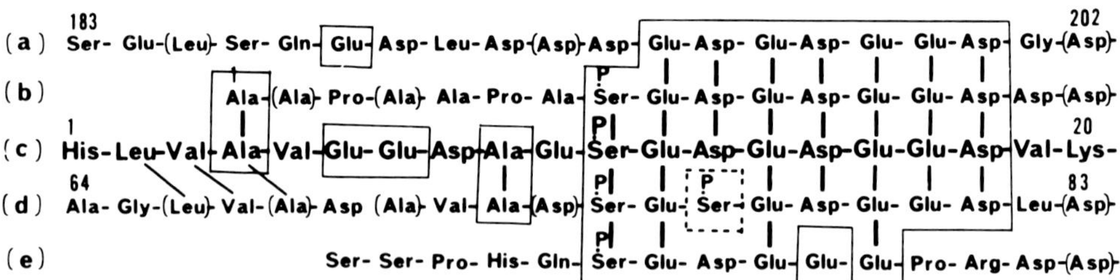

FIGURE 4.   Comparison of the amino acid sequences in phosphorylation sites of various proteins. (a) DNA-dependent RNA polymerase, sigma chain; (b) nucleolar phosphoprotein C23, peptide ClH2; (c) nucleolar phosphoprotein B23; (d) R_{II} of cAMP-dependent protein kinase; (e) glycogen synthase. □, Amino acids homologous to protein B23; ( ), amino acids homologous to R_{II} of cAMP-dependent protein kinase. (From Chan, P. K., Aldrich, M., Cook, R. G., and Busch, H., *J. Biol. Chem.*, 261, 1868, 1986. With permission.)

4. Discussion

The function of the acidic amino acid clusters in protein B23 is not known. The additional negative charge contributed by the phosphate group in the phosphopeptide may intensify the binding capacity of the protein. These acidic residues could possibly bind to positively charged molecules, such as basic proteins (histones, ribosomal proteins), small cations (polyamines), or metal ions (Mg^{++}, Ca^{++}). These glutamic and aspartic acid clusters may be essential recognition sites for kinases such as casein kinase II, which has a preference for phosphorylating acidic regions of polypeptides.[31,51]

The proteins with this phosphorylation site have different functional and chemical properties.[7,9,11,48] The reason that these proteins possess a common amino acid sequence is not clear at this moment. It is possible that this homologous amino acid sequence represents a recognition site for secondary messengers, such as kinases of a cascade system which is initiated by hormonal or growth factor stimuli. This site could serve as the protein activator or as a signal for events subsequent to protein phosphorylation.

It is not known whether phosphorylation of this homologous site in protein B23 is associated with ribosome synthesis activity. Direct studies of this relationship are possible because only one major phosphorylation site has been found in the B23 molecule. Nonetheless, other minor phosphorylation sites may also play a functional role for protein B23.

B. Antigenic Sites of Protein B23[14]

1. Identification of a Specific Antigenic Peptide of Protein B23

Figure 5 shows the antigenic activity of protein B23 after partial digestion with various proteases. About 20% of the original antigenic activity was lost after 30 min of digestion with trypsin, V8 protease, and leucine amino peptidase. Further digestion with trypsin and V8 protease resulted in the complete loss of antigenic activity. Chymotrypsin destroyed the antigenic activity of protein B23 during the first 30 min of digestion. The Western immunoblot assay of the antigenic peptide obtained after 30 min of V8 protease digestion (Figure 5 insert) showed that the molecular weight was between 3000 and 6000.

2. Purification of the Antigenic Peptide P42

Figure 6 shows the purification of the antigenic peptide by HPLC on a reverse-phase C18 column. Among the 40 components that were resolved by this column chromatography, antigenic activity was found in peaks A, B, and C (arrows), which were eluted with 42% acetonitrile. These antigenic peptides (A, B, and C) are the result of partial cleavages of V8 protease. Antigenic peptide B was further purified by rechromatography on the same reverse-phase C18 HPLC column. Figure 7 shows a single peak of the purified antigenic peptide B, which is designated as P42 and used for sequencing analysis. Amino acid composition analysis of P42[14] indicated that this peptide contains 14.7 mol% and 18.7 mol% of glutamic acid and lysine, respectively.

3. Amino Acid Sequence of the Antigenic Peptide P42

The antigenic peptide P42 was further cleaved into smaller peptides by V8 protease or trypsin. These peptides (V_a to V_h, from extensive V8 protease digestion and T_9 to T_{34}, from trypsin digestion) were separated by HPLC on a reverse phase C18 column. The amino acid sequences of these peptides were determined.[14] The complete amino acid sequence of antigenic peptide P42 is shown in Figure 8. The peptide contains 68 amino acids and is located at the carboxyl terminal of protein B23. The NH_2-terminal region of this peptide is lysine-rich and contains the twice repeated tri-peptide sequence Lys-Thr-Pro.

4. Discussion

Attempts have been made to determine the minimum amino acids required for antigenic activity by further cleavage with proteolytic enzymes. It was found that the whole peptide was required for antibody recognition. The antibody binding site (epitope) is most likely to be constructed by folding of the ploypeptide chain which is readily destroyed by proteolytic cleavages. Antibody-antigen interaction is particularly sensitive to cleavage by chymotrypsin (Phe or Tyr residues) (Figure 5). When the antigenic peptide is modified by Bolton-Hunter or fluorescamine reagents, it loses its antigenic activity. This loss of activity suggests that modification of the α-NH_2 groups of the lysines may result in blockage of the antibody binding.

The NH_2-terminal region (amino acids #1 to 15) of the antigenic peptide contains 34% lysine. The fact that the hydrophilicity in this region averages greater than 1.5 indicates the location of the antigenic determinant.[34] This region also contains three prolines and the twice-repeated tri-peptide with the amino acid sequence of Lys-Thr-Pro. The high value of hydrophilicity and the twisted structure (3 prolines) in this region suggest the possibility that the whole antigenic peptide sticks out of the protein B23 molecule, thus favoring interaction with the antibody.

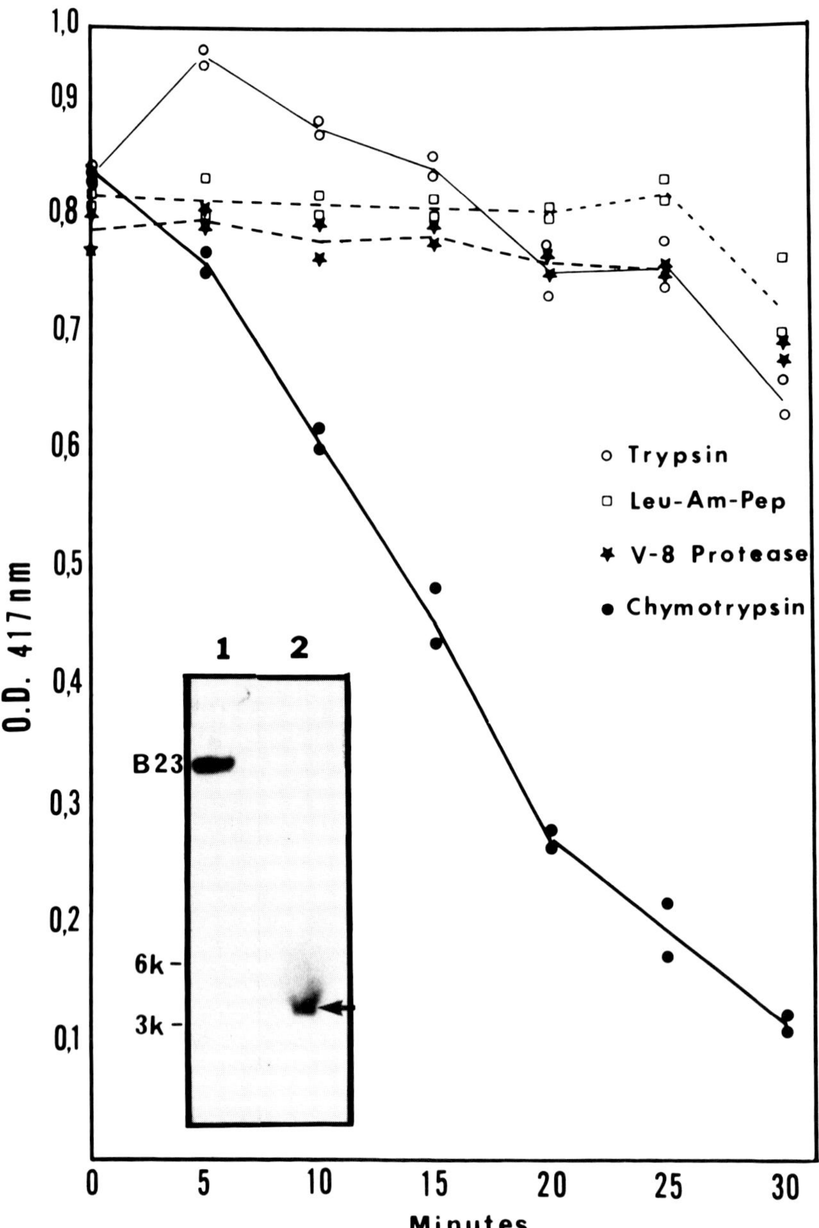

FIGURE 5.   Enzymatic digestion of protein B23. Protein B23 (1 mg/ml) was incubated with various enzymes (10 μg/ml) at 37°C. Aliquots were removed at different time intervals (abscissa) and the antigenic activity was determined by ELISA assay (ordinate). (O) Trypsin-TPCK; (□) leucine amino peptidase; (*) V8-protease; (●) chymotrypsin. Insert: immunoblot assay of protein B23 (lane 1) and antigen peptide obtained after 30 min of V8-protease digestion (lane 2). (From Chan, P. K., Chan, W.-Y., Yung, B. Y.-M., Cook, R. G., Aldrich, M. B., Ku, D., Goldknopf, I. L., and Busch, H., *J. Biol. Chem.*, 261, 14335, 1986. With permission.)

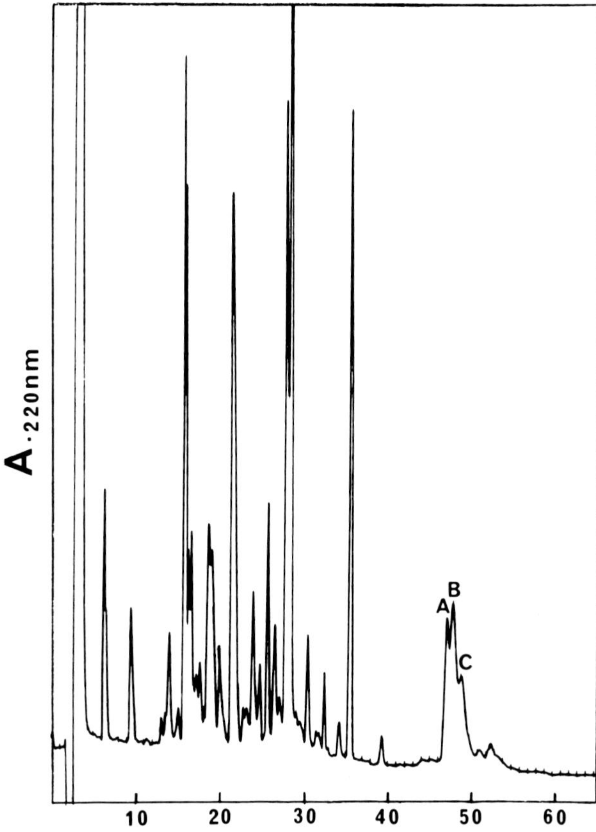

FIGURE 6. Purification of antigenic peptide by HPLC. The flow through fraction of the DEAE column (14) was applied to a reverse phase (C18) HPLC column. The column was eluted with an acetonitrile gradient in 0.1% TFA. Three antigenic peaks (A, B, C) were eluted at 42% acetonitrile as determined by ELISA assay. (From Chan, P. K., Chan, W.-Y., Yung, B. Y.-M., Cook, R. G., Aldrich, M. B., Ku, D., Goldknopf, I. L., and Busch, H. J., *J. Biol. Chem.*, 261, 14335, 1986. With permission.)

Previous studies have indicated that protein B23 may bind to the pre-rRNA,[83] particularly the spacer region of the 32S RNA. A positively charged lysine cluster such as the one in this antigenic region may be responsible for the binding.

C. Studies of cDNA Clones of Protein B23[14]

1. Identification of B23 cDNA Clones

Initial screening of the human placental expression library with the ^{125}I-labeled anti-rat B23 antibody yielded ten positive clones. Repeated screening resulted in three strongly hybridizing clones, hpB 1, hpB 2, and hpB 7. These clones were plaque-purified. All three clones contained a cDNA insert of approximately 700 bp.[14]

2. DNA Sequence Analysis

Figure 9 shows the 5' end nucleotide sequence of cloned cDNA (hpB 1) and the corresponding amino acid sequence. This sequence matches the amino acid sequence of the antigenic peptide p42 of rat B23 (Figure 8). This cloned cDNA (hpB 1) contains the C-terminal 82 amino acids, the stop codon, TAA, and the 3' nontranslating sequence. Figure

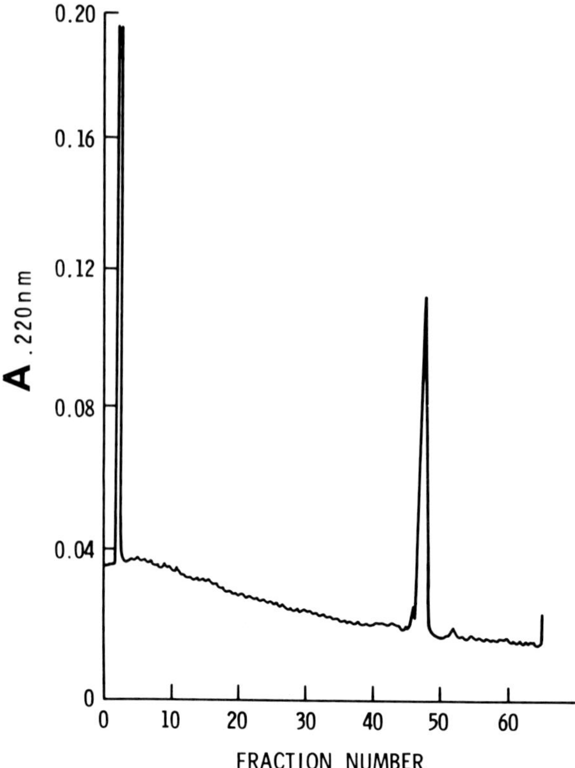

FIGURE 7.   Purity of the antigenic peptide. The antigenic peptide B was purified by rechromatography on a reverse phase (C18) HPLC column. A single peak was observed. This peptide was designated as P42 and used for sequence studies. (From Chan, P. K., Chan, W.-Y., Yung, B. Y.-M., Cook, R. G., Aldrich, M. B., Ku, D., Goldknopf, I. L., and Busch, H., *J. Biol. Chem.*, 261, 14335, 1986. With permission.)

Antigenic Peptide of Protein B 23

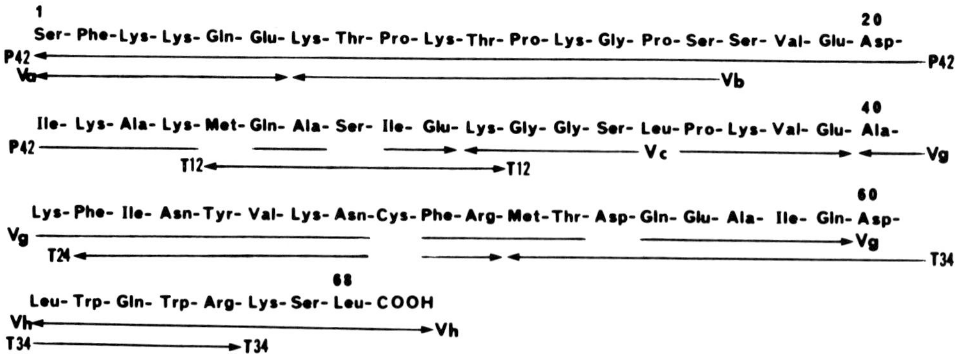

FIGURE 8.   Amino acid sequence of antigenic peptide (P42) of protein B23. $V_{peptide}$ = amino acid sequence of peptides from extensive V8 protease digest. $T_{peptide}$ = amino acid sequence of peptide from extensive tryptic digest. (From Chan, P. K., Chan, W.-Y., Yung, B. Y.-M., Cook, R. G., Aldrich, M. B., Ku, D., Goldkopf, I. L., and Busch, H., *J. Biol. Chem.*, 261, 14335, 1986. With permission.)

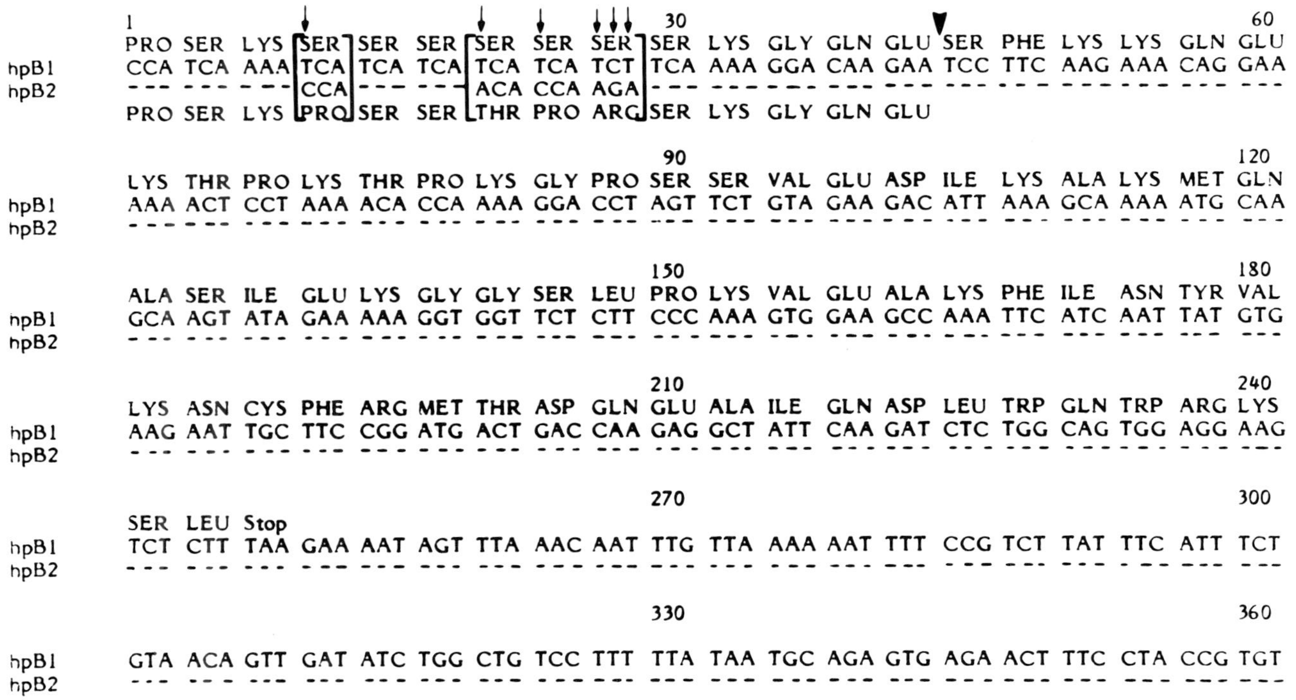

FIGURE 9.   Nucleotide sequence of B23 clones hpB1 and hpB2. Six mutated nucleotides (arrows) were observed which caused changes in four amino acids (brackets) preceding the antigenic peptide. Arrow head indicates the beginning of the antigenic peptide. (From Chan, P. K., Chan, W.-Y., Yung, B. Y.-M., Cook, R. G., Aldrich, M. B., Ku, D., Goldkopf, I. L., and Busch, H., *J. Biol. Chem.*, 261, 14335, 1986. With permission.)

9 also shows the nucleotide sequence of the second clone (hpB 2), which has an identical nucleotide sequence to that of hpB 1, except for 6 base changes just preceding the immunoreactive region (Figure 9). Alterations of these 6 nucleotides (2Ts to Cs, 3Ts to As, and 1C to G), cause changes in 4 amino acids (2 serines to prolines, a serine to threonine, and a serine to arginine) in this region. This result suggests that there are two immunologically similar but biochemically distinct proteins recognizable by the anti-B23 antibody.

3. Discussion

The results of these studies indicate that there is an evolutionary conservation of the amino acid sequence in the C-terminal portion of protein B23 between rats and humans. Studies in previous sections have shown that the amino acid sequence of the phosphorylation site of protein B23 is identical to that of several other proteins[11] and is another conserved domain of the molecule. This hypothesis has been recently confirmed by Schmidt-Zachmann et al.[69] in Franke's group, who observed identical amino acid sequences in the C-terminal antigenic and phosphorylation sites in the B23 of *Xenopus laevis*.

The cDNA sequence study reveals at least two human genes for protein B23. These results agree with the observation that there are two forms of protein B23 (α and β) in HeLa cells.[10,85] It is interesting to note that the α form of protein B23 is predominantly found in the nucleoplasm during serum starvation or actinomycin D treatment.[10,82,83] Future studies of the expression of these two B23 genes will be essential to understanding the function and control mechanisms of B23 in proliferating cells.

D. Hexameric Form of Protein B23[85]

1. Purification of Native Protein B23

HeLa cell nucleoli were suspended in 10 mM Tris-HCl, 0.5 mM MgCl$_2$, 1 mM PMSF, 1 mM leupeptin, 1 mM pCMPS, and were magnetically stirred for 15 min at 10°C. The suspension was then centrifuged at 27,000 $\times$ g for 20 min. The supernatant (Tris-extract) was then loaded onto a protein B23 immunoaffinity column. The elution profile is shown in Figure 10. Protein B23 bound to the immunoaffinity column but not to the control column.

2. Identification of the High Molecular Weight Form of Protein B23

Figure 11, lane D, shows the proteins specifically bound to the affinity column. A weak band corresponding to protein B23 (37 kDa) and a dense high molecular weight band were identified by silver staining. The proteins in the gels were transferred to nitrocellulose at 10 V for 16 h for Western blot assay. These conditions insured that the high molecular weight proteins were efficiently transferred to the nitrocellulose. Both the 37 kDa and the high molecular weight bands had protein B23 immunoactivity (Figure 11 E and F). In previous Western blot experiments, the electrophoretic transfer was performed for 1 h at 50 V.

3. Dissociation of the High Molecular Weight Form of Protein B23 into 37 kDa Monomers

Protein B23 that was purified from the immunoaffinity column (Figure 10) was treated with 7 M urea and then analyzed by SDS gel electrophoresis and immunoblot assay. Figure 12, lane B, shows that both high and low molecular weight protein B23 were in the sample. The high molecular weight form of protein B23 disappeared (Figure 12, lane C), and the intensity of the monomeric protein B23 band increased after 7 M urea treatment. To further confirm that the high molecular weight form was dissociated into monomers, the high molecular weight band of protein B23 (Figure 12, lane B) was excised and electrophoretically eluted from the polyacrylamide gel. The sample was then treated with 7 M urea and reloaded onto a one-dimensional gel (Figure 12, lane D). The high molecular weight form of protein B23 was not observed. Instead, the monomers (α and β) appeared in the one-dimensional gel. The α and β monomers were present in an approximate 2:1 ratio, as measured by densitometric scanning.

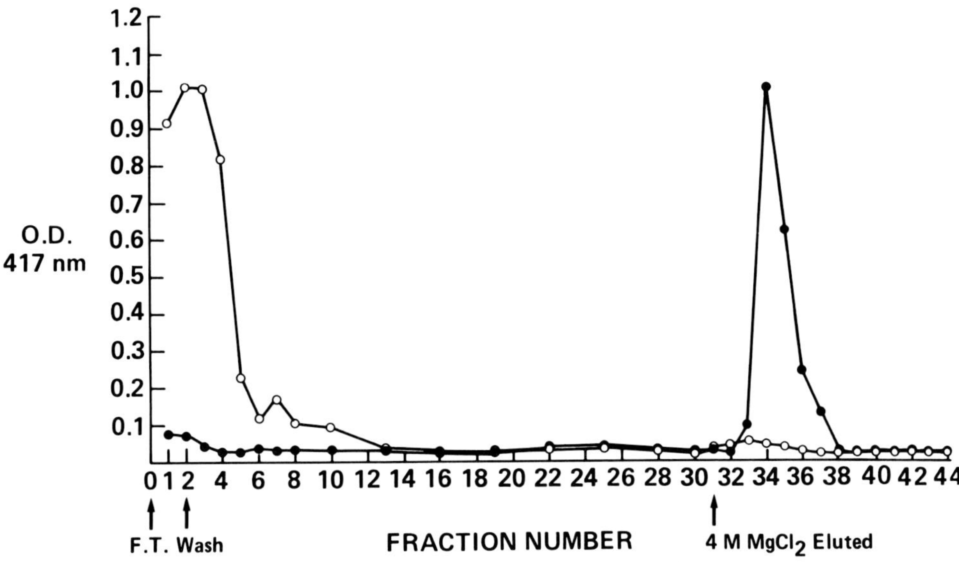

FIGURE 10. Purification of native protein B23 by immunoaffinity chromatography. The Tris extract of HeLa nucleoli was collected and applied to the protein B23 immunoaffinity column (●– – –●) or to the pre-immune control column (○ – – – ○). The antigen was eluted with 4 *M* MgCl₂ in the Tris buffer; 2-ml fractions were collected and assayed for protein B23 ELISA activities. (From Yung, B. Y.-M., Busch, H., and Chan, P. K., *Biochim. Biophys. Acta*, 826, 167, 1985. With permission.)

4. Characterization of the High Molecular Weight Form of Protein B23

The Stoke's radius and the sedimentation coefficient of the native protein B23 were determined by gel filtration (Ultrogel AcA 22) and sucrose density gradient centrifugation. Two immunoreactive peaks, corresponding to the high molecular weight form of protein B23 and the monomer of protein B23 (37 kDa), were observed.[85] The Stoke's radius of the high molecular weight form of protein B23 was estimated to be 51 Å.[85] The sedimentation coefficient of the high molecular weight protein B23, as determined by sucrose density gradient centrifugation, was about 10S.[85] The molecular weight of the high molecular weight form of protein B23 was calculated by the following equation:[72] M.W. = 6 π Nas/(1 − ūp) where M. W. = molecular weight; a = Stoke's radius; s = sedimentation coefficient; ū = partial specific volume; p = density of the medium; and N = Avogadro's number. Assuming a partial specific volume of 0.725, which is the average of most known proteins, the molecular weight of the high molecular weight form of protein B23 was calculated to be 230 kDa. Since the molecular weight of a monomer of protein B23 is 37 kDa, these results suggest that the high molecular weight form of protein B23 may be a hexamer. Previous experiments showed that the ratio of the α and β forms in the high molecular weight protein B23 is 2 to 1 (Figure 12). Accordingly, the high molecular weight form of protein B23 is composed of four α and two β monomers. Table 2 summarizes the results of the characterization of the hexameric form of protein B23.

5. Discussion

Protein B23 was previously isolated from Novikoff hepatoma cells with urea.[53] The protein may have been denatured under these conditions. In order to preserve its nativeness and understand its function, this study attempted to purify protein B23 without urea. Under ''no urea'' conditions, a high molecular weight form (M.W. = 230 kDa) of protein B23 was identified. This high molecular weight form could be detected in cellular, nuclear, and nucleolar extracts, and in pre-rRNP particles.[85]

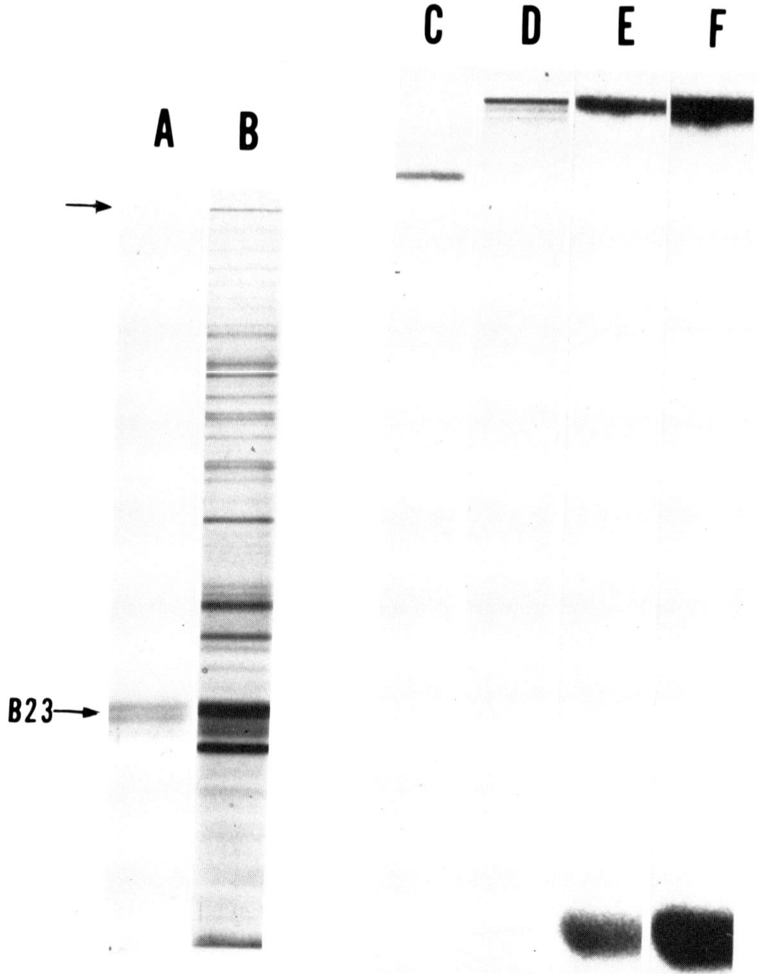

FIGURE 11.   Characterization of native protein B23 by SDS Gel electrophoresis and immunoblot analysis. (Lane A) Purified protein B23 from Novikoff hepatoma cells; (Lane B) the Tris extract of HeLa nucleoli. (Lane C) Proteins eluted from the pre-immune control column; a protein band with molecular weight of 160 kDa was observed. The identity of this band is not presently known. (Lane D) proteins eluted from the protein B23 immunoaffinity column. A dense high molecular weight band and a protein B23 band (37 kDa) were observed. (Lanes A to D) Gels were fixed and silver-stained. (Lane F) Immunoblot of the Tris extract of HeLa nucleoli. (Lane E) Immunoblot of proteins eluted from immunoaffinity column. Two dense immunobands corresponding to protein B23 and the high molecular weight form of protein B23 were observed in both Lanes E and F. (From Yung, B. Y.-M., Busch, H., and Chan, P. K., *Biochim. Biophys. Acta,* 826, 167, 1985. With permission.)

Protein B23 is associated with pre-RNPs,[65,74] and our previous studies suggest that protein B23 is bound to the spacer region of 45S rRNA.[83] It is possible that the hexameric form of protein B23 is a structural element (analogous to histones in nucleosomes) which associates with pre-RNP particles. There are approximately 6×10^6 molecules of monomeric protein B23 per HeLa cell.[10]

Recent studies indicate that protein B23 is associated with the matrix structure in the nucleoplasm. Protein B23 may be involved in the transport of large ribosomal subunits through the internal nuclear matrix network.[28]

FIGURE 12.   Dissociation of the oligomer of protein B23 into monomers by treatment with 7 *M* urea. (Lane A) Purified protein B23 from Novikoff hepatoma cells; (Lane B) proteins eluted from the affinity column without urea treatment; (Lane C) proteins eluted from the affinity column and treated with 7 *M* urea; (Lane D) the high molecular weight band of protein B23 (Lane B) was excised, electrophoretically eluted, and treated with 7 *M* urea. (From Yung, B.Y.-M., Busch, H., and Chan, P. K., *Biochim. Biophys. Acta,* 826, 167, 1985. With permission.)

E. Alternations in Cellular Localization of Protein B23 during Serum Starvation
1. Serum Starvation[10]

Systems which suppress RNA and protein synthesis have been employed to investigate the function of protein B23. The cellular localization of protein B23 in HeLa cells is affected by serum deprivation. When HeLa cells are cultured in serum-free medium for 48 h, the nucleolar fluorescence of protein B23 is diminished, and a general nuclear immunofluorescence is observed. Figure 13 shows the localization of protein B23 in HeLa cells under different conditions. Figure 13A shows the bright nucleolar fluorescence observed in the untreated cells. Figure 13B shows the decreased brightness of the nucleolar fluorescence and increased nucleoplasmic fluorescence after 48 h of incubation in a serum-free medium.

Table 2
CHARACTERIZATION OF THE
HEXAMERIC FORM OF PROTEIN
B23

Stoke's radius	51.0 Å
Sedimentation coefficient	10 S
Molecular weight	230000
Chemical composition	
Protein	97%
Nucleic acid	<3%
Subunits	4 α and 2 β
	monomers of
	protein B23
Estimated number of hex-amers per pre-rRNP	2

From Yung, B.Y.-M. and Chan, P. K., *Biochim. Biophys. Acta,* 925, 74, 1987. With permission.

When the starved cells were stained with Giemsa stain, there was no apparent change in morphology of the nucleoli by light microscopy (Figure 13G). The nucleolar immunofluorescence of another nucleolar phosphoprotein, C23, was not affected by serum deprivation (Figure 13C). Relocalization of protein B23 in nucleoli was observed 24 h after refeeding the cells with fresh medium containing 10% FCS (Figure 13D). Refeeding with fresh serum-free medium which contained glucose, amino acids, and vitamins did not stimulate the nucleolar relocalization of protein B23 (as in Figure 13B).

Refeeding the serum-starved HeLa cells with serum-supplemented medium together with 0.2 mM of cycloheximide for 24 h (less than 10% of [^{35}S]-methionine was incorporated by cells under these conditions) produced heterogeneous nucleolar fluorescence (Figure 13E). In some cells, the nucleoli were small and fluorescence was weak, while in others, nucleolar fluorescence was the same as in the controls. This result indicated that cycloheximide did not block the relocalization of protein B23 to nucleoli. However, refeeding the serum-starved cells with serum-supplemented medium together with actinomycin D (5 μg/ml or 50 ng/ml) resulted in loss of both nuclear and nucleolar fluorescence (Figure 13F). Under these conditions, the nucleoli were small and irregular as shown by the Giemsa stain (Figure 13H).

2. Quantitation

Quantitation of protein B23 in starved or fed cells was done by a competition ELISA assay. A linear competition curve was obtained between 0.12 to 1 μg of protein B23 per microtitre well.[10] The quantities in the samples were determined in the linear portion of the standard curve (between 0.4 to 0.6 OD). There were 20.8 μg of B23 per mg DNA and 11.8 μg of B23 per mg DNA in fed and starved nucleoli, respectively. Accordingly, there was a 43% reduction of protein B23/mg of DNA in nucleoli during starvation. The fed and starved nucleoplasm contained 4 μg/mg DNA and 6.9 μg/mg DNA of protein B23, respectively, which amounts to a 73% increase of protein B23 in the starved cell nucleoplasm. The total amounts of protein B23 in control (fed) and serum-starved cells were 24.8 μg/mg DNA and 18.7 μg/mg DNA, respectively. An overall decrease of 6.1 μg/mg DNA occurred during starvation. Based on the molecular weight of protein B23 (37 kDa) and 15 pg of DNA/HeLa cell, it was calculated that there are approximately 6×10^6 molecules of protein B23 per HeLa cell.

3. Discussion

This study indicates that during serum starvation, in which most cells are in a quiescent state, an increased concentration of protein B23 is found in the nucleoplasm (Figure 13).

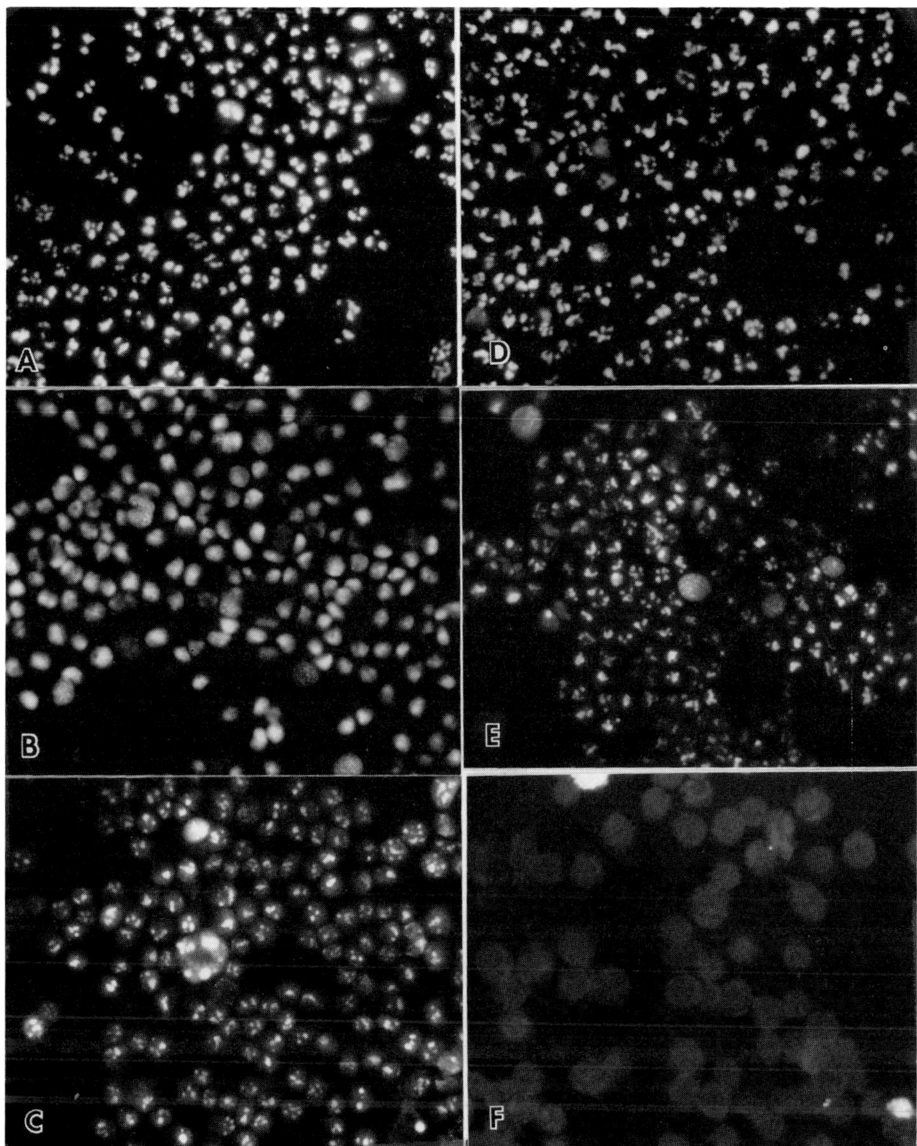

FIGURE 13.   Immunofluorescence of HeLa cells under various conditions. (A) Control HeLa cells, immunostained with anti-B23 antibody. (B,C) Starved cells, immunostained with (B) anti-B23; (C) anti-C23 antibody. (D to F) Starved cells were refed with serum (D) for 24 h, (E) and cycloheximide; (F) and actinomycin D; (D to F) immunostained with anti-B23 antibody. (G, H) Giemsa stain of (G) starved cells; (H) actinomycin D-treated cells; × 400. (From Chan, P. K., Aldrich, M., and Busch, H., *Exp. Cell Res.*, 161, 101, 1985. With permission.)

When cell growth is stimulated, protein B23 relocalizes in the nucleoli. This process appears to depend on specific serum factors, because replenishment with fresh serum-free medium (which contains glucose, vitamins, and amino acids) had no effect.

To study whether inhibition of RNA or protein synthesis affected the localization of protein B23 in nucleoli, starved HeLa cells were refed with medium containing serum and cycloheximide or actinomycin D. The results suggested that the re-entry process did not require newly synthesized protein, but that new synthesis of RNA is essential. Willems et

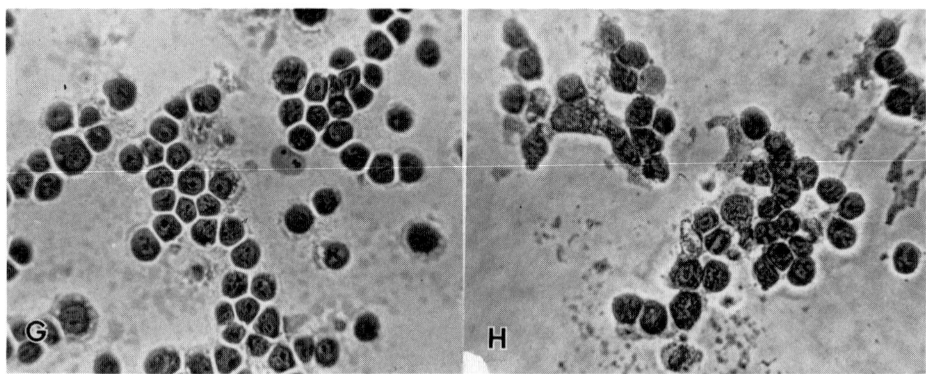

FIGURES   13G and H.

al.[80] pointed out that the efficiency of processing of 45S to 32S rRNA in cycloheximide-treated HeLa cells remained unchanged; the cells continued to produce small quantities of the large ribosomal subunit. The fact that there was no inhibition of the localization of protein B23 in nucleoli during cycloheximide treatment suggests that protein B23 may play a role in rRNA processing.

When nucleolar function is activated, essential enzymes, proteins, and cofactors are concentrated into nucleoli for ribosome synthesis. Under conditions unfavorable for cell growth, proteins such as protein B23 may lose their binding sites in nucleoli and dissipate into the nucleoplasm.

F. Effects of Antibiotics on B23 Translocation

1. Inhibition of rRNA Synthesis by Actinomycin D[82]

To further study whether "B23 translocation" is related to decreased synthesis of protein or RNA, various inhibitors have been employed. The first drug used was actinomycin D, which inhibits RNA synthesis. In control HeLa cells, bright nucleolar fluorescence, but little or no nuclear fluorescence, was observed (Figure 14a). This fluorescence pattern is defined as A. After treatment with actinomycin D (250 ng/ml) for 2 h, a uniform nucleoplasmic fluorescence was observed (Figure 14b). This fluorescence pattern is defined as C. When HeLa cells were treated with reduced amounts of actinomycin D (10 ng/ml) for 2 h, both nucleoplasmic and nucleolar fluorescence were observed (Figure 14b). This fluorescence pattern is defined as B. Table 3 summarizes the effects of actinomycin D and its analogs on B23 translocation. "B23 translocation" correlates with the inhibition of RNA synthesis and the potency of the drugs.

2. Effects of Other Antibiotics on B23 Translocation[82-84]

The inhibition of RNA polymerase activity has also been investigated using different doses of α-amanitin. A low dose of α-amanitin (5 μg/ml), which inhibits RNA polymerase II activity, has no effect on B23 nucleolar localization. A high dose of α-amanitin (200 μg/ml), which not only inhibits RNA polymerase III activity but also inhibits 5S RNA synthesis, produces a nuclear fluorescence.[83] This result is interesting because 5S RNA is one of the components of the 80S RNP. Treatment with toyocamycin, an adenosine analog which alters RNA conformation and blocks the processing of 45S RNA to 32S RNA[3] also results in nucleoplasmic fluorescence. These results are summarized in Table 4. Cycloheximide and puromycin, which are both inhibitors of protein synthesis that decrease RNA synthesis but have no direct effect on rRNA processing, do not alter the nucleolar localization of protein B23. These results indicate that interruption of rRNP formation or processing results in migration of protein B23 to the nucleoplasm.

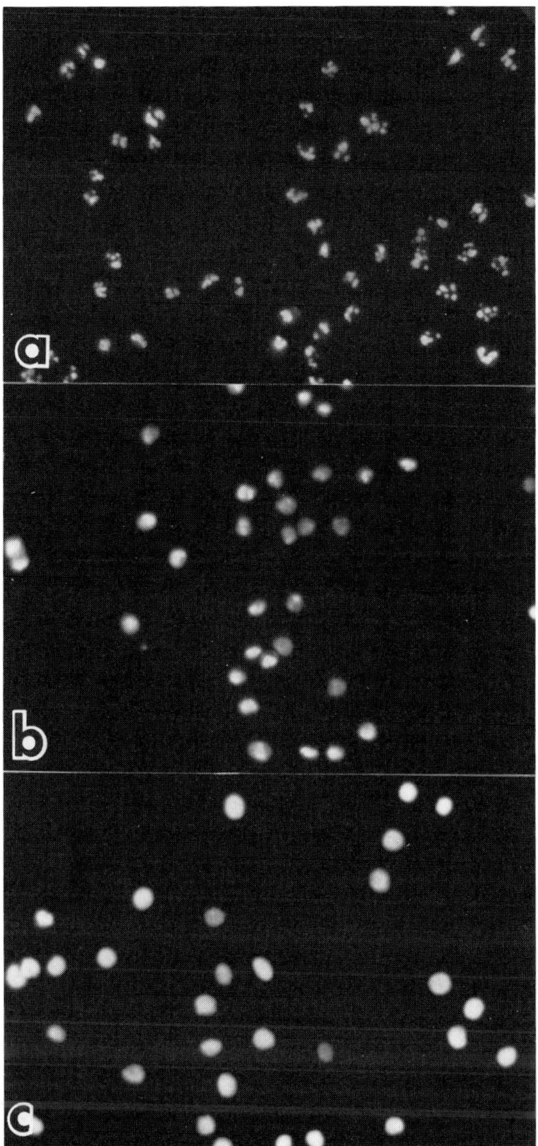

FIGURE 14.   Effects of actinomycin D on the cellular localization
of protein B23. (a) Control HeLa cells: only nucleolar fluorescence
was observed; (b) after treatment with actinomycin D (10 ng/ml,
2 h), both nuclear and nucleolar fluorescence were observed; and
(c) maximum effect after treatment with actinomycin D (250 ng/
ml, 2 h) in which only nuclear fluorescence was observed. (From
Yung, B.Y.-M., Busch, H., and Chan, F. K., *Cancer Res.*, 46,
922, 1986. With permission.)

G. Correlation of Antitumor Activity and B23 Translocation

To investigate whether B23 translocation correlates with the potency of drugs, four lu-
zopeptin analogs (BBM 928 A, B, C, and D)[37,38] with different antitumor activities[60] were
tested using this system. The results (Table 5) indicate that protein B23 translocation correlates very well with the antitumor activity of the luzopeptins.[84] To study whether B23
translocation can be used for detection of drug-resistant cells, doxorubicin-resistant and

Table 3
TIME COURSE OF THE EFFECT OF ACTINOMYCIN D AND ITS ANALOGS ON THE RNA SYNTHESIS AND LOCATION OF PROTEIN B23

Drug	Time (h)	Inhibition of RNA synthesis[a] (%)	Immuno-fluorescence[b]
Actinomycin D	0	0	A
(10 ng/ml)	1	47.5 ± 5.2[c]	B
	2	55.0 ± 4.3[d]	B
	4	88.8 ± 5.6[e]	C
Actinomycin Z5	0	0	A
(40 ng/ml)	1	46.7 ± 3.9[c]	B
	2	54.5 ± 3.8[d]	B
	4	77.1 ± 4.9[e]	C
Actinomycin K2T	0	0	A
(1400 ng/ml)	1	46.8 ± 4.7[c]	B
	2	50.7 ± 6.1[d]	B
	4	75.5 ± 3.0[e]	C

Note: HeLa cells were cultured on slides. An equivalent dose (IC_{50}) of actinomycin D or its analogs was added to the culture medium for 1, 2, and 4 h. The cellular incorporation of [3H]uridine and the localization of protein B23 were then determined. Viability of the cells was over 95% under these conditions.

[a] Percent of RNA synthesis inhibition was calculated as

$$100 \times \left[1 - \frac{[^3H]\text{uridine uptake in treated HeLa cells}}{[^3H]\text{uridine uptake in control HeLa cells}} \right]$$

The values represent the mean ± S.D. of three experiments.

[b] A = over 90% of the cells had bright nucleolar fluorescence with little or no nuclear fluorescence (Figure 14a). About 1200 cells were counted. B = over 90% of the cells had both nuclear and nucleolar flourescence (Figure 14b). About 1200 cells were counted. This is an intermediate phenomenon between A and C. C = over 95% of the cells had homogeneous nuclear fluorescence but no distinct nucleolar fluorescence was observed (Figure 14c). About 1200 cells were counted.

[c] There were about 1.0×10^6 cpm [3H]uridine incorporated/1.0 mg protein in control HeLa cells.

[d] There were about 2.0×10^6 cpm [3H]uridine incorporated/1.0 mg protein in control HeLa cells.

[e] There were about 4.0×10^6 cpm [3H]uridine incorporated/1.0 mg protein in control HeLa cells.

From Yung, B.Y.-M., Busch, R. K., Busch, H., Mauger, A. B., and Chan, P. K., *Biochem. Pharmacol.*, 34, 4059, 1985. With permission.

doxorubicin-sensitive mouse leukemia cells (P388) have been cultured in medium containing various doses of doxorubicin for 4 h, and the responsive levels of the cells to doxorubicin have been compared.[13] Table 6 shows the results of these studies, which demonstrate that identification of "B23 translocation" can be used to detect drug-resistant cells.

V. CONCLUSIONS AND HYPOTHESES

Our studies indicate that protein B23 binds to certain elements in the nucleus (rRNA, RNP, rDNA, and/or matrix proteins), and plays an essential role in ribosome assembly and transport.[56] As pointed out by Perry[62] and Warner,[78] the packaging of pre-RNP[71] is not a random process. The sequential cleavages of the 45S RNA in the RNP may relate to a sequential removal of steric constraints with consequent exposure of new cleavage or binding sites.[62] It is conceivable that temporary binding of protein B23 to the preribosomal particles in the nucleolus stabilizes the RNA so that the RNP can be processed, ribosomal proteins can be incorporated, or the matured ribosome subunit can be transported out of the nucleus.

Table 4
THE EFFECT OF ACTINOMYCIN D, TOYOCAMYCIN, PUROMYCIN, CYCLOHEXIMIDE, AND α-AMANITIN ON THE CELLULAR LOCALIZATION OF PROTEIN B23

| | Treatment time (h) | | | |
Drug	0	2	4	24
Actinomycin D (10 ng/ml)	A	B	C	
Toyocamycin (100 ng/ml)	A	A	B	
Toyocamycin (1 μg/ml)	A	B	C	
Puromycin (2 μg/ml)	A	A	A	A*
Cycloheximide (56 μg/ml)	A	A	A	A*
α-Amanitin (10 μg/ml)	A	A	A	A*
α-Amanitin (200 μg/ml)	A		C	

Note: HeLa cells were cultured on slides in minimum essential medium (Eagle's). Actinomycin D, toyocamycin, puromycin, cycloheximide, or α-amanitin was added for 2, 4, or 24 h before the cells were fixed and immunostained by protein B23 antibody. Viability of the cells was over 95% under these conditions except those with asterick (*) where it was not determined. A, over 90% of the cells showed bright nucleolar fluorescence with little or no nuclear fluorescence (Figure 14a). About 1200 cells were counted. B, over 90% of the cells showed both nuclear and nucleolar fluorescence (Figure 14B). About 1200 cells were counted. This is an intermediate phenomenon between A and C. C, over 95% of the cells showed homogeneous nuclear fluorescence. No distinct nucleolar fluorescence was observed (Figure 14c). About 1200 cells were counted.

Table 5
EFFECT OF LUZOPEPTINS ON THE CELLULAR LOCALIZATION OF PROTEIN B23

| | Immunofluorescence at following luzopeptin doses and times | | | | | | |
| | | 10 ng/ml | | 50 ng/ml | | 500 ng/ml | |
	0	2h	4h	2h	4h	2h	4h
Luzopeptin A	A	B	C	C	C	C	C
Luzopeptin B	A	A	A	B	B	C	C
Luzopeptin C	A	A	A	A	A	A	A
Luzopeptin D	A	C	C	C	C	C	C

When cells are treated with antibiotics which intercalate or alkylate the nucleolar DNA or RNA, protein B23 loses its binding site in the nucleolus and diffuses into the nucleoplasm (translocation). The phosphate group of protein B23 may enhance the binding of the protein to pre-RNPs in the nucleolus. It is feasible that interaction of the negatively charged phosphorylation site with positively charged molecules such as polyamines[43] or metal ions (Mg^{2+}, Ca^{2+}) could alter the conformation of protein B23 and modify its binding to RNP. These changes in the conformation of protein B23, therefore, could depend on the concentration of polyamines or other cations in the nucleolus.

Table 6
COMPARISON OF B23 TRANSLOCATION IN
DOXORUBICIN-SENSITIVE AND -RESISTANT
MOUSE LEUKEMIA (P388) CELLS

Drug	Dose (µg/ml)	Cells in each translocation pattern (%)[a]					
		Sensitive cells			Resistant cells		
		A	B	C	A	B	C
Doxorubicin	0.0	100	0	0	100	0	0
	1.0	100	0	0	93	7	0
	5.0	99	1	0	94	7	0
	10.0	13	87	0	87	13	0
	20.0	1	73	26	78	21	0
	50.0	2	12	86	75	16	9
	100.0	0	0	100	17	57	26
Actinomycin D	0.0	100	0	0	100	0	0
	0.01	1	74	25	98	2	0
	0.1	0	31	69	0	87	13
	1.0	0	3	97	0	6	94
	10.0	0	0	100	0	1	99

Note: Mouse leukemia cells (P388) were grown in suspension in medium containing various doses of doxorubicin or actinomycin D for 4 h. Cells were cytocentrifuged onto slides, fixed, and immunostained with B23 antibody.

[a] About 500 cells were counted. Classification of immunofluorescence is defined in the Table 3.

ACKNOWLEDGMENTS

These studies were supported by PHS grant CA 42476 and CA 10893-p9 awarded by the National Cancer Institute, DHHS. We thank Dr. H. Busch for his advice and support throughout these studies.

REFERENCES

1. **Ahmed, K., Davis, A. T., and Goueli, S. A.,** *Biochem. J.,* 209, 197, 1983.
2. **Ahmed, K. and Wilson, M. J.,** in *The Cell Nucleus,* Vol. 6., Busch, H., Ed., 1978, 409.
3. **Auger-Buendia, M.-A., Hamelin, R., and Tavitian, A.,** *Biochim. Biophys. Acta,* 521, 241, 1978.
4. **Benton, W. D. and Davis, R. W.,** *Science,* 196, 180, 1977.
5. **Busch, H. and Smetana, K.,** in *The Nucleolus,* Busch, H. and Smetana, K., Eds., Academic Press, New York, 221.
6. **Busch, H., Lischwe, M. A., Michalik, J., Chan, P. K., and Busch, R. K.,** in *The Nucleolus,* Jordan, E. G. and Cullis, C. A., Eds., Cambridge University Press, 1982, 43.
7. **Burton, Z., Burgess, R. R., Lin, J., Moorer, D., Holder, S., and Gross, C. A.,** *Nucleic Acids Res.,* 9, 2889, 1981.
8. **Busch, R. K., Chan, P. K., and Busch, H.,** *Life Sci.,* 35, 1777, 1984.
9. **Carmichael, D. F., Geahlen, R. L., Allen, S. M., and Krebs, E. G.,** *J. Biol. Chem.,* 257, 10440, 1982.
10. **Chan, P. K., Aldrich, M., and Busch, H.,** *Exp. Cell Res.,* 161, 101, 1985.
11. **Chan, P. K., Aldrich, M., Cook, R. G., and Busch, H.,** *J. Biol. Chem.,* 261, 1868, 1986.
12. **Chan, P. K., Aldrich, M., Ochs, R., and Busch, H.,** *J. Cell Biol.,* 97, 72a, 1983.
13. **Chan, P. K., Aldrich, M., and Yung, B. Y-M.,** *Cancer Res.,* 47, 3798, 1987.
14. **Chan, P. K., Chan, W-Y., Yung, B.Y-M., Cook, R. G., Aldrich, M. B., Ku, D., Goldknopf, I. L., and Busch, H.,** *J. Biol. Chem.,* 261, 14335, 1986.

15. **Chan, P. K., Feyerabend, A., Busch, R. K., and Busch, H.**, *Cancer Res.*, 40, 3194, 1980.
16. **Chan, P. K., Frakes, R., Tan, E. M., Brattain, M. G., Smetana, K., and Busch, H.**, *Cancer Res.*, 43, 3770, 1983.
17. **Chan, P. K. and Liew, C. C.**, *Biochem. J.*, 183, 147, 1979.
18. **Chan, P. K. and Liew, C. C.**, *Can. J. Biochem.*, 57, 666, 1979.
19. **Chan, P. K., Liu, Q. R., Yung, B.Y-M., Durban, E., and Aldrich, M.**, *J. Cell Biol.*, 103, 45a, 1986.
20. **Cohen, P., Yellowlees, D., Aitken, A., Donella-Deana, A., Hemmings, B. A., and Parker, P. J.**, *Eur. J. Biochem.*, 124, 21, 1982.
21. **Constantinous, A. I., Squinto, S. P., and Jungmann, R. A.**, *Cell* 42, 429, 1985.
22. **Craigen, W. J., Cook, R. G., Tate, W. P., and Caskey, C. T.**, *Proc. Natl. Acad. Sci.*, 82, 3616, 1985.
23. **Dasgupta, J. D. and Garbers, D. L.**, *J. Biol. Chem.*, 258, 6174, 1983.
24. **Degen, S. J. F., MacGillivray, R. T. A., and Davie, E. W.**, *Biochemistry*, 22, 2087, 1983.
25. **Dills, W. L., Goodwin, C. D., Lincoln, T. M., Beavo, J. A., Bechtel, P. J., Corbin, J. D., and Krebs, E. G.**, *Adv. Cyclic Nucl. Res.*, 10, 199, 1979.
26. **Edman, P. and Begg, G.**, *Eur. J. Biochem.*, 1, 80, 1967.
27. **Erikson, R. L., Purchio, A. F., Erikson, E., Collett, M. S., and Brugge, J. S.**, *J. Cell Biol.*, 87, 319, 1980.
28. **Fields, A. P., Kaufman, S. H., and Shaper, J. H.**, *Exp. Cell Res.*, 164, 139, 1986.
29. **Foster, D. and Davis, E. W.**, *Proc. Natl. Acad. Sci. U.S.A.*, 81, 4766, 1984.
30. **Halikowski, M. J. and Liew, C. C.**, *Biochem. J.*, 241, 693, 1987.
31. **Hathaway, G. M. and Traugh, J. A.**, *Curr. Top. Cell. Regul.*, 21, 101, 1982.
32. **Hemmings, B. A., Aitken, A., Cohen, P., Raymond, M., and Hofmann, F.**, *Eur. J. Biochem.*, 127, 473, 1982.
33. **Hentzen, P. C. and Bekhor, I.**, in *Progress in Nonhistone Protein Research*, Vol. 1, Bekhor, I., Ed., CRC Press, Boca Raton, FL, 1985, 75.
34. **Hopp, T. P. and Woods, K. R.**, *Proc. Natl. Acad. Sci. U.S.A.*, 78, 3824, 1981.
35. **Hoppe, J.**, *Trends Biochem. Sci.*, 1, 29, 1985.
36. **Houghten, R. A.**, *Proc. Natl. Acad. Sci.*, 82, 5131, 1985.
37. **Huang, C. H., and Crooke, S. T.**, *Cancer Res.*, 45, 3768, 1985.
38. **Huang, C. H., Mirabelli, C. K., Mong, S., and Crooke, S. T.**, *Cancer Res.*, 43, 2718, 1983.
39. **Inoue, A., Tei, Y., Qi, S-L., Higashi, Y., Yukioka, M., and Morisawa, S.**, *Biochem. Biophys. Res. Commun.*, 123, 398, 1984.
40. **Jungmann, R. A. and Schweppe, J. S.**, *J. Biol. Chem.*, 247, 5535, 1972.
41. **Kleinsmith, L. J.**, *J. Cell Physiol.*, 85, 459, 1975.
42. **Krebs, E. G. and Beavo, J. A.**, *Annu. Rev. Biochem.*, 48, 923, 1979.
43. **Kuehn, G. D., Affolter, H.-U., Atmar, V. J., Seebeck, T., Gubler, U., and Braun, R.**, *Proc. Natl. Acad. Sci.*, 76, 2541, 1979.
44. **Laemmli, U. K.**, *Nature (London)*, 227, 680, 1970.
45. **Larson, R. and Messing, J.**, *Nucleic Acids Res.*, 10, 39, 1982.
46. **Liew, C.C. and Chan, P. K.**, *Proc. Natl. Acad. Sci.*, 73, 3458, 1976.
47. **Liew, C. C., Hentzen, P. C., and Bekhor, I.**, *Can. J. Biochem. Cell Biol.*, 63, 824, 1985.
48. **Mamrack, M. D., Olson, M. O. J., and Busch, H.**, *Biochemistry*, 15, 3381, 1979.
49. **Maniatis, T., Fritsch, E. F., and Sambrook, J.**, in *Molecular Cloning: A Laboratory Manual*, Cold Spring Harbor Laboratory, Cold Spring Harbor, N.Y., 1982, 371.
50. **Mathews, H. R. and Huebner, V. D.**, *Mol. Cell. Biochem.*, 59, 81, 1984.
51. **Meggio, F., Deana, A. D., and Pinna, L. A.**, *Biochim. Biophys. Acta*, 662, 1, 1981.
52. **Messing, J., Crea, R., and Seeburg, P.**, *Nucleic Acids Res.*, 9, 309, 1981.
53. **Michalik, J., Yeoman, L. C., and Busch, H.**, *Life Sci.*, 20, 1371, 1981.
54. **Miller, M. R., Ulrich, R. G., Wang, T. S., and Korn, D.**, *J. Biol. Chem.*, 260, 134, 1985.
55. **Murray, S. A. and Fletcher, W. H.**, *J. Cell Biol.*, 98, 1710, 1984.
56. **Nomura, M., Gourse, R., and Baughman, G.**, *Annu. Rev. Biochem.*, 53, 75, 1984.
57. **Norman, G. L. and Bekhor, I.**, *Biochemistry*, 20, 3568, 1981.
58. **Norrander, J., Kempe, T., and Messing, J.**, *Gene*, 26, 101, 1983.
59. **Ochs, R., Lischwe, M., O'Leary, P., and Busch, H.**, *Exp. Cell Res.*, 146, 139, 1983.
60. **Okhuma, H., Sakai, F., Nishiyama, Y., Ohbayashi, M., Imanishi, H., Konishi, M., Miyaki, T., Koshiyama, H., and Kawaguchi, H. J.**, *Antibiot. (Tokyo)*, 33, 1087, 1980.
61. **Olson, M. O. J., Orrick, L. R., Jones, C., and Busch, H.**, *J. Biol. Chem.*, 249, 283, 1974.
62. **Perry, R. P.**, *Annu. Rev. Biochem.*, 45, 605, 1976.
63. **Phillips, I. R., Shephard, E. A., Stein, J. L., Kleinsmith, L. J., and Stein, G. S.**, *Biochim. Biophys. Acta*, 565, 3267, 1979.
64. **Prentice, D. A., Taylor, S. E., Newmark, M. Z., and Kitos, P. A.**, *Biochem. Biophys. Res. Commun.*, 85, 541, 1978.

65. **Prestayko, A. W., Klomp, G. R., Schmoll, D. J., and Busch, H.,** *Biochemistry,* 13, 1945, 1974.
66. **Putney, S. D., Benkovic, S. J., and Schimmel, P. R.,** *Proc. Natl. Acad. Sci.,* 78, 7350, 1981.
67. **Saffer, J. D. and Coleman, J. E.,** *Biochemistry,* 19, 5874, 1980.
68. **Sanger, F., Nicklen, S., and Coulson, A. R.,** *Proc. Natl. Acad. Sci. U.S.A.,* 74, 5463, 1977.
69. **Schmidt-Zachmann, M. S., Hugle-Dorr, B., and Franke, W. W.,** *EMBO. J.,* 6, 1881, 1987.
70. **Sheorain, V. S., Ramakrishna, S., Benjamin, W. B., and Soderling, T. R.,** *J. Biol. Chem.,* 260, 12287, 1985.
71. **Shepherd, J. and Maden, B. E. H.,** *Nature,* 236, 211, 1972.
72. **Siegel, L. M. and Monty, K. J.,** *Biochim. Biophys. Acta,* 112, 346, 1966.
73. **Smith, C. J., Wejksnora, P. J., Warner, J. R., Rubin, C. S., and Rosen, O. M.,** *Proc. Natl. Acad. Sci.,* 76, 2725, 1979.
74. **Spector, D. L., Ochs, R. L., and Busch, H.,** *Chromosoma,* 90, 139, 1984.
75. **Takio, K., Smith, S. B., Krebs, E. G., Walsh, K. A., and Titani, K.,** *PNAS,* 79, 2544, 1982.
76. **Teng, C. S., Teng, C. T., and Allfrey, V. J.,** *J. Biol. Chem.,* 246, 3597, 1971.
77. **Walton, G. M., Spiess, J., and Gill, G. N.,** *J. Biol. Chem.,* 260, 4745, 1985.
78. **Warner, J. R.,** *J. Cell Biol.,* 80, 767, 1979.
79. **Weinert, T. A., Schaus, N. A., and Grindley, N. D. F.,** *Science,* 222, 755, 1983.
80. **Willems, M., Penman, M., and Penman, S.,** *J. Cell Biol.,* 41, 177, 1969.
81. **Young, R. A. and Davis, R. W.,** *Proc. Natl. Acad. Sci. U.S.A.,* 80, 1194, 1983.
82. **Yung, B.Y-M., Busch, R. K., Busch, H., Mauger, A. B., and Chan, P. K.,** *Biochem. Pharmacol.,* 34, 4059, 1985.
83. **Yung, B.Y-M., Busch, H. and Chan, P. K.,** *Biochim. Biophys. Acta,* 826, 167, 1985.
84. **Yung, B.Y-M., Busch, H., and Chan, P. K.,** *Cancer Res.,* 46, 922, 1986.
85. **Yung, B.Y-M. and Chan, P. K.,** *Biochim. Biophys. Acta,* 925, 74, 1987.
86. **Zweig, S. E.,** *J. Biol. Chem.,* 256, 11847, 1981.

Chapter 3

DNA-BOUND NONHISTONE CHROMOSOMAL PROTEINS AND LOOP ORGANIZATION OF CHROMATIN

R. Tsanev, Z. Avramova, and B. Tasheva

TABLE OF CONTENTS

I. CLASSIFICATION OF NONHISTONE CHROMOSOMAL PROTEINS (NHCPs)

Nonhistone chromosomal proteins — strictly speaking nonprotamine proteins too — are an extremely large group whose important role in the functions and organization of chromatin, although unanimously recognized, is far from being elucidated. When dealing with such a numerous population of different molecules, it would be useful to select distinct groups on the basis of some general characteristics. Such a selection requires a rational basis of classification. In our studies, we have classified NHCPs into several groups on the basis of three different criteria: (1) their possible role; (2) their metabolic behavior; and (3) the stability of their association with chromatin. Using these criteria we can consider the following groups:

1. According to their role:
    1.1 *Chromosomal enzymes* involved in the functional activities of chromatin (replication, transcription, structural reorganizations, and rearrangements)
    1.2. *Regulatory proteins* involved in the control of chromatin functions
    1.3. *Structural proteins* which may fix some specific structures, as for example chromatin loops, important for the functioning of DNA
2. According to their metabolic behavior:
    2.1. *Metabolically labile* proteins which turn over during the cell cycle with different half-lives
    2.2. *Metabolically stable* proteins which are conserved and transmitted to the progeny
3. According to the stability of their association with chromatin, NHCPs are often divided into "easily extractable" and "tightly bound", but the criteria used for such a distinction have been very different. We have adopted the following criteria which seem to us more rational (under conditions reducing disulfide bonds):
    3.1. *Easily extractable proteins* which are held in chromatin by ionic interactions or hydrogen bonding and can be extracted with solutions of high salt and urea
    3.2. *Tightly bound proteins* which are held by strong hydrophobic interactions and can be extracted only by using detergents like SDS, sarkosyl, some times hot phenol, or digestion of DNA with nucleases
    3.3. *Covalently DNA-bound proteins* which cannot be dissociated from DNA by the above procedures.

We have started a study on the NHCPs with the speculative idea that the organization of DNA into permanently repressed (blocked) and potentially active (deblocked) regions could be determined by some inheritable DNA-protein structures.[1,2] This hypothesis prompted us to look for NHCPs which should have a structural role (group 1.3), would be most probably metabolically stable (group 2.2), and presumably tightly bound to DNA (group 3.2 and/or 3.3).

With this idea our interest was focused on the structures and mechanisms which maintain eukaryotic DNA in the form of loops equivalent to circular DNA molecules.[3-6] Some data indicate that these loops are supercoiled and the torsional constraint imposed on them might be an important factor in the control of transcription.[2,5,6] Thus, the chromatin loops seem to represent separately controlled domains,[3,7] and it would be of great interest to reveal the structure and the protein composition of their anchorage sites in the nucleus. This problem is being studied by several groups[8,9] and it appears that both functional and permanent DNA binding sites exist. In our studies we were interested in the permanent sites which determine a constitutive loop organization of DNA.

II. LOOP ORGANIZATION OF EUKARYOTIC CHROMATIN

The loop organization of eukaryotic DNA has been observed by electron microscopy in metaphase chromosomes[3] and in interphase nuclei[4] after high salt treatment. Arguments have been presented that such a type of DNA organization might be an artifact of high salt-induced protein aggregation and trapping at random DNA fibers.[10]

It is very probable that such artifacts really occur. An artifactual association of actively transcribed DNA sequences with the so-called nuclear matrix was found after high salt treatment of nuclei.[11] However, this does not exclude the possibility that real DNA anchorage sites do exist *in vivo*. Even under low and physiological ionic conditions, nuclei spread according to the Miller's procedure[12] show well-formed loops organized in nucleosomes and attached to a nuclear skeleton.[13-15] The same holds true for mitotic chromosomes — during both prophase and metaphase the loop organization of the chromatin fibers could be visualized under low ionic conditions.[15]

Since in all different cellular types and different eukaryotic species studied, DNA was always found organized in loops, the latter seems to be a general feature of eukaryotic cells. The finding of similar proteins[16] and DNA sequences[17] at the DNA attachment sites in mitotic and in interphase chromatin supports the electron microscopic observations that this organization is maintained during the cell cycle.

It was of interest to see whether the loop organization of DNA is also preserved in the mature spermatozoa where DNA was considered to be packed only by interaction with sperm-specific basic proteins. We have studied by electron microscopy the mature spermatozoa of three different species — a mammalian (ram),[18] a fish (trout),[19] and a mollusk (mussel).[20] The compact sperm nuclei could be decondensed and spread for electron microscopy only after extraction of the sperm-specific basic proteins. Electron microscopy has revealed that, in all three species, DNA of the sperm is attached to a nuclear skeleton consisting of nonhistone proteins resisting the salt-urea extraction procedure.

The absence of RNP network in the mature sperm presented the great advantage of permitting the observation of the structures which hold the DNA fibers. These structures appear with similar characteristics in the three evolutionary distant species studied: (1) they consist of small granules (about 15-nm large) which associate to form ring-shaped bodies with variable size between 30 and 100 nm (Figure 1); (2) several DNA fibers are radially attached to these granular bodies; (3) they are resistant to high salt and urea even in the presence of 2-mercaptoethanol and are partially resistant to some detergents; (4) in response to increased salt concentrations, these bodies show a tendency to aggregate into dense masses from which DNA fibers emerge; (5) they contain nonhistone proteins of two groups — one which could be released by detergent treatment and a second group which was resistant to this treatment and contained the covalently linked proteins.

These data strongly indicate that some specific small granules represent the DNA anchorage structures of the sperm. The stability of these granules, their ability to form ring-shaped bodies of different size, and to aggregate into larger complexes give reason to speculate that these granules, being the attachment sites of DNA, are also the basis for the formation of different DNA-containing structures during the cell cycle. It seems most probable that these granules are identical to those described by another group of authors who suggested also that they have a basic and universal role in the formation of DNA-related structures.[21]

All these data show that the loop organization of DNA is preserved not only during the cell cycle but also during spermiogenesis. Although it is not clear whether the same or different attachment sites are used in somatic and in sperm nuclei, the persistance of this type of DNA organization in both cellular types and in evolutionary-so-distant species suggests that it is critical for the functioning of the genetic apparatus.

This strengthens the need of elucidating the protein composition and the organization of the structures maintaining this third level of DNA folding.

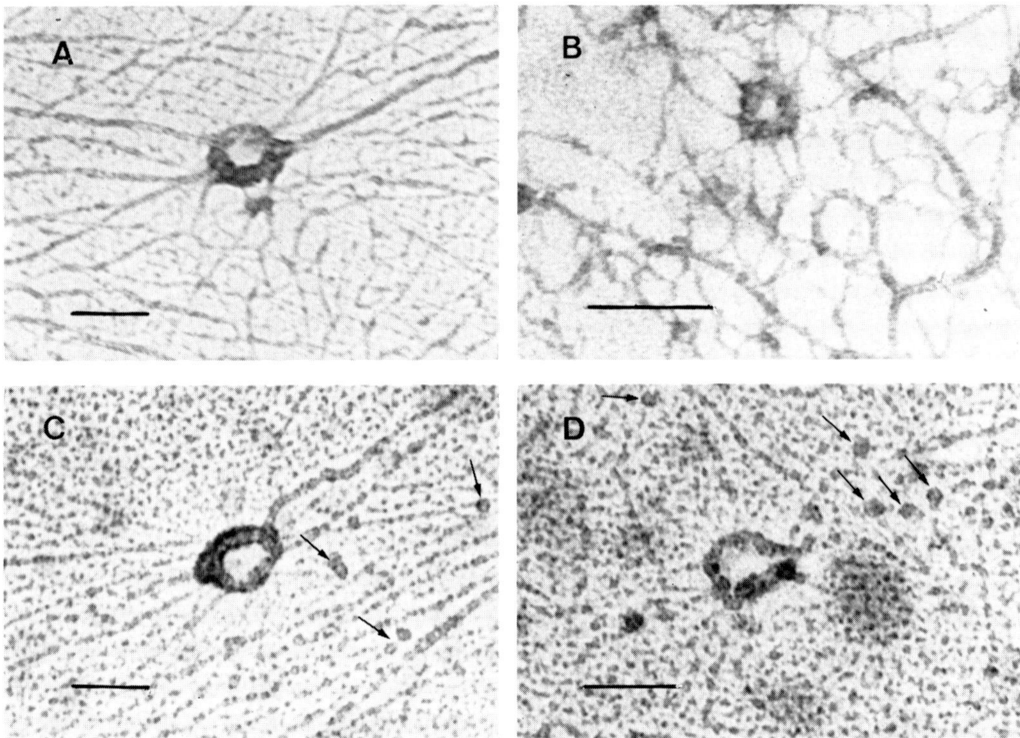

FIGURE 1.   Electron micrographs of the granular ring-shaped bodies holding DNA fibers in the spread high-salt urea extracted sperm nucleus of mussel (A), trout (B), and ram (C and D). Arrows show granules dissociated from the rings. Bars are 100 nm.

III. NONHISTONE PROTEINS OF THE SPERM NUCLEUS

Spermiogenesis is accompanied by drastic changes in the proteins associated with DNA. The basic proteins of the sperm nucleus become species-specific to such a high extent that there are no two species, even related, with identical sperm-specific proteins.[22,23] In spite of the diversity of basic sperm nuclear proteins, there are some general features characterizing the nuclear changes during spermiogenesis which are common to all species: (1) RNA synthesis being switched off, there is no RNP network in the sperm nucleus; (2) the somatic histones are partially or almost totally replaced by other basic sperm-specific proteins (sperm histones, different protamines, etc.) and (3) a small amount of nonhistone and nonprotamine proteins resisting high-salt urea extraction is present. Thus, there are two classes of proteins only in the sperm nucleus — the sperm-specific basic proteins representing the bulk of proteins and a minor fraction of tightly bound nonhistones.

The role of the first group evidently is to tightly pack DNA,[23] thus ensuring a safe transfer of genetic information. The great diversity of this class of proteins in different species is in striking contrast to the highly conserved structures of the somatic histones. This could be interpreted[22] as a reflection of the great mutational freedom of DNA-associated proteins when the latter are involved in one function only — the packaging of functionally inactive DNA.

The second group of sperm proteins which are tightly bound to DNA and resist salt-urea extraction has been detected in ram,[24] mouse,[25,26] trout,[27] and mussel.[20] Thus, these proteins seem to have a general occurrence but have not yet been well studied. In the three species that we have investigated (a mollusk, a fish, and a mammal) they amount to 4 to 5 μg/100 μg DNA.

We have found that these proteins, which are the main component of the skeleton of the sperm nucleus, can be extracted either with SDS or after digestion with DNase I.[24] We have preferred the second procedure in order to work under conditions similar to those used for the isolation of somatic nuclear matrices. Our data have shown that both kinds of treatment release the bulk of the nonhistone proteins of the sperm nuclear skeleton, but a minor protein fraction resists even these treatments and remains associated with DNA. Thus, the nuclear skeleton of the sperm contains two types of DNA-associated proteins: a major fraction which is tightly bound but can be released by detergents or after digestion of DNA, and a minor fraction which is so firmly bound to DNA that it resists all extraction procedures.

A. Extractable Tightly Bound Proteins of the Sperm Nuclear Skeleton

We have studied this protein fraction in the mature sperm nuclei of two species — the ram[28] and the mussel.[29]

Ram sperm nuclei isolated as described[24] were extracted with 2 M NaCl, 5 M urea, 0.2 M 2-mercaptoethanol for 2 h at 4°C to eliminate all protamines. The remaining proteins were precipitated together with DNA by low speed centrifugation (3800 $\times$ g for 15 min) through 1 M sucrose. To release the proteins, DNA was digested with DNase I (Sigma), 1 mg/ml, 6 h at 37°C in 10 mM Tris-HCl, pH 7.6, 5 mM Mg^{2+} buffer.

Sperm nuclei obtained from mature gonads of the mussel *Mytilus galloprovincialis*[20] were treated with 2 M NaCl to extract protamines and histones. The tightly bound proteins were pelleted together with DNA by centrifugation through 2 M sucrose containing 2 M NaCl and recovered after digestion of DNA as above.

The proteins thus obtained from the two species were dispersed in sample buffer, boiled for 3 min, saturated with urea, and fractionated electrophoretically in 10% polyacrylamide gel.[30] Selected protein fractions were cut off from the gel, radiolabeled with ^{125}I, and subjected to two-dimensional tryptic peptide mapping.[31]

The electrophoretic fractionation of the tightly bound sperm proteins from the two species is shown in Figure 2. The presence of many fractions would suggest a great heterogeneity of this protein class. However, the peptide maps of individual fractions have revealed that most of the proteins, although displaying different mobilities, exhibit similar peptide maps (Figures 3 and 4). The general tendency is seen that with increasing mobility some peptide spots disappear, indicating that some low-molecular weight fractions might be the result of proteolytic splitting.

Although the tightly bound sperm proteins of these two evolutionary-quite-distant species are different as judged from their peptide maps, a common principle emerges in both cases — a class of homologous nuclear matrix proteins displaying a size heterogeneity. The homology of the proteins within each species is supported also by the identical peptide fractions obtained by their hydrolysis with the V$_8$ protease of *Staphylococcus aureus* (not shown).

The heterogeneity in molecular masses of a set of homologous proteins may be due to one or several of the following reasons: (1) homologous products of a multigenic family; (2) differential processing of a single gene; (3) regular proteolytic splitting of a precursor protein; (4) binding of carbohydrate residues differing in nature and length. A test for carbohydrates[32] was found positive with this protein fraction.[28]

An interesting question is whether a relationship exists between this class of sperm matrix proteins and the proteins of the somatic nuclear matrix. The results reported in the literature concerning this point are rather controversial. Meiotic spermatocytes and spermatids have not revealed the presence of lamina by immunofluorescence and electron microscopic techniques,[33] while homology between lamins A and C and mouse sperm residual proteins was found on the basis of electrophoretic mobility, isoelectric points, and immunocrossreactivity.[26] An opposite conclusion — that lamins A and C were not present during gamete

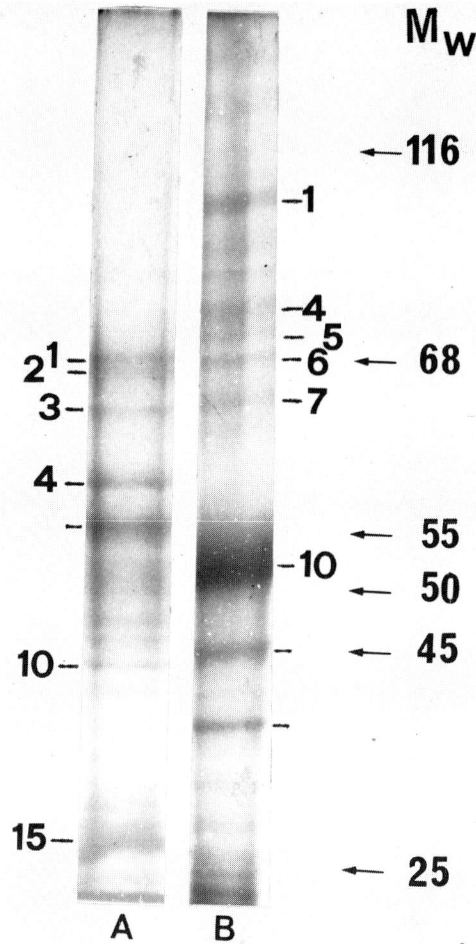

FIGURE 2.   Fractionation of the extractable tightly bound nonhistone proteins of ram sperm (A) and of mussel sperm (B) in 10% polyacrylamide gels.

development, while lamin B was a constituent of all developmental stages — was drawn from studies of mouse spermatogenesis.[34] In our studies we could not find a similarity between the two-dimensional peptide maps of ram tightly bound sperm proteins and of lamb liver lamins, although a homology between them has been suggested on the basis of immunological crossreactivity.[26] A common feature between the two sets of proteins is the presence of carbohydrate residues.

The interesting fact that the extractable tightly bound sperm proteins, although closely related, display considerable differences in their molecular masses has been reported earlier for a subfamily of cytokeratin polypeptides.[35] The major nuclear envelope polypeptides have been shown to be closely related also,[36,37] and a high degree of homology has been reported for the whole class of intermediate filaments.[38] On the other hand, it has been recently demonstrated that nuclear lamins reveal a striking homology with the entire protein family of the intermediate filaments.[36] It may be tentatively speculated that common principles might govern the origin and diversity of both nuclear and cytoplasmic skeletal proteins.

The localization of the tightly bound sperm proteins in skeletal structures associated with DNA suggests that these proteins might be part of the structures to which DNA is anchored, but this does not show whether they are directly responsible for the DNA binding or are only structural elements of an anchorage protein complex.

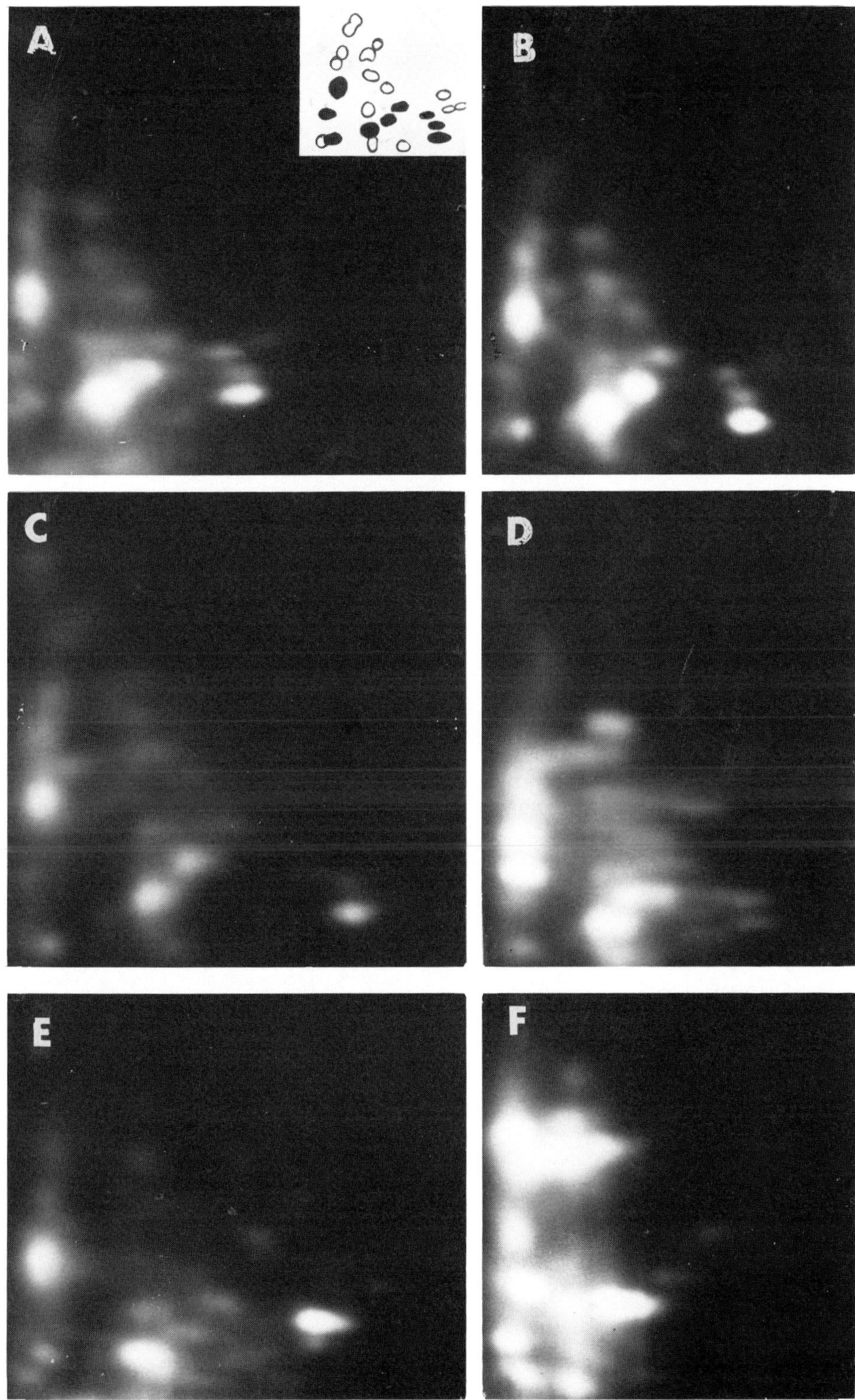

FIGURE 3. Two-dimensional tryptic peptide maps of the ram sperm tightly bound nonhistone proteins. A to F correspond to fractions 1 to 4, 10, and 15 indicated in Figure 2A. Insert in A shows a typical spot pattern. (From Avramova, Z. and Tasheva, B., *Mol. Cell. Biochem.*, 74, 67, 1987. With permission.)

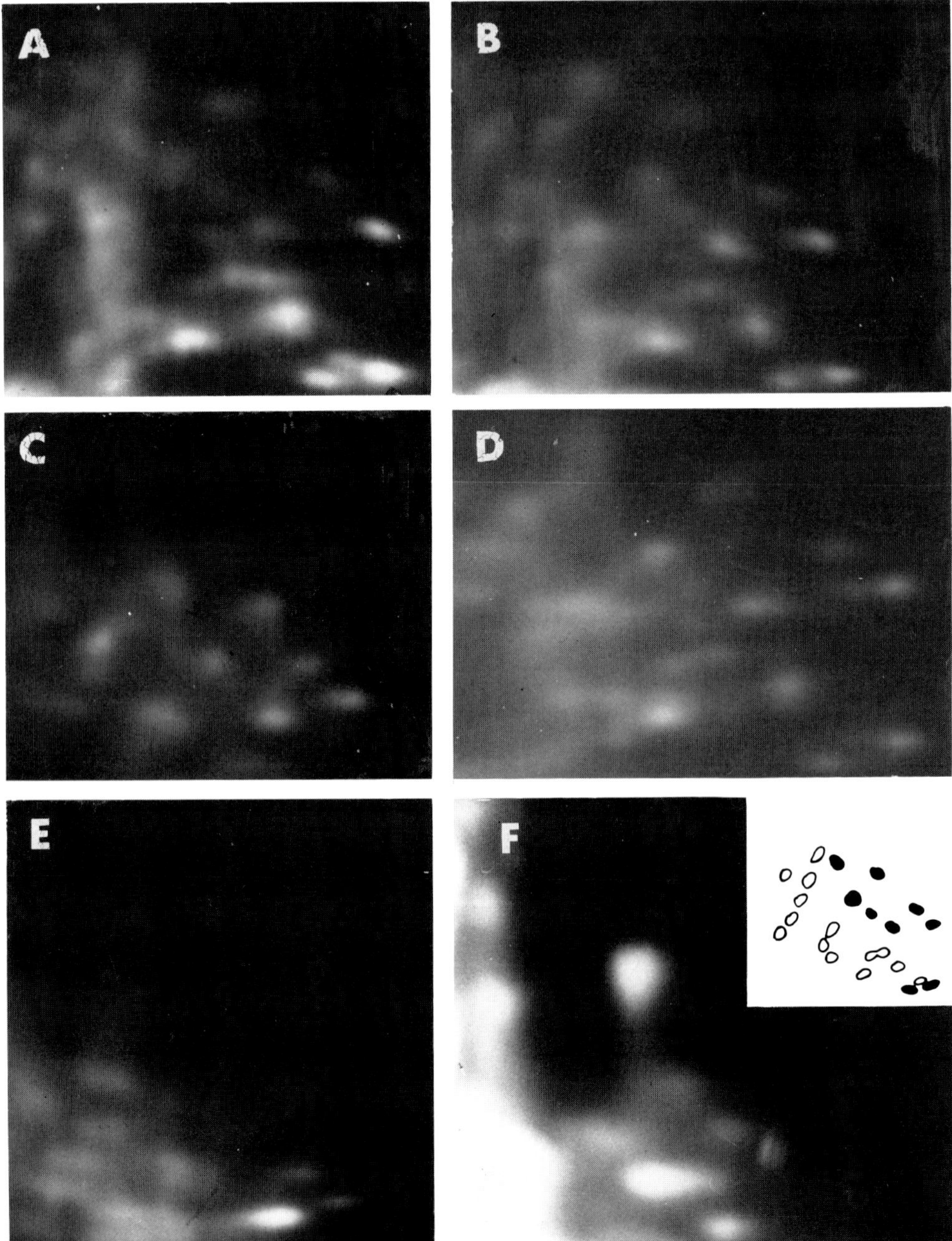

FIGURE 4.   Two-dimensional tryptic peptide maps of the mussel sperm tightly bound nonhistone proteins. A to F correspond to fractions 1, 4 to 7, and 10 indicated in Figure 2B. Insert in F shows a typical spot pattern.

B. Covalently DNA-Bound Sperm Proteins

An astonishing observation was that even after the removal of the tightly bound proteins from the sperm DNA, a small amount of proteinaceous material still remained associated with DNA, resisting all treatments disrupting noncovalent linkages. In order to study the nature of the force binding this material, the DNA tightly bound protein complex of ram sperm was radiolabeled under conditions when the radiolabel was exclusively introduced

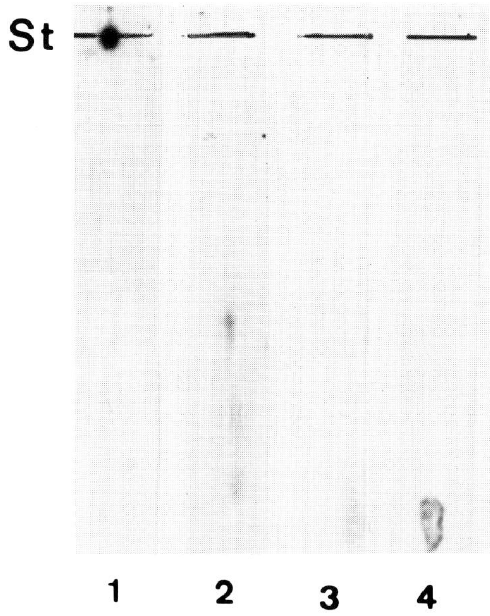

FIGURE 5.   Autoradiographs of the high-voltage electrophoregrams of the residual DNA-peptide complex after digestion with different enzymes: (1) The ^{125}I-labeled complex digested with DNase I; (2) as in (1) but additionally digested with α-chymotrypsin; (3) as in (1) but additionally digested with alkaline protease; (4) iodinated tyrosine marker (ninhydrine staining).

into the protein component.[39,40] The labeled DNA-protein complex was repeatedly precipitated with cold ethanol until no more radioactivity could be washed out. This DNA-protein complex was subjected to two digestion treatments — first, an extensive digestion with pronase E (Merck, 1 mg/ml in 50 mM Tris-HCl, pH 8.0, 1 mM EDTA, 0.5% SDS, overnight at 37°C) to eliminate all proteins, and second, following the pronase step, an extensive digesion with DNase I to avoid aggregation and coprecipitation. The nature of the oligodeoxynucleotides obtained was investigated.

After the first step, which was supposed to eliminate bound proteins, it was found invariably[41] that still about 10% of the initial radioactivity remained associated with DNA. This suggested that a residual peptide had remained complexed with DNA, resisting the pronase digestion.

To check whether the DNA-associated labeled material represented single amino acids or oligopeptides, the residual DNA-peptide complex was additionally digested with alkaline protease or with α-chymotrypsin. Under our conditions, the alkaline protease from *Bacillus subtilis* (1 mg/ml in 50 mM Tris-HCl, pH 6.0; 0.025 M CaCl for 16 h at 37°C) converted model di-, tri-, and tetrapeptides into free amino acids, while α-chymotrypsin (Sigma, 1 mg/ml in 50 mM Tris-HCl, pH 8.0) hydrolysed predominantly the peptide bonds on the C-terminal side of tyrosine.

The products obtained after this additional proteolytic treatment of the residual DNA-peptide complex were subjected to high-voltage electrophoresis on cellulose-coated glass plates (Merck) in acetic acid:formic acid:water = 15:5:80 for 30 min at 70 V/cm.[31] The autoradiographs of the electrophoregrams are shown in Figure 5. It is seen that the untreated complex remains at the start, while alkaline protease releases all radiolabel as free tyrosine. α-Chymotrypsin, however, produces a heterogeneous material, indicating the presence of oligopeptides, rather than of single amino acids. These results show that the bulk of the label (more than 90%) has been incorporated into the peptide moiety of the complex.

To draw some conclusions on the nature of the chemical bonds linking the protein to DNA, the residual DNA-peptide complex was treated with different chemical reagents by

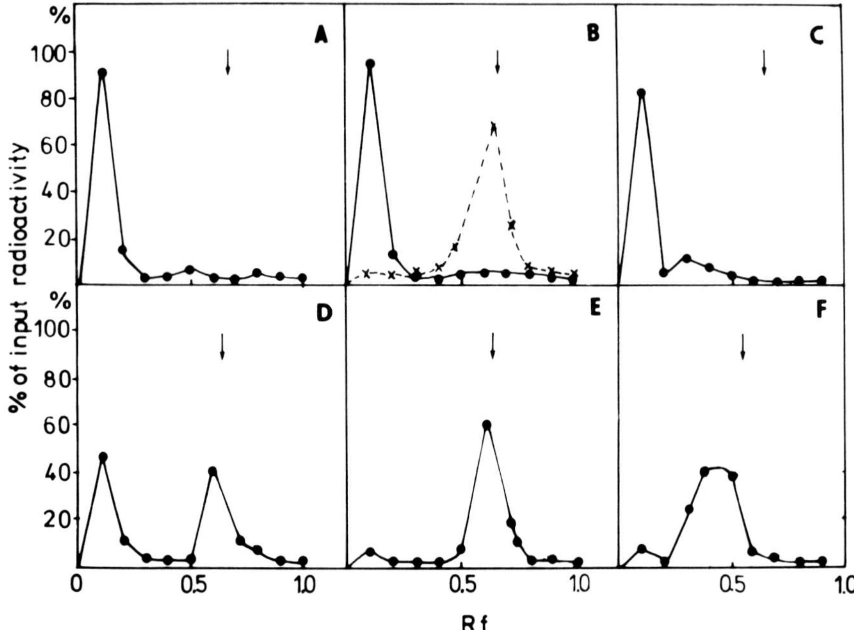

FIGURE 6.   Thin-layer chromatography (phenol-citrate system) of the products from different chemical treatments of the ^{125}I-labeled residual DNA-peptide ram sperm complex: (A) untreated complex; (B) 1 N HCl for 2 h at 37°C (full line) or 6 N HCl for 6 h at 110°C (broken line); (C) 3.8 M hydroxylamine for 2 h at 37°C, pH 4.8; (D) 0.1 N NaOH for 2 h at 60°C; (E) 10% aqueous piperidine for 2 h at 95°C; (F) dioxan/0.1 N NaOH for 4 h at room temperature. Arrows indicate the position of marker iodinated tyrosine. (From Avramova, Z. and Tsanev, R., *J. Mol. Biol.*, 196, 437, 1987. With permission.)

incubation in the following solutions: (1) 1 N HCl for 2 h at 37°C; (2) 0.1 N NaOH for 2 h at 60°C; (3) 3.8 M NH$_2$OH, pH 4.8 for 2 h at 37°C; (4) 6 N HCl for 6 h at 110°C; (5) 10% aqueous (v/v) piperidine for 2 h at 95°C and (6) dioxan: 0.1 N NaON = 1:4 for 4 h at room temperature. At the end of the reaction the samples were dried under vacuum, dispersed in a small volume of water, and subjected to radiochromatography on cellulose-coated plastic sheets (Merck) in a solvent consisting of phenol saturated with 0.22 M sodium citrate, 0.27 M KH$_2$PO$_4$, pH 6.3.[40] The chromatograms were cut into 1-cm strips and counted in a well-type crystal γ-counter (RFT 20046).

As seen in Figure 6A, the residual DNA-peptide complex remained at the start in the phenol-citrate system used. The complex was stable in 1 N HCl and in 3.8 M hydroxylamine (Figure 6 B, C). This excludes the possibility of a linkage of the aminoacyl ester type and makes less probable a phosphoamidate type. The high sensitivity of N-phosphohistidine to hydroxylamine, to acid, and its stability to alkali makes histidine also an improbable candidate for the linking amino acid.[42-44] Alkaline hydrolysis of our complex showed that it was only partially stable — incubation for 2 h at 60°C in 0.1 N NaOH or at 37°C in 0.3 N NaOH released almost half of the radioactivity (Figure 6D). An almost complete hydrolysis of the linkage was obtained in piperidine (Figure 6E) and in alkaline dioxan (Figure 6F). On the basis of these data, a preferred possibility seems to be an ester bond between a DNA phosphate and some hydroxyaminoacid. Such a bond has been shown to be easily hydrolysed by piperidine[40] as is also our complex. Thus, it may be suggested that one or more of the four hydroxyaminoacids — tyrosine, serine, threonine, and hydroxyproline — could be involved in the formation of a covalent bond with the sperm DNA.

To further study the nature of this bond, the residual DNA-peptide complex was subjected

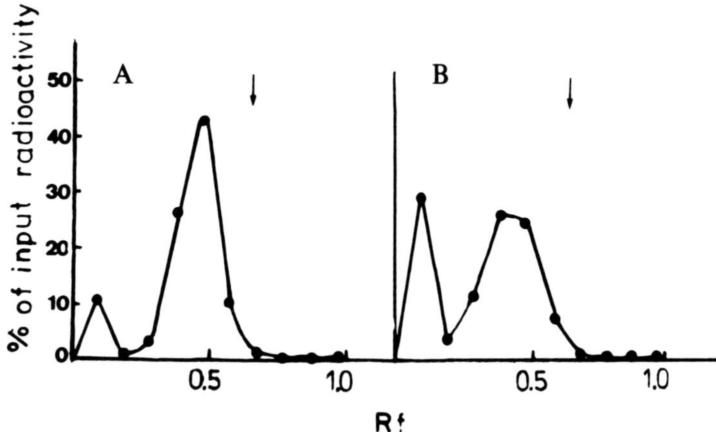

FIGURE 7.   Thin-layer chromatography (phenol-citrate system) of the products of
the enzymatic hydrolysis of the residual DNA-peptide ram sperm complex with snake
venom phosphodiesterase (A) and with spleen phosphodiesterase (B).

to a second digestion step — with DNase I (BDH Biochemicals, 150 μg/ml in 10 mM Tris-
HCl, pH 7.6, 1 mM Mg^{2+}, for 6 h at 37°C). This step resulted in products representing
small oligodeoxynucleotides with phosphate at their 5′-ends and free 3′-OH ends. As shown
by the above presented data, some of these small DNA fragments should be bound to the
residual oligopeptides. The suggestion that this association was due to a covalent linkage
was confirmed by our experiments showing the release of peptides upon digestion with
spleen and snake venom phosphodiesterases.

The oligodeoxynucleotides were digested with spleen phosphodiesterase after removal of
the 5′-phosphates with alkaline phosphatase and the products were chromatographed in the
phenol-citrate system.[40] As seen in Figure 7B, about 60% of the radioactivity was released
by the enzyme, a result reproducibly obtained in several experiments.

However, further studies have shown that the effect on the chromatographic behavior of
the residual oligonucleotide-peptide complex observed after treatment with spleen phosphodiesterase (Figures 7 and 8) is in fact due to the action of the alkaline phosphatase used to
remove 5′- phosphatase from DNA.[41] Although the nature of the product has not been
identified yet, this effect may well be due to the altered hydrophobicity of the dephosphorylated complex. Thus, the linkage of the peptide to the 3′-end of DNA remains unproven
and appears highly improbable.

A very strong effect was obtained with the 3′-end specific snake venom phosphodiesterase
which was able to release more than 80% of the total radioactivity (Figure 7A). It should
be noted that the radioactive material released moved as a broad peak with a mobility
somewhat lower than that of free tyrosine (Figure 7A).

The release of the oligopeptides from their association with DNA fragments upon their
hydrolysis with exonucleases is consistent with the suggestion that the linkage between the
two components represents a covalent ester bond, most probably involving a DNA phosphate
and a hydroxyaminoacid. Our data do not permit us to establish which hydroxyaminoacid
is engaged in phosphoester bonds. It should be noted only that tyrosine involved in such a
bond could not be iodinated.[40] It is possible that some oligopeptides covalently linked to
DNA remain undetected due to the absence of a tyrosine accessible to the labeling reaction.

In summary, all these experiments demonstrate the presence of a DNA-protein complex
in the ram sperm, which resists all treatments known to split S-S bonds, to dissociate hydrogen
bonds, ionic and hydrophobic interactions. When degraded by pronase E and DNase I to

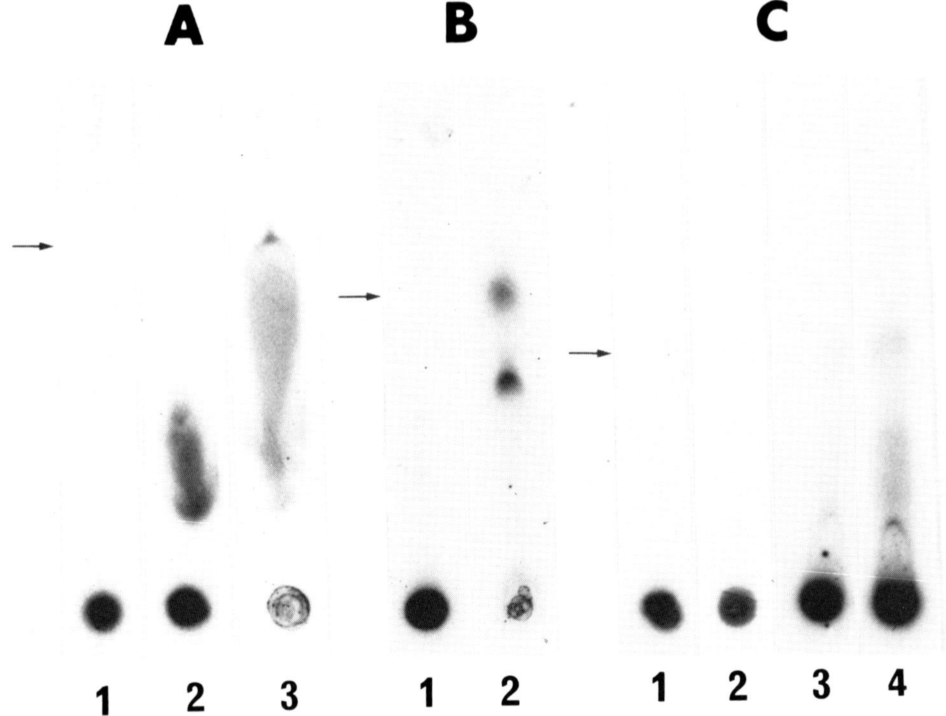

FIGURE 8.   Thin-layer chromatography (phenol-citrate system) of the products of the enzymatic and chemical treatments of the oligodeoxynucleotide-peptide complex from rat liver: A1, B1, and C1 — untreated samples; A2 — snake venom phosphodiesterase; A3 — spleen phosphodiesterase; B2 — alkaline protease; C2 — 1 N HCl for 2 h at 37°C; C3 — 3.8 M hydroxylamine for 2 h at 37°C, pH 4.8; C3 — 0.3 N NaOH for 2 h at 37°C. Arrows indicate the position of marker iodinated tyrosine.

an oligodeoxynucleotide-oligopeptide complex, the linkage between the two components behaves as a covalent bond. It can be also enzymatically split by the 3′-end-specific phosphodiesterase. All these data strongly indicate the presence in the sperm nucleus of a protein covalently linked to DNA by a phosphoester bond.

An important question is the site of the covalent attachment of the protein to DNA. A model has been proposed, suggesting the presence of staggered protein linkers in the two antiparallel strands of DNA.[40,45,46] There is an important prediction of this model — an increased sensitivity of pronase-treated DNA to S1 nuclease.

This prediction is based on the fact that treatment of such a complex with pronase will digest the protein linkers creating single-strand gaps in DNA, thus making the latter highly sensitive to S1 nuclease. Our experiments have shown that this is not the case: pronase-digested and control DNA from ram sperm were only to an equal extent affected by S1 nuclease. In both cases, identical high-molecular-weight fragments were detected by gel electrophoresis (not shown).

Another possibility is a side attachment of the proteins through a phosphoester bond. This may be the third hydroxyl of a DNA phosphate thus forming a phosphotriester compound. The plausibility of such a model would depend on the ability of a phosphodiesterase to attack a phosphotriester.

Experiments with model triester oligonucleopeptides have shown that neither spleen, nor snake venom phosphodiesterase degrades phosphotriester bonds (Juodka, personal communication).

All these new data indicate that most probably the linkage of the residual peptide to DNA is realized via its 5′-end only.

Table 1
PERCENTAGE OF ACID-INSOLUBLE
RADIOACTIVITY AFTER ENZYMATIC
TREATMENT OF THE RAT LIVER
^{125}I-LABELED DNA-PROTEIN COMPLEX

	1st Expt	2nd Expt
Untreated sample	100	100
DNase I (1 mg/ml, 16 h, 37°C)	94	100
Pronase E (1 mg/ml, 1 h, 37°C)	36	25
Alkaline protease after pronase E (1 mg/ml, 1 h, 37°C)	7	5

From Avramova, Z. and Tsanev, R., *J. Mol. Biol.*, 196, 437, 1987. With permission.

IV. COVALENTLY BOUND NONHISTONE CHROMOSOMAL PROTEINS IN SOMATIC NUCLEI

The existence of proteins covalently linked to nucleic acids has been reported for various prokaryotic and viral systems, while their presence in eukaryotic cells is still open to discussion.[43] In this connection, it is intersting to note an early observation that the majority of the phosphorus associated with highly purified nonhistone chromatin proteins of HeLa cells originates from nucleic acids.[48] This fact could be easily explained by the existence of phosphoester bonds covalently linking some proteins to DNA. Recently, the presence of proteins covalently bound to DNA was reported for EAT cells.[40]

The complex isolated from mammalian sperm in our experiments showed the same stability in acid and in hydroxylamine as the complex from EAT cells, but it differed from it in its partial alkaline lability. However, it cannot be excluded that in the case of the EAT-complex, which was obtained under protein-denaturing conditions, additionally trapped topoisomerases could be present.

To check whether covalently DNA-linked proteins are present in somatic cells, we used rat liver nuclei. In a first step, all chromatin proteins held by ionic interactions or hydrogen bonds were dissociated by 2 M NaCl and 5 M urea in the presence of 0.2 M 2-mercaptoethanol. In a second step, the material was treated with SDS and in some cases with hot phenol. Then the material was dispersed in 5% sarcosyl, 1% Triton X100 and CsCl (initial density 1.57 to 1.60 g/ml) containing 5 μg/ml ethidium bromide. The mixture was centrifuged in the vertical Ti60 rotor of the Beckman L60 ultracentrifuge for 16 h at 50,000 rpm at 20°C.

A band of DNA was obtained which had the spectral characteristics of pure DNA (E_{260}/E_{280} = 1.80 to 1.83 and E_{260}/E_{230} = 2.2 to 2.3). However, upon iodination with ^{125}I this DNA incorporated the label as in the case of the sperm complex and this label was released by the sequential action of pronase E and alkaline protease (Table 1), thus indicating the presence of a protein component incorporating the label. The residual oligodeoxynucleotide-oligopeptide complex (obtained after extensive digestion of the complex with pronase E and DNase I) was also attacked by snake-venom phosphodiesterase. The complex showed the same behavior towards different chemical reagents (Figure 8) as the sperm complex. It may be concluded, therefore, that rat liver contains also a DNA-peptide complex held by a covalent linkage.

V. PEPTIDE MAPPING OF THE COVALENTLY DNA-BOUND PROTEINS

A point of great interest is the nature of the proteins covalently bound to DNA. Attempts to fractionate the DNA-protein complex after its digestion with DNase I were not successful, since the complex did not enter even the 3% stacking polyacrylamide gel. This is probably the reason why these covalently linked proteins have not been described till now, although protein material remaining at the gel start has been observed by other authors as well.[40,49] Under similar conditions of isolation and digestion of the DNA-protein complex Razin et al.[16] obtained several nonhistone proteins upon fractionation in polyacrylamide gels. It seems that our procedure was more efficient in dissociating all noncovalently tightly bound proteins coprecipitating with DNA in CsCl gradient. It was possible, however, to characterize the protein moiety of the complex by two-dimensional tryptic peptide mapping of the ^{125}I-labeled complex. To this end, the complex was isolated from the sperm of two mammalian species (rat and ram) and from two somatic tissues (rat liver and Guerin ascites tumor cells). In some experiments, the complex was additionally treated with hot phenol after CsCl gradient centrifugation with the same final results.

As seen in Figure 9A and B, the two-dimensional tryptic peptide maps displayed eight iodinated spots forming a typical pattern, identical for the sperm chromatins of the two mammalian species.

The identity of these two peptide maps shows definitely that the complex does not represent a random association of proteins with DNA. This conclusion is reinforced by the important fact that the eight-spot motif in the peptide map can be detected in the peptide maps of the two somatic chromatins as well (Figure 9C and D).

Analysis of the peptide maps of the DNA-linked protein component in the somatic chromatins of evolutionarily remote species (plants,[41] Drosophila, amphibian, fish, bird,[65] and mammals[50]) have revealed its remarkable preservation in evolution.

An important question is whether these proteins are topoisomerases or not. It is known that when denatured these enzymes could be covalently trapped to the ends of the cleaved DNA.[50] In our experiments we have deliberately avoided treatment of the nuclei with alkali or other denaturing agents before the high salt-urea extraction. On the other hand, peptide maps of the two topoisomerases were not identical to those of the covalently bound proteins.[50a] However, the possibility could not be excluded that this protein may be a part of topoisomerase II or its chemical modification.

VI. METABOLIC BEHAVIOR OF THE NONHISTONE CHROMOSOMAL PROTEINS

In connection with the role of the different classes of nonhistone chromosomal proteins, it was of interest to study their turnover and to see whether there are proteins which are conserved and transmitted to the progeny. This seems especially important when studying proteins which might have a structural role. This problem was studied in our laboratory on various cell systems using the following approach. The cellular DNA was labeled with ^{3}H-TdR (or with ^{14}C-TdR) for 24 h or more, then the nonhistone chromosomal proteins were pulse-labeled with [14]C-(respectively [3]H-) tryptophan and the ratio ^{14}C/[3]H (or ^{3}H/^{14}C, respectively) of labeled proteins to labeled DNA was followed during a long chase period with cold tryptophan.

This experimental design revealed that there are two main metabolic types of nonhistone proteins in chromatin — a metabolically labile group representing a fraction turned over during the cell cycle in about 24 h, and a metabolically stable fraction conserved during many cell generations. The bulk of the metabolically stable proteins were found localized in a small chromatin fraction which had the characteristics of active chromatin and comprised proteins with different turnover rates, including also proteins of the hnRNP particles.[51,52]

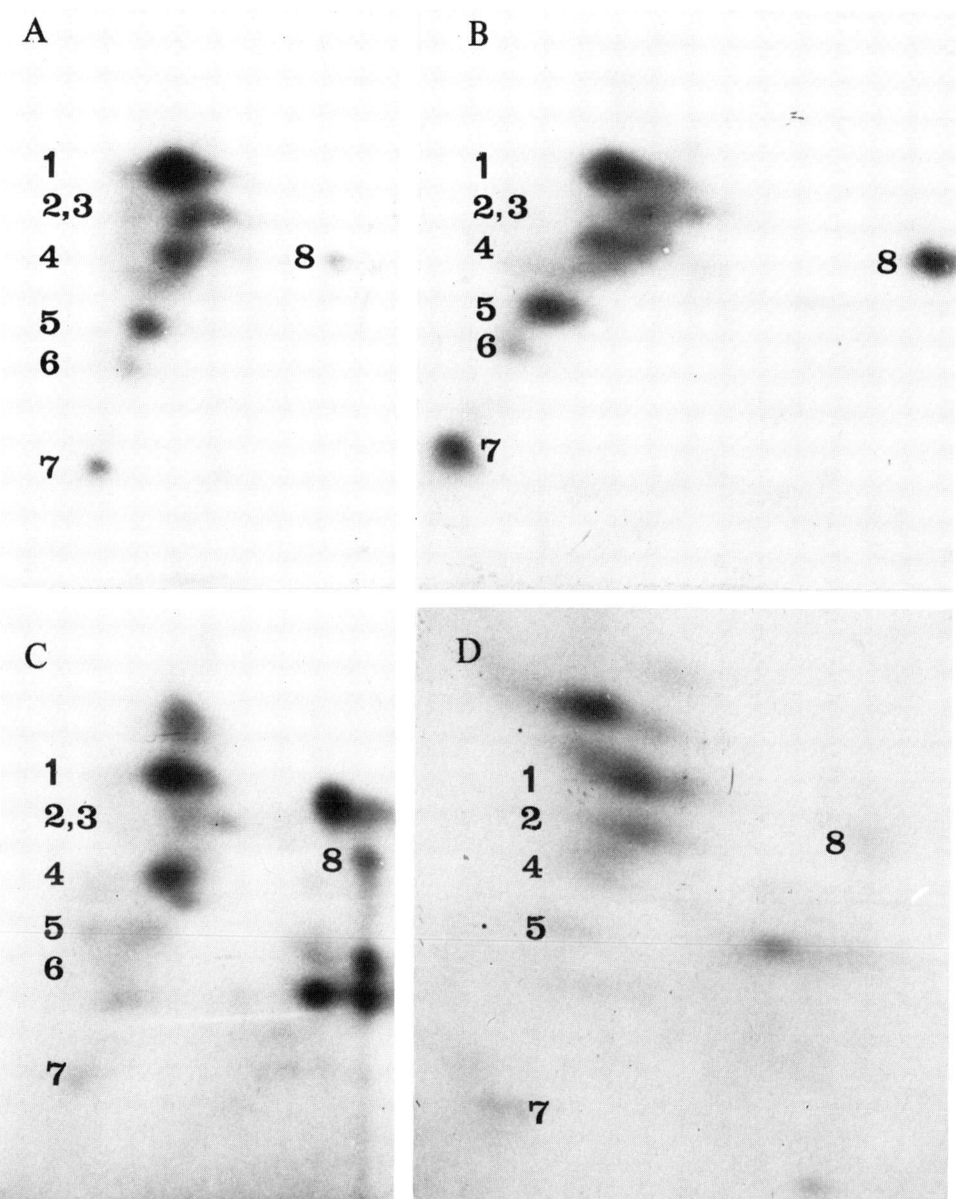

FIGURE 9.   Two-dimensional tryptic peptide maps of the covalently DNA-bound nonhistone proteins from ram sperm (A), rat sperm (B), rat liver (C), and Guerin ascites tumor cells (D). (From Avramova, Z and Tsanev, R., *J. Mol. Biol.*, 196, 437, 1987. With permission.)

The metabolically stable nonhistone proteins were found to have a universal occurrence since we have detected them in all different normal and malignant cellular types studied: Chinese hamster fibroblasts,[51,53,54] human diploid fibroblasts,[54] rat hepatocytes,[55] rat brain cells,[56,57] mouse erythroleukemia cells,[51] EAT cells,[58] plant cells.[59] The metabolic stability of this group of proteins is illustrated in Figure 10A.

To follow the metabolic behavior of the covalently linked nonhistone chromosomal proteins, we have used[60] mouse erythroleukemia cells cultured in MEM supplemented with 10% fetal calf serum. DNA of these cells was labeled with ^{3}H-TdR and the proteins with ^{14}C-tryptophan for 24 h. The DNA-protein complex was isolated from cellular nuclei as

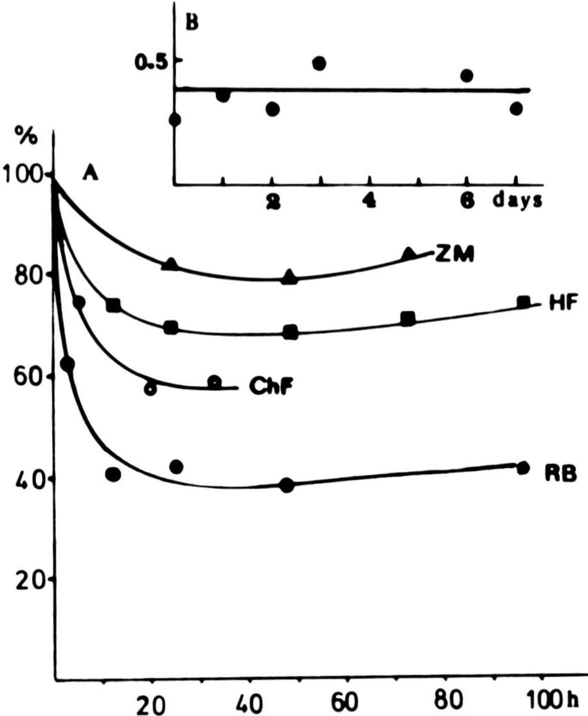

FIGURE 10.   Metabolic behavior of the nonhistone chromosomal proteins. (A) Relative changes of the ratio of ^{3}H-tryptophan labeled total nonhistone proteins to ^{14}C-TdR-labeled DNA in Zea mays nuclei (ZM), in human diploid fibroblasts (HF), in Chinese hamster fibroblasts (ChF), and in rat brain (RB) during a chase period of 100 h.; (B) ratio of ^{14}C-tryptophan-labeled covalently linked nonhistone proteins to ^{3}H-TdR-labeled DNA in mouse erythroleukemia cells during a 7-d chase period.

described for the somatic cells (see above). When the fraction of ^{3}H-labeled "pure" DNA was isolated, it contained also ^{14}C-labeled tryptophan. Thus, these experiments confirmed once more the presence in somatic cells of tryptophan-containing proteins associated with DNA in a complex resisting all reagents disrupting noncovalent associations. The ^{14}C/^{3}H ratio of this complex did not decrease during a chase period of 7d (Figure 10B). These data show that the proteins covalently associated with DNA in the somatic cells belong to the metabolically stable group, i.e., they are transmitted to the progeny as DNA itself.

VII. LOCALIZATION OF THE COVALENTLY BOUND NONHISTONE CHROMOSOMAL PROTEINS

We have obtained two different sets of data which could indicate the localization of the covalently linked nonhistone proteins — electron microscopic observations on spread sperm nuclei and biochemical data on somatic chromatin gradually digested with DNase II.

As already mentioned, when high salt-urea-treated sperm nuclei were spread for electron microscopy, it was found that the DNA fibers were attached to nuclear skeletal structures as in somatic cells. The presence of tightly bound sperm proteins in the DNA-holding structures shows that these proteins are most probably localized at the DNA anchorage sites. It is attractive to speculate that the covalently linked nonhistones play the main role in this respect.

To check this point we have used a somatic cell line — the Guerin ascites tumor cells

Table 2
PROTEIN/DNA RATIO (^{14}C/^{3}H) OF
THE MATRIX-BOUND COVALENT
COMPLEX

DNase II units per 1 A_{260}	Matrix-DNA (%)	^{14}C/^{3}H
1	44	0.02
2	28	0.03
5	13	0.28
20	2.5	2.53

— to follow the localization of the covalently linked proteins on the DNA fibers.[61] The cells (3 to 4 ml of the ascites fluid) were grown for 3 h at 37°C in 40 ml serum-free medium 199, pH 7.2, containing 0.5 U/ml of heparin, 100 μCi H-TdR (Amersham), and 220 μCi ^{14}C-protein hydrolysate (UVVVR, ČSSR).

Chromatin was isolated as described[62] and dispersed in the digestion buffer (10% sucrose, 2 mM Tris-HCl, 0.2 mM EDTA, pH 6.8) at a concentration of 5 A_{260} U/ml. Samples were digested for 45 min at 28°C with different concentrations of DNase II (Sigma) in combination with 10 μg/ml of RNase A (Reanal). DNase II was used in order to work with less condensed nuclei and to avoid precipitation of digested material by divalent cations.[63] Digestion was carried out before dehistonization to avoid precipitation of insoluble DNA-protein complexes.

After digestion, 4-ml samples in 2 M NaCl were layered on top of 10 ml 10% sucrose-2 M NaCl and centrifuged for 30 min at 5000 rpm in the K23 refrigerating centrifuge (GDR). The pellets containing the nuclear matrices and different amount of matrix-associated DNA were collected and the ^{14}C and ^{3}H radioactivities measured in a Triton X100-scintilation cocktail. As expected, with the increase of the amount of digested DNA, the nuclear pellet became gradually enriched in proteins.

To see the localization of the covalently linked proteins, the pellets were deproteinized and fractionated in a CsCl density gradient as described above. DNA was recovered and the ^{14}C/^{3}H ratio measured. As seen in Table 2, increasing the degree of digestion resulted in a gradual enrichment of the "matrix" DNA in labeled proteins. When only 2.5% of it remained associated with the matrix, the amount of DNA-linked ^{14}C-labeled proteins increased more than 100 times as compared to the mildly digested DNA when about half of it was matrix associated. Figure 9D shows that the peptide map of the proteins, remaining covalently bound to "matrix" DNA is similar, although not identical, to that of rat liver. These data show that the covalently-linked proteins are predominantly localized at the attachment sites of DNA to a nuclear structure which in additional experiments was found' to be the lamina. It should be mentioned that part of the ^{14}C-label was found associated also with the DNA digested by DNase II and released into the supernatant. However, we cannot say at present whether part of these proteins are localized at other sites as well.

Thus, two independent sets of data obtained by two different methods and on two different cellular types — sperm and somatic cells — indicate that there are covalently DNA-linked nonhistones which most probably are localized at the basis of the DNA loops and are part of the anchorage structures.

VIII. CONCLUSIONS

The data presented show that, in both sperm and somatic nuclei, nonhistone proteins can be detected which resist extraction with all reagents known to disrupt noncovalent associations. The chemical behavior of the DNA-protein linkage and its sensitivity to phosphodiesterases strongly suggest the presence of proteins covalently linked to DNA via phosphoester bonds. These proteins are metabolically stable and predominantly localized in the lamina at the bases of the chromatin loops. Tryptic peptide mapping has shown the identity of one of these proteins in the sperm of two mammalian species and its presence also in some somatic tissues.

On the basis of these results, the following tentative model of the loop organization of eukaryotic DNA could be proposed: Some specific nonhistone proteins are normally attached to DNA via covalent bonds. These proteins associate with some other nonhistone proteins by hydrophobic interactions, thus forming protein granules to which DNA fibers are attached. These granules can associate to form different DNA-holding structures which organize the DNA loops in different arrangements in the interphase nucleus, in the mitotic chromosome and in the sperm nucleus.

The metabolic stability of the covalently DNA-linked proteins shows that they are transmitted to the progeny. This is to be expected if the loop organization is preserved during the whole cell cycle and during spermiogenesis. This type of DNA organization seems to be preserved in evolution which suggests that it represents a basic principle of DNA folding in eukaryotes indispensable for its functioning.

Independently of the exact nature of the bond, the finding of covalently DNA-linked proteins in eukaryotic cells raises a number of important questions — the exact role of these proteins in the loop organization of DNA, their tissue specificity, their presence in different species, their localization with respect to different genes, etc. — questions which we hope will be answered in the near future.

REFERENCES

1. **Tsanev, R. and Sendov, B.,** Possible molecular mechanism for cell differentiation in multicellular organisms, *J. Mol. Biol.,* 30, 337, 1971.
2. **Weintraub, H.,** Assembly and propagation of repressed and depressed chromosomal sites, *Cell,* 42, 705, 1985.
3. **Paulson, J. R.,** Chromatin and chromosomal proteins, in *Electron Microscopy of Proteins,* Vol. 3, Harris, J. R., Ed., Academic Press, London, 1982, 78.
4. **Hancock, R.,** Topological organization of interphase DNA: the nuclear matrix and other skeletal structures, *Biol. Cell,* 46, 105, 1982.
5. **Luchnik, A. N., Bakayev, V. V., and Glassev, V. M.,** DNA supercoiling: changes during cellular differentiation and activation of chromatin transcription, *Cold Spring Harbor Symp. Quant. Biol.,* 47, 793, 1983.
6. **Stirdivant, S. M., Crossland, L. M., and Bogorad, L.,** DNA supercoiling affects in vitro transcription of two maize chloroplast genes differentially, *Proc. Natl. Acad.Sci. U.S.A.,* 82, 4886, 1985.
7. **Igo-Kemenes, T., Horz, W., and Zachau, H. G.,** Chromatin, *Anal. Rev. Biochem.,* 51, 89, 1982.
8. **Razin, S. V., Chernokhvostov, V. V., Yarovaya, O. V.,and Georgiev, G. P.,** Organization of the sites for DNA attachment to the nonhistone proteinaceous nuclear skeleton, in *Progress in Nonhistone Protein Research,* Vol. II, Bekhor, I., Ed., CRC Press, Boca Raton, FL, 1985, 91.
9. **Mirkovitch, J., Mirault, M.-E., and Laemmli, U. K.,** Organization of the higher-order chromatin loop: specific DNA-attachment sites on nuclear scaffold, *Cell,* 39, 223, 1984.
10. **Hadlaczky, G., Sumner, A. T., and Ross, A.,** Protein depleted chromosomes. I. Structure of isolated protein-depleted chromosomes, *Chromosoma,* 81, 537, 1981.
11. **Kirov, N., Djondjurov, L., and Tsanev, R.,** Nuclear matrix and transcription activity of the mouse globin gene, *J. Mol. Biol.,* 180, 601, 1985.
12. **Miller, O. L. and Bakken, A. H.,** Morphological studies on trancrition, *Acta Endocrinol.,* 168, 155, 1972.
13. **Rattner, J. B. and Hamkalo, B. A.,** Nucleosome packing in interphase chromatin, *J. Cell Biol.,* 81, 453, 1979.
14. **Earnshaw, W. C. and Laemmli, U. K.,** Architecture of chromsomes and chromosome scaffold, *J. Cell Biol.,* 96, 84, 1983.

15. **Tsanev, R. and Tsaneva, I.,** Molecular organization of chromatin as revealed by electron microscopy, in *Methods and Achievements in Experimental Pathology,* Vol. 12, Jasmin, G. and Simard, R., Eds., Karger, Basel 1986, 63.
16. **Razin, S. V., Chernokhvostov, V. V., Roodin, A. V., Zbarsky, I. B., and Georgiev, G. P.,** Proteins tightly bound to DNA in the regions of DNA attachment to the skeletal structures of interphase nuclei and interphase chromosomes, *Cell,* 28, 99, 1981.
17. **Razin, S. V., Mantieva, V. L., and Georgiev, G. P.,** The similarity of DNA sequences remaining bound to scaffold upon nuclease treatment of interphase nuclei and metaphase chromosomes, *Nucleic Acids Res.,* 7, 1713, 1979.
18. **Tsanev, R. and Avramova, Z.,** Nonprotamine nucleoprotein ultrastructures in mature ram sperm nuclei, *Eur. J. Cell Biol.,* 24, 139, 1981.
19. **Tsanev, R. and Avramova, Z.,** Trout sperm chromatin. II. Ultrastructural aspects after salt dissociation of proteins, *Eur. J. Cell Biol.,* 31, 143, 1983.
20. **Avramova, Z., Zalensky, A., and Tsanev, R.,** Biochemical and ultrastructural study of the sperm chromatin from Mytilus galloprovincialis, *Exp. Cell Res.,* 152, 231, 1984.
21. **Engelhardt, P., Plagens, U., Zbarsky, I. B., and Filatova, L. S.,** Granules 25-30 nm in diameter — a basic constituent of the nuclear matrix, nuclear envelope and chromosome scaffold, *Proc. Natl. Acad. Sci. U.S.A.,* 79, 6937, 1982.
22. **Tsanev, R.,** Role of histones in cell differentiation, in *Eukaryotic Gene Regulation,* Vol. 2, Kolodny, G. M., Ed., CRC Press, Boca Raton, FL, 1980, 57.
23. **Bellvé, A. R.,** The molecular biology of mammalian spermatogenesis, in *Oxford Reviews of Reproductive Biology,* Vol. 1, Finn, C. A., Ed., Oxford University Press, Oxford, 1979, 159.
24. **Avramova, Z., Dessev, G., and Tsanev, R.,** DNA-associated proteins of ram sperm nuclei, *FEBS Lett.,* 118, 58, 1980.
25. **O'Brien, D. A. and Bellvé, A.,** Protein constituents of the mouse spermatozoon. An electrophoretic characterization, *Dev. Biol.,* 75, 386, 1980.
26. **Pruslin, F. M. and Rodman, T. C.,** Proteins of demembraned protamine-depleted mouse sperm. Homology with the proteins of somatic cell nuclear envelope matrix, *Exp. Cell Res.,* 144, 115, 1983.
27. **Avramova, Z., Uschewa, A., Stephanova, E., and Tsanev, R.,** Trout sperm chromatin. I. Biochemical and immunological study of the protein composition, *Eur. J. Cell. Biol.,* 31, 137, 1983.
28. **Avramova, Z. and Tasheva, B.,** Tightly bound nonprotamine proteins from ram sperm nuclei studied by one — and two-dimensional peptide mapping, *Mol. Cell. Biochem.,* 74, 67, 1987.
29. **Avramova, Z. and Tasheva, B.,** Proteins of the sperm nuclear matrix of a bivalvia mollusc; analysis by peptide mapping, *C. R. Acad. Bulg. Sci.,* 40, 101, 1987.
30. **Laemmli, U. K.,** Cleavage of the structural proteins during the assembly of the head of bacteriophage T4, *Nature (London),* 227, 680, 1970.
31. **Elder, J. H., Pickett, R. A., Hampton, J., and Lerner, R. A.,** Radioiodination of proteins in single polyacrylamide gel slices. Tryptic peptide analysis of all major members of a complex multicomponent system using microgram quantities of total protein, *J. Biol. Chem.,* 252, 6510, 1977.
32. **Walborg, E. F. and Christensson, L.,** A colorimetric method for the quantitative determination of monosaccharides, *Anal. Biochem.,* 13, 186, 1965.
33. **Stick, R. and Schwartz, H.,** The disappearance of the nuclear lamina during spermatogenesis: an electron microscopic and immunofluorescent study, *Cell Differ.,* 11, 235, 1982.
34. **Moss, S. B., Burnham, B. L., and Bellvé, A. R.,** Differential occurrence of lamina proteins during spermatogenesis, *J. Cell. Biol.,* 99pB, 463, 1984.
35. **Schiller, D. L., Franke, W. W., and Geiger, B.,** A subfamily of a relatively large and basic cytokeratin polypeptides as defined by peptide mapping is represented by one or several polypeptides in epithelial cells, *EMBO J.,* 1, 761, 1982.
36. **Mc Keon, D., Kirschner, M. W., and Caput, D.,** Homologies in both primary and secondary structure between nuclear envelope and intermediate filament proteins, *Nature (London),* 319, 463, 1986.
37. **Shelton, K. R., Egle, P. M., and Cochran, D. L.,** In *The Nuclear Envelope and the Nuclear Matrix,* Alan R. Liss, New York, 1982, 158.
38. **Quax, W., Egberts, W. V., Hendrix, W., Quax-Jenken, Y., and Bloemendal, H.,** The structure of vimentin gene, *Cell,* 35, 215, 1983.
39. **Krohn, K. A., Knight, L. C., Harwig, J. F., and Welch, M. J.,** Differences in the sites of iodination of proteins following four methods of radioiodination, *Biochim. Biophys. Acta,* 490, 497, 1977.
40. **Neuer, B., Plagens, V., and Werner, D.,** Phosphodiester bonds between polypeptides and chromosomal DNA, *J. Mol. Biol.,* 164, 213, 1983.
41. **Avramova, Z., Ivanchenko, M., and Tsanev, R.,** A protein fraction stably linked to DNA in plant chromatin, *Plant Mol. Biol.,* 11, 401, 1988.
42. **Shabarova, Z. A.,** Synthetic nucleotide-peptides, in *Progr. Nucl. Acids Res. Mol. Biol.,* 10, 145, 1970.

43. **Juodka, B. A.,** Covalent interaction of proteins and nucleic acids. Synthetic and natural nucleotide-peptides, *Nucleosides Nucleotides,* 3, 445, 1984.

44. **Juodka, B. A.,** *Covalent Nuleo - Protein Structures and their Chemical Modeling,* Lichvar, Yu., Ed., Mokslas Press, Vilnius, 1985 (in Russian).

45. **Hershey, H. V. and Werner, D.,** Evidence for non-deoxynucleotide linkers in Ehrlich ascites tumor cell DNA, *Nature (London),* 262, 148, 1976.

46. **Werner, D., Krauth, W., and Hershey, H. V.,** Internucleotide protein linkers in Ehrilich ascites cell DNA, *Biochim. Biophys. Acta,* 608, 243, 1980.

47. **Bhorjee, J. S. and Pederson, T.,** Chromosomal proteins: tightly bound nucleic acids and its bearing on the measurement of nonhistone protein phosphorylation, *Anal. Biochem.,* 71, 393, 1976.

48. **Bodnar, J. W., Jones, C. J., Coombs, D. H., Pearson, G. D., and Ward, D. C.,** Proteins tightly bound to HeLa cell DNA at nuclear matrix attachment sites, *Mol. Cell. Biol.,* 3, 1567, 1983.

49. **Gellert, M.,** DNA topoisomerases, *Annu. Rev. Biochem.,* 50, 879, 1981.

50. **Avramova, Z. and Tsanev, R.,** Stable DNA-protein complexes in eukaryotic chromatin, *J. Mol. Biol.,* 196, 437, 1987.

51. **Djondjurov, L, Ivanova, E., and Tsanev, R.,** Two chromatin fractions with different metabolic properties of nonhistone proteins and of newly synthesized RNA, *Eur. J. Biochem.,* 97, 133, 1979.

52. **Djondjurov, L, Ivanova, E., Pironcheva, V. G., and Tsanev, R.,** Metabolically labile nonhistone proteins in chromatin, *Eur. J. Biochem.,* 107, 105, 1980.

53. **Tsanev, R., Djondjurov, L., and Ivanova, E.,** Cell cycle dependent turnover of nonhistone proteins in chromatin, *Exp. Cell Res.,* 84, 137, 1974.

54. **Djondurov, L, Ivanova, E., and Tsanev, R.,** Metabolic behaviour of nonhistone chromosomal proteins in proliferating and resting fibroblasts, *Eur. J. Biochem.,* 77, 545, 1977.

55. **Russev, G., Anachkova, B., and Tsanev, R.,** Fractionation of rat liver chromatin nonhistone proteins into two groups with different metabolic rates, *Eur. J. Biochem.,* 58, 253, 1975.

56. **Stambolova, M., Angelova, A., and Tsanev, R.,** Metabolic stability of nonhistone chromosomal proteins in two different classes of developing rat brain nuclei, *Cell Differ.,* 8, 195, 1979.

57. **Stambolova, M., Angelova, A., and Tsanev, R.,** Metabolic behaviour and heterogeneity of nonhistone chromosomal proteins in fractionated rat brain chromatin, *Int. J. Biochem.,* 15, 395, 1983.

58. **Srebreva, L., Russev, G., and Tsanev, R.,** Metabolic stability of nonhistone chromosomal proteins in Ehrlich ascites tumor cells, *Int. J. Biochem.,* 10, 691, 1979.

59. **Koleva, S. and Tsanev, R.,** Metabolically stable nonhistone chromosomal proteins in growing maize roots, *Cell Differ.,* 7, 83, 1978.

60. **Avramova, A., Mikhailov, I., and Tsanev, R.,** Metabolic behavior of a stable DNA-protein complex, *Int. J. Biochem.,* 20, 61, 1988.

61. **Krachmarov, Ch., Avramova, Z., and Tsanev, R.,** unpublished results.

62. **Yaneva, M. and Dessev, G.,** Isolation and properties of structured chromatin from Guerin ascites tumor and rat liver, *Eur. J. Biochem.,* 66, 535, 1976.

63. **Galcheva-Gargova, Z., Petrov, P., and Dessev, G.,** Effect of chromatin decondensation on the intranuclear matrix, *Eur. J. Cell Biol.,* 28, 155, 1982.

64. **Bradford, M.,** A rapid and sensitive method for the quantitation of microgram quantities of protein utilizing the principle of protein dye-binding, *Anal. Biochem.,* 72, 248, 1976.

65. **Avramova, Z., Mikhailov, I., and Tsanev, R.,** An evolutionarily conserved protein fraction stably linked to DNA, *Biochim. Biophys. Acta,* in press.

Chapter 4

NONHISTONE PROTEINS AND THE ORGANIZATION OF EUKARYOTIC DNA

Ronald Hancock and George Dessev

TABLE OF CONTENTS

I. INTRODUCTION

The chromosomal DNA within the interphase nucleus is folded in a heirarchy of higher order structures, and current evidence shows that nonhistone proteins play a role in at least some levels of this organization. The DNA of each chromosome, a giant DNA molecule several centimeters in length, is folded into a series of loops of the order of 100-kb long. It is intuitively attractive that each loop could represent an independent unit of replication (a replicon), of transcription (for example, a coordinately expressed gene family), and of folding into higher order structure. Within the nucleus, individual chromosomes occupy restricted territorial domains; this localization, suggested by earlier electron microscopical evidence, has recently been demonstrated directly. In mitotic chromosomes, DNA is organized into loops in a manner similar, and perhaps identical, to the first level of organization in the interphase nucleus. Skeletal structures which could be responsible for these levels of topological organization have been studied by several different approaches. We discuss here our recent work on the roles of nonhistone proteins in this organization.

II. STRUCTURAL ORGANIZATION OF INTERPHASE CHROMOSOMES

Chromosomal DNA molecules within the interphase nucleus behave, by several criteria, as though they are subdivided into a series of topologically independent domains.[1,2] We have visualized and studied directly the loops of DNA which form these domains, using interphase chromatin isolated gently in structured form under low ionic strength conditions in the absence of divalent cations,[3] and subsequently dissociating nucleosomes from the DNA. The measured mean length of the DNA loops in mouse cells is 53 kb, with a range of 10 to 180 kb (Figure 1).[4] These values coincide with those (34 to 140 kb) estimated earlier by indirect methods.[2]

The folding of the chromosomal DNA into loops is maintained by nonhistone proteins, since the DNA is unfolded by digestion of proteins but not when histones alone are removed.[4] We discuss below some artifacts which we have observed during fractionation of nuclei, which can lead to an apparent tight binding of DNA to the nuclear lamina induced by Mg^{++} ions, and we note that Mg^{++} is absent during the preparation of the structures shown in Figure 1. Further evidence that the attachment of DNA to this skeletal structure is not artifactual includes the criteria of their cosedimentation through sucrose gradients,[4] cobanding in equilibrium density gradients of metrizamide,[5] and cofractionation in two-phase polymer separation systems.[4] The predominant polypeptides in the skeletal structure to which the DNA loops are attached (Figure 1) are the lamins; other polypeptides of 42 and 175 kDa are present (Figure 2). Chromatin fibers in the interphase nucleus, both as extended nucleosome chains and in compact higher order structures, have also been visualized in a looped conformation.[6]

Both replication and transcription of the DNA take place on DNA regions located on the loops, as demonstrated by several criteria. Nascent RNA transcripts are visualized on the loops by electron microscopy[4] (Figure 3). Pulse-labeled DNA and RNA, and transcribing RNA polymerase II titrated with $(^3H)\alpha$-amanitin, are localized on the DNA fragments released from the loops by digestion with restriction nucleases[4] (Table 1). In some other experimental systems, replication and transcription have been observed to occur on closely adjacent regions of DNA.[6,7]

It is well established that replication and transcription sites are localized on DNA regions which show no evidence of attachment to a structure when chromatin is prepared for electron microscopy by Miller's[7,8] procedure; this procedure, like that which we have employed, uses low ionic strength conditions in the absence of divalent cations. A fundamentally different model, which proposes that replication and transcription take place on DNA regions

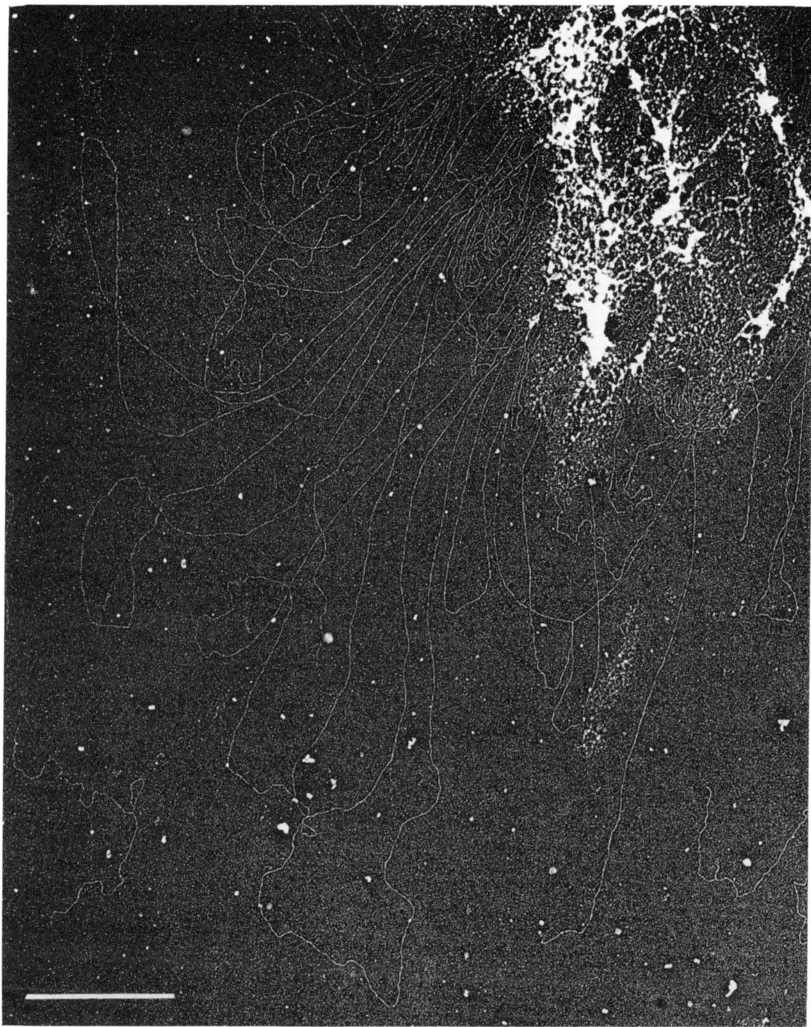

FIGURE 1.   Chromsomal DNA in the interphase nucleus is attached as a series of loops to a nonhistone protein skeletal structure. This organization is visualized after dissociating nucleosomes from interphase chromatin structures of cultured mouse (P815) cells.[4] The DNA loops have a mean length of 53 kb. Bar = 1 μm.

attached to an intranuclear structural element, has been deduced from work on nuclear matrix preparations.[9] Recent work suggests that at least some of the molecular interactions forming the internal structural elements of the nuclear matrix may be generated during its isolation, by formation of disulfide bonds between proteins[9,10] and by aggregation of proteins derived from RNP particles.[11] We have observed an artifactual, irreversible binding of DNA to the nuclear lamina induced in the presence of Mg^{++} ions; single-stranded and pulse-labeled DNA are preferentially bound *in vitro* by bonds which are subsequently resistant to 2 *M* NaCl, suggesting caution in the interpretation of studies on the apparent attachment of nascent DNA to the nuclear matrix (Section III). The generation of such new interactions between nuclear macromolecules which are in close proximity *in vivo* may explain the marked discrepancies in published findings on the association of transcription, replication, and other functions with the nuclear matrix.[2,6,9]

A theoretical advantage of the association of processes such as replication and transcription

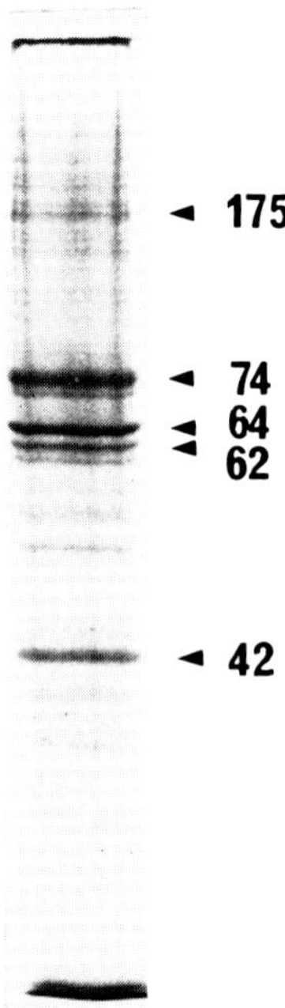

FIGURE 2.   The polypeptide components of the skeletal structure in Figure 1. The 62- to 74- kDa polypeptides have been identified as lamins by peptide mapping.[5]

with a structural element could be that this structure may attract enzymes and substrates to their sites of action, thus reducing from 3 to 2 the dimensions involved in the reactions; this concept was first discussed for membrane-bound enzyme systems.[12] However, careful analysis of this concept has shown that the result of such a situation can equally well be a diminution rather than an enhancement of reaction rates, depending on the two- and three-dimensional diffusions rates, reactant concentrations, and affinity constants.[13] Thus this justification, for kinetic reasons, of models in which transcription and replication take place on a structural support is not supported by theoretical arguments.

Electron microscope studies of sectioned interphase nuclei show that regions of compact chromatin are localized in contact with the nuclear envelope,[1] and where these regions may be counted their number is equal to that of the mitotic chromosomes.[14] These observations suggested that single chromosomes occupy a defined domain in contact with the nuclear envelope, and this model has been confirmed directly by localizing specific chromosomes in the nuclei of hybrid cells.[15,16] This arrangement of individual chromosomes must be accomplished through components which attach them to the nuclear envelope at appropriately located sites. Nonhistone proteins which are located in the centromeric regions of mitotic chromosomes remain detectable in the interphase nucleus,[17] and are potential candidates for the attachment of individual chromosomes to a nuclear structure.

III. BINDING INTERACTIONS BETWEEN THE NUCLEAR LAMINA AND DNA

The lamins are the best characterized nonhistone proteins of the nucleus. Ultrastructural studies established that the lamina is in intimate contact with the peripheral chromatin.[1] Contact at the molecular level is demonstrated by the crosslinking of lamins to DNA *in vivo* by K_2CrO_4 and by *cis*-diamminedichloroplatinum(II).[18] The assembly of the lamina from its component polypeptides after mitosis takes place onto the chromosome surface, and can

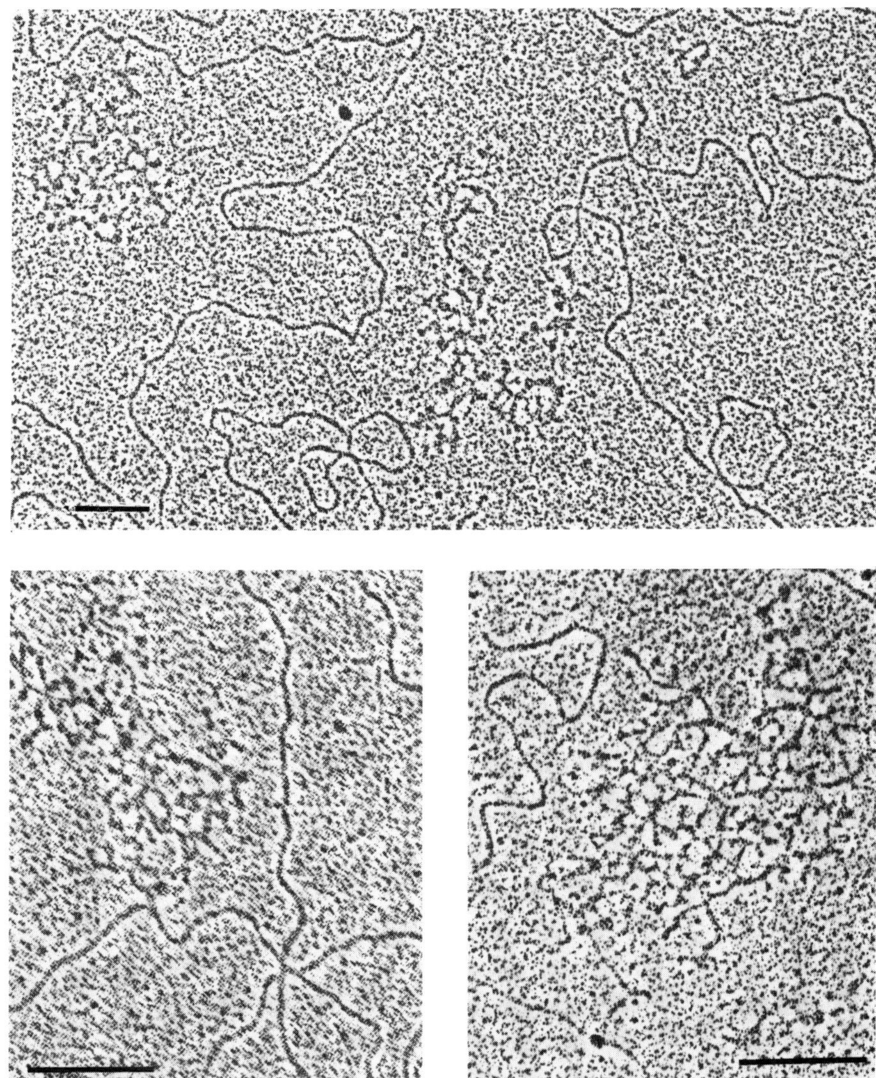

FIGURE 3.   Nascent RNA transcripts localized on DNA regions in the loops in a preparation similar to that in Figure 1. This localization of transcription sites is supported by the localization of both pulse-labeled RNA and of transcribing RNA polymerase II (titrated with (^{3}H) α-amanitin) on the DNA fragments detached from the skeletal structure by restriction nucleases[4] (Table 1). Bar = 0.2 μm.

also occur onto foreign DNAs.[19] In order to study the nature and significance of these interactions, we have investigated lamina-DNA binding interactions *in vivo* and *in vitro*.

We have recently described a new procedure for isolation of the nuclear lamina, which employs low ionic strength conditions in the absence of divalent cations during digestion of chromatin structures[3] with DNAase II together with RNAase.[20] The lamina prepared from mouse cells by this procedure is an empty shell-like structure containing about 6% of the protein of the initial chromatin, and which contains no internal elements (Figure 4). The predominant polypeptides in these lamina preparations are the lamins, identified by peptide mapping and by immunoblotting[20] (Figure 5).

The lamina purified in this way is able to bind exogenous DNAs *in vitro*, as detected by a specific filter binding assay and by sedimentation of the product in sucrose gradients[21] (Table 2). This binding process does not show a preference for eukaryotic DNA. The binding

Table 1
**LOCALIZATION OF SITES OF REPLICATION
AND TRANSCRIPTION IN DNA LOOPS**

	Percent of total in loop DNA fraction
Nascent RNA ((^{3}H)uridine, 3 min)	92
RNA polymerase II ((^{3}H)α-amanitin binding)	98
Nascent DNA ((^{3}H)thymidine, 20 s)	92

Note: Skeletal structures with attached DNA loops (Figure 1) were prepared from interphase chromatin of mouse P815 cells after dissociation of nucleosomes followed by centrifugation.[4] The DNA in the loop regions was cleaved off with EcoRI, and the skeletal structures with remaining attached DNA fragments separated by centrifugation. Radioactivity from pulse-labeled RNA or DNA was measured in these two fractions. Transcribing RNA polymerase II was quantitated by binding of (^{3}H)α-amanitin.

is essentially abolished by digestion of DNA with nuclease S1 and thus appears to be predominantly to single-stranded DNA regions. The product of the binding reaction is resistant to dissociation by 2 *M* NaCl and to EDTA. The skeletal structure shown in Figure 1 is also able to bind DNA in a manner resistant to 2 *M* NaCl, when dissociated in 6 *M* urea, mixed with DNA, and dialyzed; binding is detected by the two criteria employed above and also by analysis of the products of the binding reaction in metrizamide gradients. Lamin A is the principal polypeptide binding to DNA (Figure 6).

It is not yet clear whether the observation that the lamins are strong DNA binding proteins *in vitro* has any biological significance. However, this property of the lamina proteins entails the danger that artifactual lamina-DNA complexes may be formed during fractionation of nuclei or chromatin. To avoid this possibility, we have taken advantage of the fact that complex formation between lamina proteins and DNA *in vitro* is minimized if the reaction is carried out in high ionic strength conditions (Table 2). If chromatin structures prepared at low ionic strength are digested only slightly with DNAase II prior to dissociation of histones in 2 *M* NaCl, some DNA appears to remain attached to the lamina as shown by two criteria, cosedimentation and electron microscopy.[21] As digestion progresses, DNA undergoes a size-dependent release from the lamina. The more rigorous criterion of centrifugation to equilibrium in density gradients of metrizamide containing 1.5 *M* NaCl shows, however, that DNA is not attached to the lamina by bonds stable to EDTA or to high NaCl concentrations (Figure 7A). In contrast, after exposure to Mg^{++} ions at low ionic strength, for digestion with DNAase I or restriction nucleases in place of DNAase II, DNA becomes *irreversibly* bound to the lamina as shown by this criterion (Figure 7B). Further, after digestion by EcoRI or Hae III, to which mouse satellite DNA is relatively resistant, this bound DNA is enriched in satellite sequences[21] (Figure 8). The same effect is observed when chromatin is digested with DNAase I in the presence of Mg^{++} ions,[21] conditions which maintain centromeric chromatin in a highly condensed state.

These experiments thus demonstrate several artifacts which occur during this type of study. Two of these appear to be caused by the presence of Mg^{++} ions during enzymatic digestion with DNAase I or with restriction nucleases, and are *irreversible* by EDTA. The first is an irreversible stabilization of DNA-lamina associations leading to spurious NaCl-resistant binding; the second is a preferential binding of DNA regions enriched in satellite DNA sequences and resistant to the restriction enzymes used to remove DNA, leading to a biased population of DNA fragments associated with the lamina. We have observed that upon exposure to Mg^{++} ions, the lamina contracts;[22] this may explain the generation of

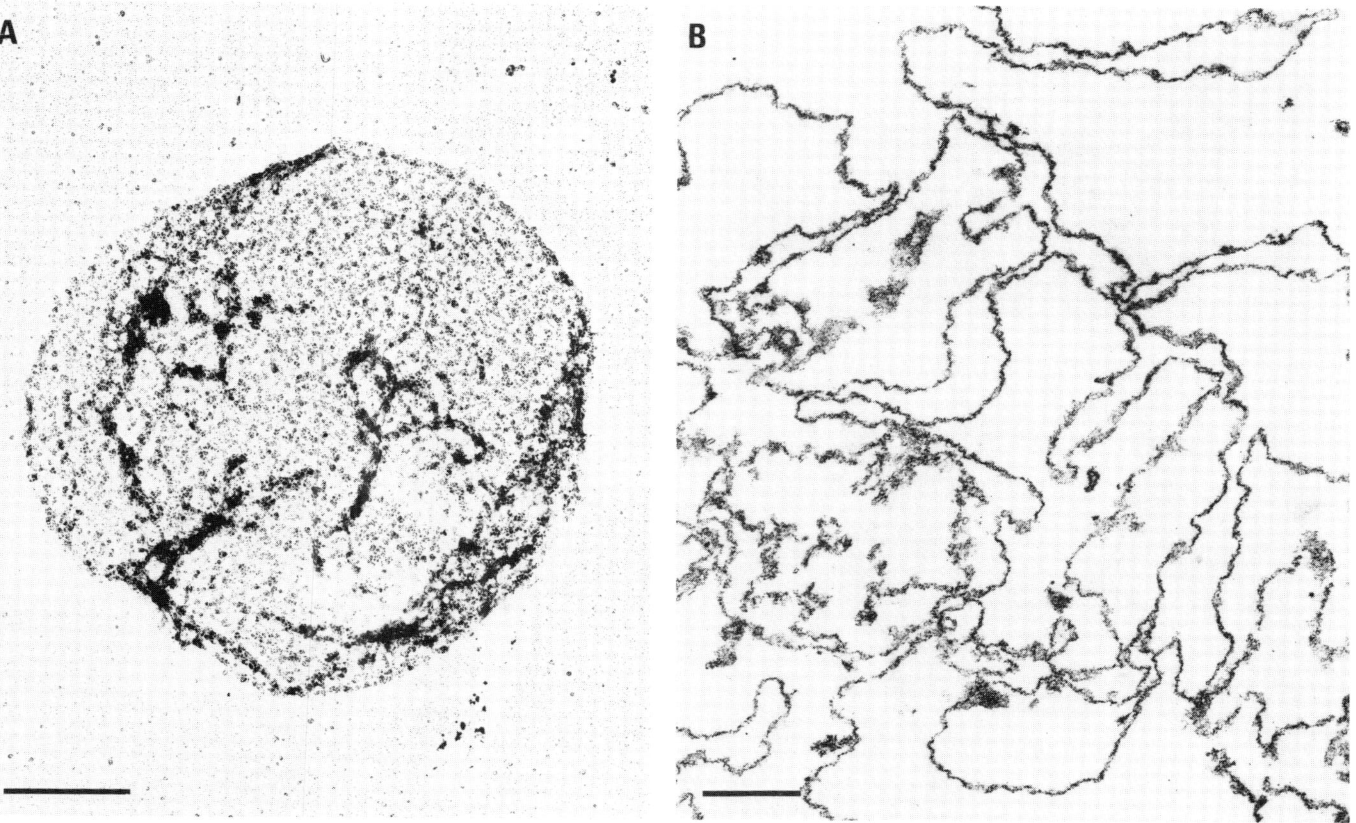

FIGURE 4. The nuclear lamina, purified from interphase chromatin structures[3] of mouse (Ehrlich ascites) cells by digestion with DNAase II and RNAase in the absence of divalent cations. (A) whole mount preparation, bar = 1 μm; B, section, bar = 0.5 μm. (From Krachmarov, C., Tasheva, B., Markov, D., Hancock, R., and Dessev, G., *J. Cell Biochem.*, 30, 351, 1986. With permission.)

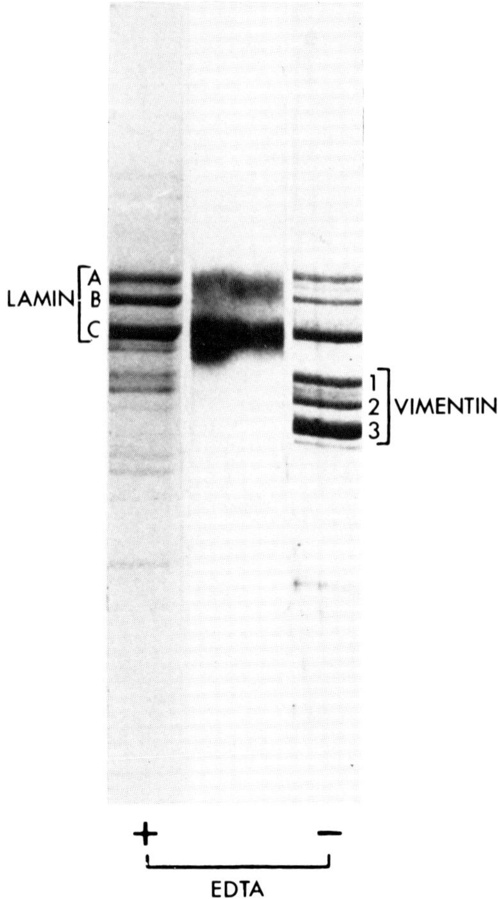

FIGURE 5.   The polypeptides of the purified nuclear lamina shown in Figure 4. The sample in lane 1 was prepared under standard conditions[20] in the presence of EDTA, and that in lane 3 in the absence of EDTA; the additional bands in the 50- to 58-kDa range have been identified as vimentins by peptide mapping.[22] The central lane shows the identification of lamins A, B, and C by probing lane 3, after transfer to nitrocellulose, with a monoclonal antibody against the 3 lamins. This identification has been confirmed by peptide mapping. (From Krachmarov, C., Tasheva, B., Markov, D., Hancock, R., and Dessev, G., *J. Cell Biochem.*, 30, 351, 1986. With permission.)

these new interactions of the lamina with DNA which are irreversible by EDTA. A further potential artifact arises from the marked preference of the nuclear lamina to bind single-stranded DNA regions *in vitro* in a manner resistant to $2M$ NaCl; such binding could make difficult the interpretation of studies on the localization of nascent DNA. The existence of such artifacts emphasizes the great caution which must be exercised in interpreting studies on intermolecular interactions in the nucleus. Similar phenomena may also occur in work on structural elements in mitotic chromosomes, as discussed in Section V.

Our observations are compatible with a model in which the DNA in contact with the lamina is loosely interwoven into the lamina *in vivo*[21] (Figure 9). Our data suggest that lamin A may provide the principal sites of contact with regions of single-stranded DNA. The observation that the lamina preferentially binds single-stranded regions of DNA *in vitro* explains in a satisfactory manner why, *in vivo*, the reformation of the lamina from lamin monomers after mitosis is prevented if potentially single-stranded regions of the DNA have been crosslinked by psoralen.[23]

Table 2
BINDING OF DNA TO NUCLEAR LAMINA
PREPARATIONS *IN VITRO*

Exogenous (³H)DNA	Percent of DNA bound	
	Low salt	**High salt (2 *M* NaCl)**
Mouse, total	22.4	0.7
Mouse, satellite	16.8	—
E. coli	15.4	0.7
Mouse, single-stranded (96°, 10 min)	96.2	0.5
E. coli, single-stranded (96°, 10 min)	97.0	0.8
Mouse, S1 nuclease digested	3.6	—
Mouse, pulse-labeled 30 s	71.8	—
Mouse, pulse-labeled 30 s, S1 nuclease	8.4	— -

Note: Samples of nuclear lamina (100 μg protein) were mixed with (³H)DNA (5 μg) which had been fragmented by Hae III (or BstN1 for satellite DNA) in 1 ml of 10 m*M* Tris-HCl, 1 m*M* EDTA (pH 7.5). After incubation for 2 h at 37° EDTA was added to 20 m*M* and NaCl to 4 *M*. The samples were filtered on glass-fiber filters under conditions specifically retaining DNA bound to proteins[21] and washed with 1.5 *M* NaCl in the same buffer. Retained (³H)DNA was determined by scintillation counting. Complete experimental details are given in Reference 21.

IV. INTERACTION OF INTERMEDIATE FILAMENT PROTEINS WITH THE LAMINA AND DNA

Purified preparations of the nuclear lamina from mouse (Ehrlich ascites) cells always contain small amounts of polypeptides characteristic of intermediate filaments, which appear on gels as bands of M_r in the 56 to 58-kDa range (Figure 5). When the lamina is isolated in the absence of EDTA, the relative amount of these polypyptides is much greater (Figure 5), and in such preparations filaments of about 100-A diameter are observed attached to one face of the lamina.[20] Most of this material has been identified as vimentin by peptide mapping and immunoblotting.[24] These results are in agreement with earlier observations showing that in many types of cells, intermediate filaments are associated with the nucleus.[25,26] Inversely, lamins have been found to coisolate with intermediate filaments of BHK cells.[27]

The close proximity of intermediate filament proteins to the lamina can explain several observations which show that these proteins are located *in vivo* within crosslinking distance of chromosomal DNA. Cytokeratins may be crosslinked to DNA by K_2CrO_4 *in vivo*,[18] and a protein which we have tentatively identified as vimentin becomes crosslinked to DNA after UV irradiation of whole cells.[28] It may be recalled that vimentin binds to DNA *in vitro* with a preference for single-stranded regions.[29]

These proteins may be derived from the terminal regions of intermediate filaments which are attached to the nuclear envelope; alternatively, they may be real components of the nuclear lamina, which would perhaps not be surprising in view of the primary and secondary structure homologies between intermediate filament proteins and lamins.[30,31] While there is no evidence that the association of intermediate filaments with the nucleus is directly related to the structural organization of DNA, it is tempting to speculate that the interactions intermediate filaments (nuclear lamina) chromatin may provide an important communication system between nucleus and cytoplasm, complementing that based on the transport of molecules through the nuclear pores.

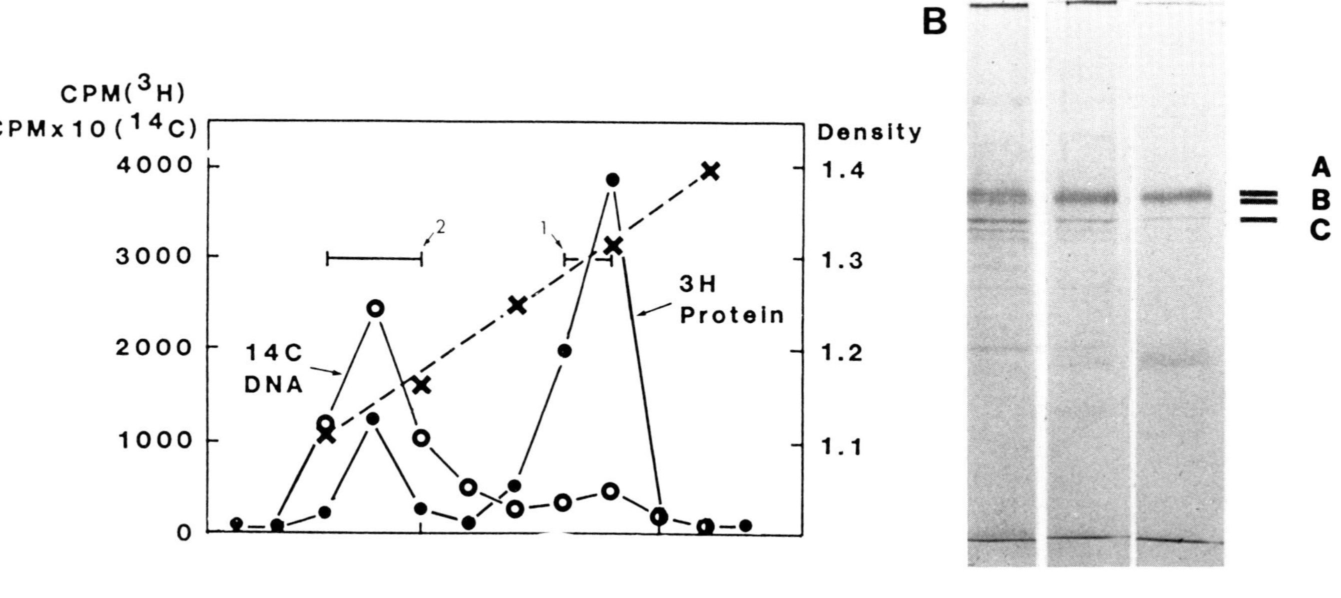

FIGURE 6.   Binding of DNA *in vitro* by the skeletal structure from interphase chromatin (Figure 1). This structure, containing (^{3}H) proteins, was exhaustively digested with DNAase I, disassembled in 6 M urea, and allowed to bind sheared (^{14}C) mouse DNA during dialysis into 10 mM Tris-HCl (pH 7.6), 1 mM EDTA. (A) The product was analyzed by centrifugation to equilibrium in a metrizamide gradient (19 to 46%) containing 1.5 M NaCl. (B) The peak 2 from this gradient (DNA carrying bound proteins) contains lamin A as the principal DNA-bound polypeptide; this identification has been confirmed by peptide mapping.[5]

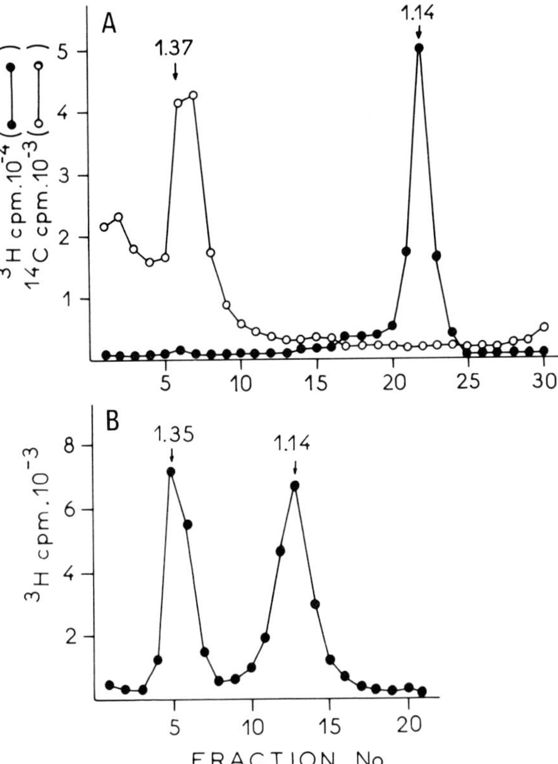

FIGURE 7.   Generation of a tight, NaCl-resistant DNA-lamina association in the presence of Mg^{++} ions. The lamina was prepared by digestion of interphase chromatin structures, containing (^{3}H) DNA (●) and (^{14}C) proteins (○), with either (A) DNAase II without Mg^{++}, or (B) EcoRI with 10 mM Mg^{++} followed by addition of EDTA to 20 mM. In both cases about 5% of the initial DNA sedimented with the lamina. The lamina preparations were then analyzed by centrifugation to equilibrium in gradients of metrizamide (19 to 46%) containing 1.5 M NaCl. Without exposure to Mg^{++} ions (A), the DNA (density 1.14) separated from the lamina (density 1.36), while after exposure to Mg^{++} ions (B) about 50% was bound to the lamina.[21]

V. STRUCTURAL ELEMENTS IN MITOTIC CHROMOSOMES

Mitotic chromosomes are commonly isolated and studied in media containing divalent cations or polyamines, in order to maintain their stability during experimental manipulations. In these conditions, an apparently rigid linear axial structure or scaffold, containing non-histone proteins, may be visualized in each chromatid.[32] We have studied chromosomes liberated from mitotic cells into a low ionic strength buffer containing no divalent cations and no EDTA, by lysis with a nonionic detergent (Nonidet P40), a procedure which we developed to isolate chromatin from interphase cells.[3] These chromosomes remain stable and retain their microscopic structure for several hours, but are fragile upon centrifugation. Earlier studies of chromosomes prepared in this way concluded that they are ''stabilized mainly by fiber-fiber contacts and give no evidence for a shape-maintaining backbone or scaffold''.[33]

Chromosomes prepared in this way are characterized by the great extensibility of their chromatids; these may be observed stretched to many times their length while conserving an identifiable morphology (Figure 10). It could be postulated that a skeletal element has dissociated in the absence of divalent cations, but it then follows that this element cannot be essential to the integrity of the chromosome, since this integrity is still conserved. This

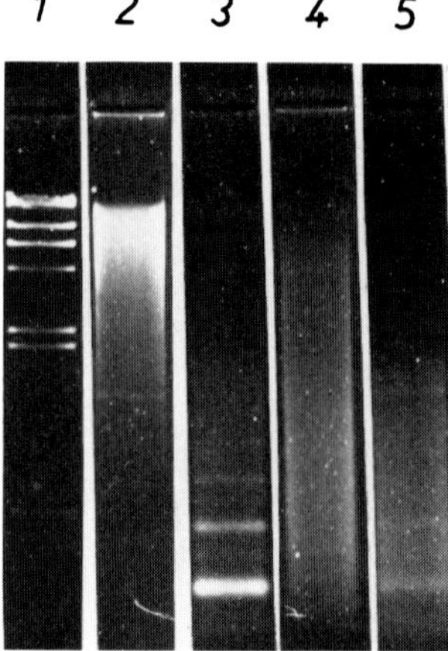

FIGURE 8.   The apparent preferential attachment of satellite DNA to the lamina after digestion of chromatin structures with EcoRI in presence of Mg^{++} ions. DNA was isolated from the two peaks of the metrizamide gradient shown in Figure 7B, and digested with BstN1 to detect the 234 bp repeated fragment, and its multimers, derived from satellite sequences. Lane 1, size markers of λ DNA digested with Hind III; lanes 2 and 3, DNA from the peak of density 1.35 (lamina-bound DNA) before and after digestion with BstN1; lanes 4 and 5, DNA from the peak of density 1.14 (free DNA) before and after digestion with BstN1. (From Krachmarov, C., Iovcheva, C., Hancock, R., and Dessev, G., *J. Cell. Biochem.*, 31, 59, 1986. With permission.)

extensibility is clearly not compatible with the existence of a rigid, linear skeletal element within each chromatid.

We have obtained further evidence supporting this conclusion, and which shows that chromatids are not linear but are formed of a helically wound fiber whose length is thus several times the apparent length of the chromatid. This conformation is visualized in polyamine-stabilized chromosomes incubated with protein-dissociating agents such as deoxycholate or diiodosalicylate (Figure 11). These observations support the conclusion quoted above[33] that the winding of this fiber is determined by protein-protein interactions between neighboring gyres. We and others[34] have visualized similar gyres in chromosomes from cells grown in the presence of intercalating agents. It follows that any skeletal element in the chromatin fiber in mitotic chromosomes must be wound helically, following the path of the chromatid gyres.

A further phenomenon with which models of chromosome structure must be compatible, and which is not predicted by the model of a rigid, structure-determining scaffold, is the existence of major rearrangements of the chromosomal DNA such as translocations and fused chromosomes. These occur spontaneously, and their frequency is greatly increased in cells grown in the presence of intercalating agents[35] (Figure 12). From our knowledge of the action of intercalating agents, it appears certain that these chromosome rearrangements must be consequences of intercalation on DNA topology, and could not be the result of direct effects on a nonhistone protein skeletal element. Thus in the fused chromosomes, shown in Figure 12, which contain fused DNA molecules, it would be necessary to postulate that the increased length of DNA has imposed the formation of a longer skeletal element. It must, therefore, be concluded that if a skeletal element does exist in mitotic chromosomes, it does not dictate the dimensions of the chromosome; rather, the form and dimensions of the skeletal element must be determined by the length of the DNA with which it is associated.

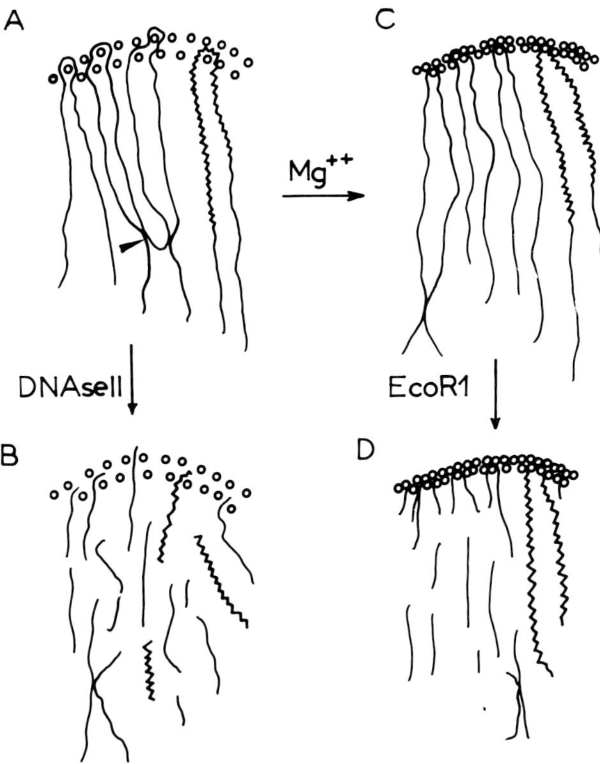

FIGURE 9.   A model of the association of DNA with the nuclear lamina *in vivo* which is compatible with the observations presented here and with most published results. (A) DNA in the form of chromatin is loosely interwoven with the lamina *in vivo;* (B) when digested with DNAase II in the absence of Mg^{++} followed by extraction of histones in 2 M NaCl, DNA fragments are completely released; (C) in the presence of Mg^{++} ions, for example during digestion with DNAase I or restriction nucleases, the lamina (and other nuclear structures) are compacted so that (D) undigested DNA fragments, including satellite sequences (zigzag lines), remain tightly bound to the lamina and resist extraction in high concentrations of NaCl. (From Krachmarov, C., Iovcheva, C., Hancock, R., and Dessev, G., *J. Cell. Biochem.*, 31, 59, 1986. With permisson.)

ACKNOWLEDGMENTS

This work was supported in part by grants from the Medical Research Council of Canada and the National Cancer Institute of Canada to Ronald Hancock.

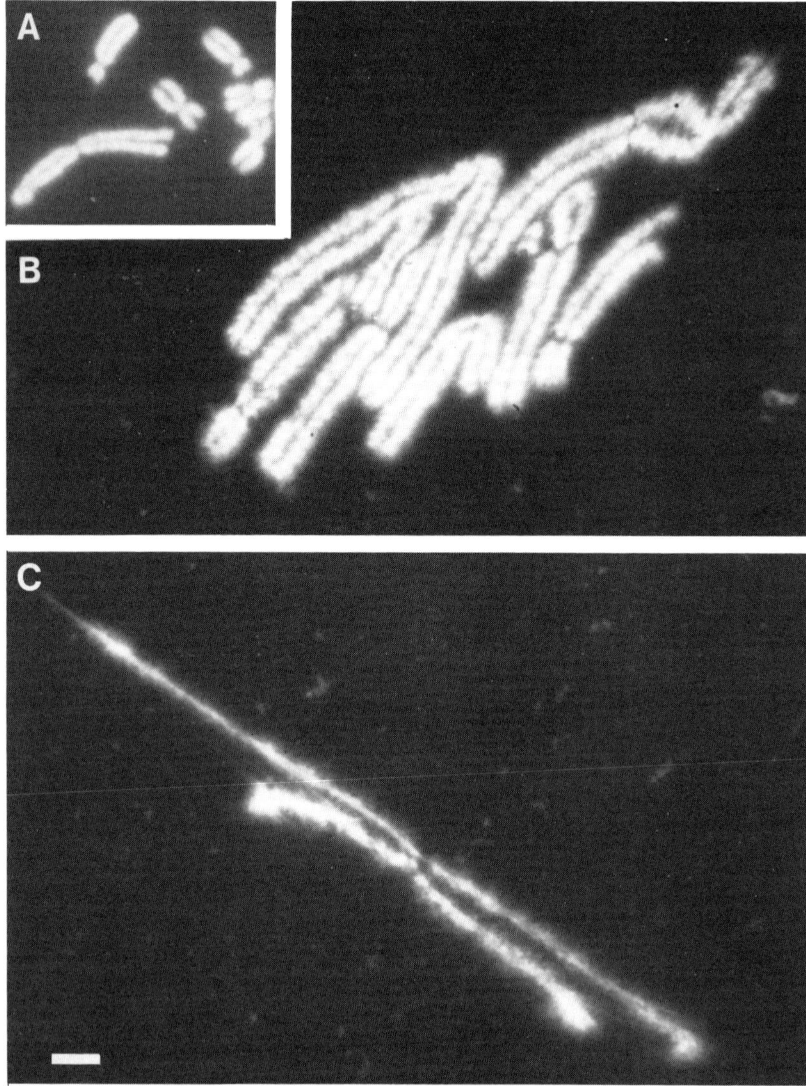

FIGURE 10.   Mitotic chromosomes are highly extensible in media without divalent cations. Chromosomes were prepared from mitotic hamster cells (line CHO) by lysis with Nonidet P40 (0.25%) in 0.2 mM phosphate pH 7.5, 0.15 M sucrose.[3] They were stained in the lysate with Hoechst 33258 and examined in suspension by fluorescence microscopy. Upon incubation the chromosomes extend greatly (B, C) from their initial length (A). Bar = 10 μm.

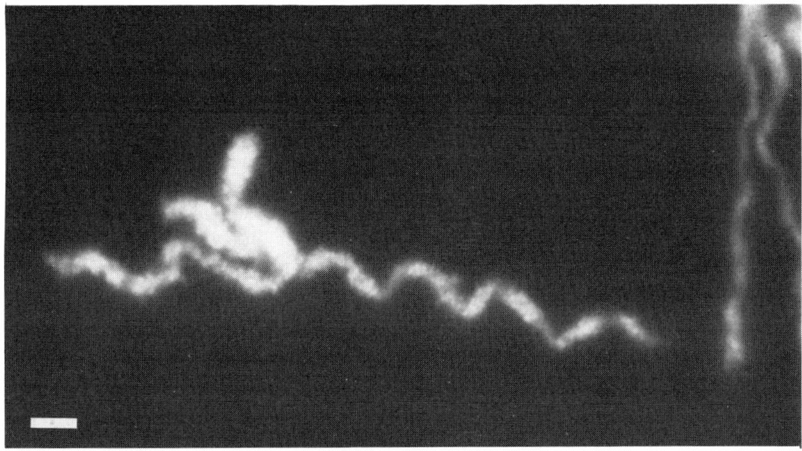

FIGURE 11.   Chromatids are constructed from a chromatin fiber wound in helical gyres. Chromosomes were prepared from mitotic CHO cells in a polyamine containing medium[36] by lysis with Nonidet P40. Lithium diiodosalicylate was added to 25 mM and the chromosomes immediately examined by fluorescent staining with Hoechst 33258. Bar = 10 μm.

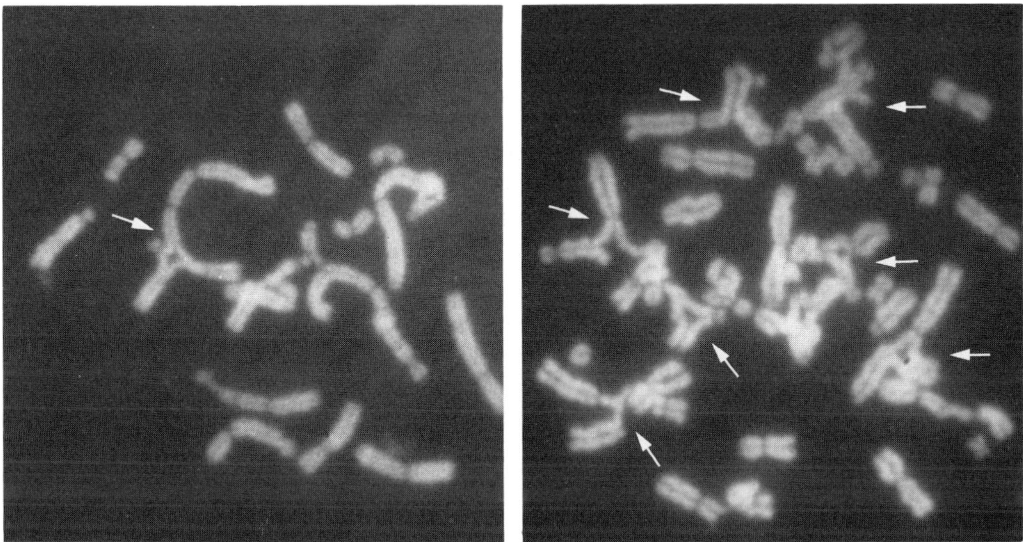

FIGURE 12.   Fused and rearranged chromosomes in CHO cells grown for one replication cycle in the presence of the intercalating agent mAMSA[35] (0.04 μg/ml). Chromosomes were prepared by standard methods for karyotyping. The induction of these rearrangements by an agent which interacts only with DNA suggests that the length and form of any skeletal element which exists in these chromosomes must be determined by the length of DNA with which it is associated, rather than *vice versa*.

REFERENCES

1. **Hancock, R. and Boulikas, T.,** Functional organisation in the nucleus, in *International Review of Cytology,* Vol. 79, Prescott, D. M., Ed., Academic Press, New York, 1982, 165.
2. **Hancock, R.,** Topological organisation of interphase DNA: the nuclear matrix and other skeletal structures, *Biol. Cell.,* 46, 105, 1982.
3. **Hancock, R., Faber, A. J., and Fakan, S.,** Isolation of interphase chromatin structures from cultured cells, in *Methods of Cell Biology,* Vol. 15, Prescott, D. M., Ed., Academic Press, New York, 1977, 127.

 4. **Hancock, R. and Hughes, M.,** Organisation of DNA in the interphase nucleus, *Biol. Cell.,* 44, 201, 1982.
 5. **Hancock, R.,** unpublished data, 1986.
 6. **Tsanev, R. and Tsaneva, I.,** Molecular organisation of chromatin as revealed by electron microscopy, *Methods Achiev. Exp. Pathol.,* 12, 63, 1986.
 7. **Pierron, G. and Sauer, H. W.,** Physical relationship between replicons and transcription units in Physarum polycephalum, *Eur. J. Cell Biol.,* 29, 104, 1982.
 8. **Miller, O. J. and Bakken, A. H.,** Morphological studies of transcription, *Acta Endocrinol. (Copenhagen) Suppl.,* 168, 155, 1972.
 9. **Kaufmann, S. H., Fields, A. P., and Shaper, J. H.,** The nuclear matrix: current concepts and unanswered questions, *Methods Achiev. Exp. Pathol.,* 12, 141, 1986.
10. **Kaufmann, S. H., Okret, S., Wikstrom, A., Gustafsson, J., and Shaper, J. H.,** Binding of the glucocorticoid receptor to the rat liver nuclear matrix: the role of disulfide bond formation, *J. Biol. Chem.,* 261, 11962, 1986.
11. **Arenstorf, H. P., Conway, G. C., Wooley, J. C., and LeSturgeon, W. M.,** Nuclear matrix-like filaments form through artefactual rearrangement of hnRNP proteins, *J. Cell Biol.,* 99, 233a, 1984.
12. **Adam, G. and Delbruck, M.,** Reduction of dimensionality in biological diffusion processes, in *Structural Chemistry and Molecular Biology,* Rich, A. and Davidson, N., Eds., W. H. Freeman, New York, 1968, 198.
13. **McCloskey, M. A. and Poo, M.,** Rates of membrane-associated reactions: reduction of dimensionality revisited, *J. Cell Biol.,* 102, 88, 1986.
14. **Murray, A. B. and Davies, H. G.,** Three-dimensional resonstruction of the chromatin bodies in the nuclei of mature erythrocytes of *Triturus cristatus:* the number of nuclear envelope attachment sites, *J. Cell Sci.,* 35, 59, 1979.
15. **Schardin, M., Cremer, T., Hager, H. D., and Lang, M.,** Specific staining of human chromosomes in chinese hamster × man hybrid cell lines demonstrates interphase chromosome territories, *Hum. Genet.,* 71, 281, 1985.
16. **Manuelidis, L.,** Individual interphase chromosome domains revealed by in situ hybridisation, *Hum. Genet.,* 71, 288, 1985.
17. **Moroi, Y., Hartman, A. L., Nakane, P. K., and Tan, E. M.,** Distribution of kinetochore (centromere) antigen in mammalian cell nuclei, *J. Cell Biol.,* 90, 254, 1981.
18. **Wedrychowski, A, Schmidt, W. N., Ward, S. W.,and Hnilica, L. S.,** Crosslinking of cytokeratins to DNA in vivo by chromium salt and *cis*diamminedichloroplatinum(II), *J. Biol. Chem.,* 260, 7150, 1985.
19. **Forbes, D. J., Kirschner, M. W., and Newport, J. W.,** Spontaneous formation of nucleus-like structures around bacteriophage DNA microinjected into Xenopus eggs, *Cell,* 34, 13, 1985.
20. **Krachmarov, C., Tasheva, B., Markov, D., Hancock, R., and Dessev, G.,** Isolation and characterisation of nuclear lamina from Ehrlich ascites tumor cells, *J. Cell. Biochem.,* 30, 351, 1986.
21. **Krachmarov, C., Iovcheva, C., Hancock, R.,and Dessev, G.,** Association of DNA with the nuclear lamina in Ehrlich ascites tumor cells, *J. Cell. Biochem.,* 31, 59, 1986.
22. **Krachmarov, C. and Dessev, G.,** unpublished data, 1986.
23. **Peterson, S. P. and Berns, M. W.,** Chromatin influence on the function and formation of the nuclear envelope shown by laser-induced psoralen photoreactivation, *J. Cell Sci.,* 32, 197, 1978.
24. **Dessev, G.,** unpublished data, 1986.
25. **Jones, J. C. R., Goldman, A. E., Yang, H., and Goldman, R. D.,** The organisational fate of intermediate filament networks in two epithelial cell types during mitosis, *J. Cell Biol.,* 100, 93, 1985.
26. **Goldman, R., Goldman, A. E., Green, K., Jones, J., Lieska, N., and Yang, H.,** Intermediate filaments: cytoskeletal connecting links between the plasma membrane and nuclear surface, *Ann. N.Y. Acad. Sci.,* 455, 1, 1985.
27. **Goldman, A. E., Maul, G., Steinert, P. M., Yang, H., and Goldman, R.,** Keratin-like proteins that coisolate with intermediate filaments of BHK-21 cells are nuclear lamins, *Proc. Natl. Acad. Sci. U.S.A.,* 83, 3839, 1986.
28. **Galcheva-Gargova, Z. and Dessev, G.,** unpublished data, 1986.
29. **Traub, P., Nelson, W. J., Kuhn, S., and Vorgias, C. E.,** The interaction in vitro of the intermediate filament protein vimentin with naturally occurring RNAs and DNAs, *J. Biol. Chem.,* 258, 1456, 1983.
30. **McKeon, F. D., Kirschner, M. W., and Caput, D.,** Homology in both primary and secondary structure between nuclear envelope and intermediate filament proteins, *Nature,* 319, 463, 1986.
31. **Fisher, D. Z., Chaudhary, N., and Blobel, G.,** cDNA sequencing of nuclear lamins A and C reveals primary and secondary structural homology to intermediate filament proteins, *Proc. Natl. Acad. Sci. U.S.A.,* 83, 6450, 1986.
32. **Paulson, J. R. and Laemmli, U. K.,** The structure of histone-depleted metaphase chromosomes, *Cell,* 12, 817, 1977.
33. **Labhart, P., Koller, T., and Wunderli, H.,** Involvement of higher order chromatin structures in metaphase chromosome organisation, *Cell,* 30, 115, 1982.

34. **Rattner, J. B. and Lin, C. C.,** Radial loops and helical coils coexist in metaphase chromosomes, *Cell,* 47, 291, 85.
35. **Deaven, L., Oka, M. S., and Tobey, R. A.,** Cell cycle-specific chromosome damage following treatment of cultured CHO cells with mAMSA, *J. Natl. Cancer Inst.,* 60, 1155, 1978.
36. **Lewis, C. D. and Laemmli, U. K.,** Higher order metaphase chromosome structure: evidence for metalloprotein interactions, *Cell,* 29, 171, 1982.

Chapter 5

CHARACTERIZATION OF NUCLEOSOME AND NUCLEOSOME-LIKE SUBPOPULATIONS DIFFERENTLY INVOLVED IN GENE EXPRESSION

Paola Caiafa

TABLE OF CONTENTS

I. INTRODUCTION

For many years, the study of the nucleosome structure has been investigated independently from the biological activity of the chromatin fraction to which they belong. The resulting "average" nucleosomal core particle has been represented[1-3] as a 145-bp DNA wrapped around the histone octamer, the latter being composed of two molecules of each of the following histones: H_{2a}, H_{2b}, H_3, H_4.

Recent investigations on the presence of nucleosome or nucleosome-like structures in actively transcribing chromatin fractions have provided evidence of the existence of "active nucleosomes". In fact, their characteristics indicate their origin from chromatin fractions involved in gene expression.[4-12] The term active nucleosome is due to the fact that, through the use of cDNA as gene probe, this kind of nucleosomes has proved to be heavily enriched in transcribed DNA sequences.

Prior et al.,[6] by using the sulfhydryl reagent iodoacetamidofluorescein and electron-microscopic observations, demonstrated an unfolding of the nucleosome core as part of the transcription process of ribosomal genes and called this transient core particle "lexosome".

Nucleosome core particles from transcribing genes with ability to bind eukaryotic RNA polymerase II have been prepared by Baer and Rhodes.[7] This nucleosome subpopulation, enriched in transcribing DNA sequences, showed a lower level of histones H_{2a} and H_{2b}. The authors suggest that the absence of an H_{2a}-H_{2b} dimer is related to the bond with the above mentioned enzyme, and to the gene transcription activity present in the chromatin fraction used. By using DNAase I to study the accessibility of DNA in the nucleosome structure from different nucleosome subpopulations, they showed an apparent unfolding in the active nucleosome.

Sorgen and Butterworth[13] have indicated that both RNA polymerases I and II may be retained on the nucleosome-like particles *in vivo*. HMG_{14} and HMG_{17} have also been found to be specifically associated with active nucleosomes,[15] and Weisbrod and Weintraub[9] have described a method to obtain active nucleosomes using an HMG proteins affinity column. A rather thorough and detailed description of these active nucleosomes could, therefore, be carried out.[10]

Active nucleosome-like core particles, as described in the literature, have revealed the same DNA size as the bulk of nucleosomes, but have been found to contain different amounts of histone proteins,[7-12] somewhat modified by postsynthetic enzymatic reactions,[14] and to be enriched in the nonhistone proteins HMG_{14} and HMG_{17}.[15]

Such conformational alterations of the nucleosome structure[6,7] can be visualized as allowing the chromatin to perform its functions in a dynamic way.

In the present paper, we describe a method for obtaining as many as three different nucleosome or nucleosome-like subpopulations, which can be comparatively investigated, in the aim of correlating their origin from chromatin fractions differently involved in gene expression.

II. PREPARATION OF NUCLEOSOME OR NUCLEOSOME-LIKE PARTICLES AND METHODS USED FOR THEIR CHARACTERIZATION

A. Preparation of Nucleosome Core Particles

Chromatin was prepared from pig kidney, according to Richwood et al.,[16] protease(s) being irreversibly inhibited by 1 mM phenylmethylsulfonyl fluoride (PMSF, Fluka), 0.1 mM 1.2 epoxy-3(p-nitrophenoxyl-propane (EPMP, Kodak), and 30 U aprotinin per mg DNA (Boehringer).

Nucleosomes were obtained by digestion of chromatin with staphylococcal nuclease (Boehringer), 150 U/mg DNA, digestion being then stopped after 13 min at 37°C by addition of

10 mM Na-EDTA.[19] Digested chromatin was layered on a linear 5 to 20% sucrose gradient in 10 mM Tris HCl (pH 7.4) buffer containing 1 mM Na-EDTA and 1 mM PMSF *plus* 0.1 mM EPNP, *plus* aprotinin (30 U/mg DNA) and centrifuged in a Beckman L 5 ultracentrifuge at 25,000 rpm, 4°C, for 17 h, using a SW 27 rotor. The nucleosome subpopulation obtained under these conditions at density ranging between 1.058 and 1.066 was called "N$_1$".

The pellet was resuspended and dialysed against 10 mM Tris/HCl *plus* 1 mM PMSF (pH 7.4) *plus* 0.1 mM EPMP *plus* 30 u aprotinin per mg DNA for 17 h and re-layered on a linear 5 to 20% sucrose gradient in the same Na-EDTA-free dialysis buffer. By centrifugation at 25,000 rpm, 4°C, for 17 h, a nucleosome subpopulation was obtained, at density ranging between 1.024 and 1.029, that was called "N$_2$". The pellet was resuspended in 0.6 M NaCl and then re-layered on a linear 5 to 20% sucrose gradient in 10 mM Tris/HCl *plus* 0.6 M NaCl *plus* 1 mM PMSF (pH 7.4) *plus* 0.1 mM EPHP *plus* 30 U aprotinin per mg DNA centrifuged at 25,000 rpm, 4°C for 17 h. At density ranging between 1.063 and 1.086, a nucleosome subpopulation was obtained that was called "N$_3$".[12]

B. Isolation, Qualitative, and Quantitative Determination of DNA

DNA was purified from whole chromatin or from different nucleosome subpopulations by incubation with 50 µg/ml of proteinase K (Boehringer) for 3 h at 37°C, followed by two extractions with freshly distilled phenol saturated with 100 mM Tris/HCl pH 8, two extractions with chloroform/isoamyl alcohol (24/1, v/v), and precipitation with 2.5 volumes of absolute ethanol in the presence of 0.3 M Na-acetate. DNA was then dissolved in buffer (10 mM Tris/HCl; pH 7.5) and incubated with 25 µg/ml of RNAase (Boehringer) (previously boiled for 10 min to inactivate any contaminating DNAase) for 30 min at 37°C, extracted again with chloroform/isoamyl alcohol, precipitated with 2.5 volumes of absolute ethanol and finally washed with 80% ethanol in the presence of 0.3 M Na-acetate. Determination of the DNA size was performed by agarose gel electrophoresis, using a 40 mM Tris/HCl, 20 mM Na-acetate, 2 mM Na-EDTA (pH 7.9) buffer and, as standards, comigrating Hae III restriction endonuclease fragments of φX 174 DNA or pBR Hin FI.

DNA content was estimated according to Burton,[18] using native calf thymus DNA as standard.

C. Analysis of 5-Methylcytosine in DNA

The content of 5-methylcytosine in various DNA samples was measured using a gas chromatographic-mass spectrometric procedure. Samples of DNA (25 to 100 µg) were dried under vacuum, dissolved in 0.2-ml formic acid (98%) in a pyrolysis tube, heated to 180°C for 1 h so as to hydrolyse the DNA into its component bases, and transported to microvials (total capacity 1 ml). Formic acid was then evaporated by a stream of pure nitrogen at 50°C, and each sample was additioned with 20 µg of phenantrene, dissolved in 20 µl acetone, as internal standard. After evaporation of acetone by flushing with pure nitrogen, the samples were dissolved into 0.1 ml acetonitrile, silylated[19] with 30 µl of bis (trimethyl-silyl)-tri-fluoroacetamide (BSTFA) by heating for 1 h at 150°C. The silyl derivatives were analyzed by GLC, using a Dani model 800 gas chromatograph equipped with a FID unit and PTV injector programmed from 100° to 240°C in total mode. To this purpose, measured aliquots (0.2 to 0.8 µl) of the reaction mixture were injected into a 25-m long, 0.25 mm i.d. silica column coated with methylsilicone fluid, operated at 170°C for 6 min, then programmed at the rate of 30°C min^{-1} to the final temperature of 230°C. The identity of the products was established by comparison of their retention volumes with those of authentic samples, obtained from the silylation of pure bases, and further confirmed by gas chromatography/mass spectrometer or a 2AB-2F mass instrument (Micromass Ltd.) operated in the electron-impact mode, with an electron energy of 70 eV.

Under the experimental conditions used, each base gave only one derivative, both the amino and the hydroxyl groups undergoing silylation. The relative yields of thymine, cytosine, and 5-methylcytosine were determined from the heights of the correspondent elution peaks. Appropriate calibration curves, obtained by plotting the ratios of the heights of each peak vs. the relative amounts of the silylated pyrimidine bases used as standards (either as free bases or as hydrolysis products of DNA with known base composition) showed that the peak heights reflected faithfully the base content.

D. Extraction and Characterization of Histones and of Loosely-Bound Nonhistone Proteins with High or Low Electrophoretic Mobility

Histone proteins were obtained from the different nucleosome subpopulations by extraction in 0.4 M H_2SO_4.[22]

In order to obtain loosely-bound nonhistone proteins (which can be characterized in terms of their electrophoretic mobility), 1.4 mg (expressed as DNA) of each nucleosome population was dialysed against 10 mM Tris/HCl (pH 7.2) buffer and adsorbed on hydroxyapatite[12] in the same buffer (1 g hydroxyapatite per mg DNA). After centrifugation, the pelleted nucleosome-hydroxyapatite complex was resuspended in 10 mM Tris/HCl (pH 7.2) buffer additioned with 0.35 M NaCl, kept for 30 min at 4°C and then centrifuged at 7000 rpm. The HMG (high mobility group) protein fraction was prepared from the supernatant by fractional precipitation between 2 and 20% (w/v) thrichloroacetic acid.[21] The LMG (low mobility group) fraction was in the supernatant of 20% TCA.

One-dimensional slab gel electrophoresis was performed according to Laemmli[22] in 15 or 18% polyacrylamide gels for HMG and 10% for LMG. Gel electrophoresis of histone protein standards (Sigma), or of marker proteins with molecular masses ranging between 68 kDa and 21.2 kDa allowed a comparison among the protein fractions.

E. RNA Polymerase Activity

Presence of endogenous RNA polymerase was assayed, as previously described,[23] with a constant amount of nucleosomes (9 μg DNA) and a constant amount of thymus DNA (3.5 μg DNA). The standard reaction mixture (150 μl) contained: 50 mM Tris/HCl (pH 8) buffer, 10% glycerol (v/v) 5 mM $MgCl_2$, 0.01 mM Na-EDTA, 0.5 mM each of GTP, CTP, and ATP, 0.08 mM UTP, 1 μCi of ^{3}H UTP, the final specific activity of UTP being 83.3 μCi/mmol. After incubation for 13 min at 28°C, a 100 μl aliquot was withdrawn from each reaction mixture and applied directly to a 24 mm Whatman DE 81 filter disc. The filters were washed five times with 5% (w/v) Na_2HPO_4, twice with water, twice with absolute ethanol, twice with acetone, and dried. Radioactivity was estimated in a liquid scintillation spectrometer. The fungal toxin α-amanitin (Boehringer) was used as a specific inhibitor for a qualitative characterization of the enzyme.[24]

F. Extraction and Characterization of Tightly-Bound Nonhistone Proteins

Nonhistone chromatin proteins that need 5 M urea *plus* 4 M guanidine-HCl to be released from DNA are designated as tightly-bound nonhistone chromatin proteins (tbNHCp). To obtain this protein class from whole chromatin or nucleosome subpopulations, two alternative methods were used (Figure 1).

As shown in Figure 2, whole chromatin or nucleosome subpopulations were pooled and dialyzed against 10 mM Tris/HCl pH 7.2 buffer (buffer A) and adsorbed on hydroxyapatite[25] (HTP, 1 g/mg DNA) in conditions of low ionic strength. Proteins were dissociated from DNA by batchwise treatment with the following buffers of increasing ionic strengths: buffer A *plus* 0.35 M NaCl; buffer A *plus* 3 M NaCl; buffer A *plus* 3 M NaCl *plus* 5 M urea *plus* 4 M guanidine-HCl. The last buffer released from DNA the tightly bound nonhistone proteins.[26]

To prove that the retention on hydroxyapatite of proteins released from chromatin fractions,

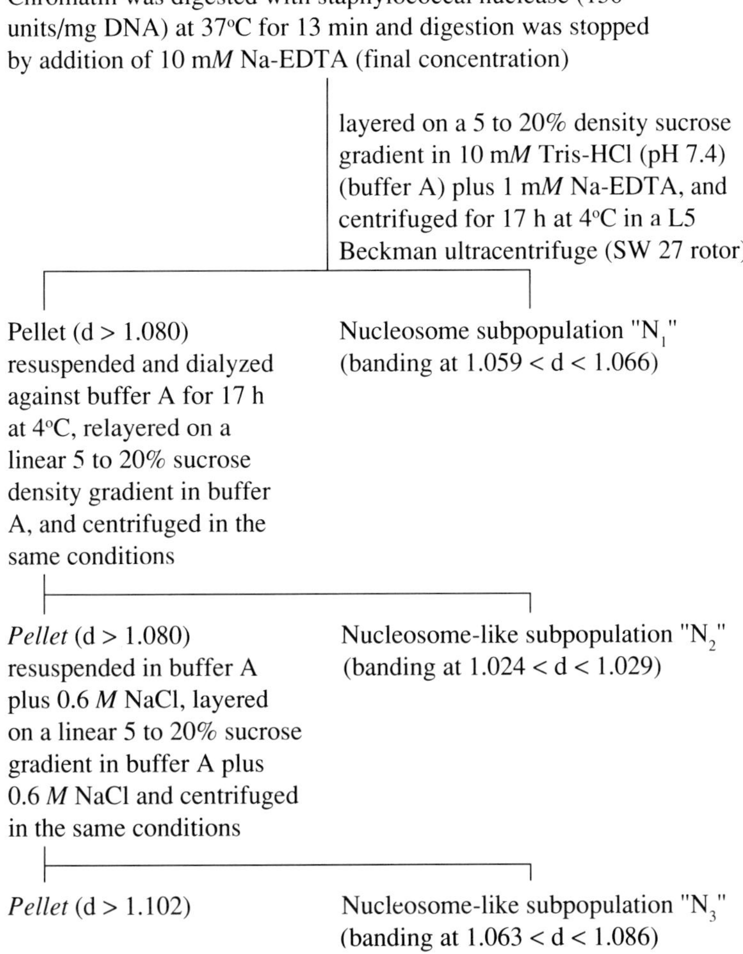

FIGURE 1. Scheme for the preparation of nucleosome or nucleosome-like subpopulations.

but not eluted by 3 M NaCl, did not result in any contamination of the final extract containing the tightly-bound nonhistone proteins, an alternative method was used (Figure 3), essentially analogous to the procedure proposed by Bekhor and Mirell.[21] To this purpose, whole chromatin, diluted so as to have A_{260} = 0.5, was layered, after dissociation in buffer A *plus* 3 M NaCl *plus* 20 mM Na-EDTA for 1 h, over a 10-ml cushion of 1 M sucrose in the same dissociating buffer and centrifuged for 20 h at 58,000 rpm, at 4°C, in a Beckman-type 70 Ti rotor. The presence of 20 mM Na-EDTA in the buffer during the chromatin dissociation and ultracentrifugation step was found to be essential in preventing the nuclease digestion of DNA. The resulting pellet was considered as being the tbNHCp-DNA complex from whole chromatin or "stripped chromatin".

In order to obtain the tbNHCp-DNA complex from N_1 nucleosomes, the latter, purified by the usual linear sucrose gradient, were dissociated in buffer A *plus* 3 M NaCl and 20 mM Na-EDTA for 1 h, and centrifuged again for 6 h at 58,000 rpm, 4°C in a linear 5 to 12.5% sucrose gradient in the same dissociating buffer.[26]

The tbNHCp-DNA complexes from whole chromatin or from the core particles were then adsorbed on hydroxyapaptite in buffer A *plus* 3 M NaCl, and the tbNHCp released from DNA by means of urea plus guanidine-HCl. We found that, for a given chromatin or N_1 nucleosome preparation, the same quantitative and qualitative tbNHCp components were

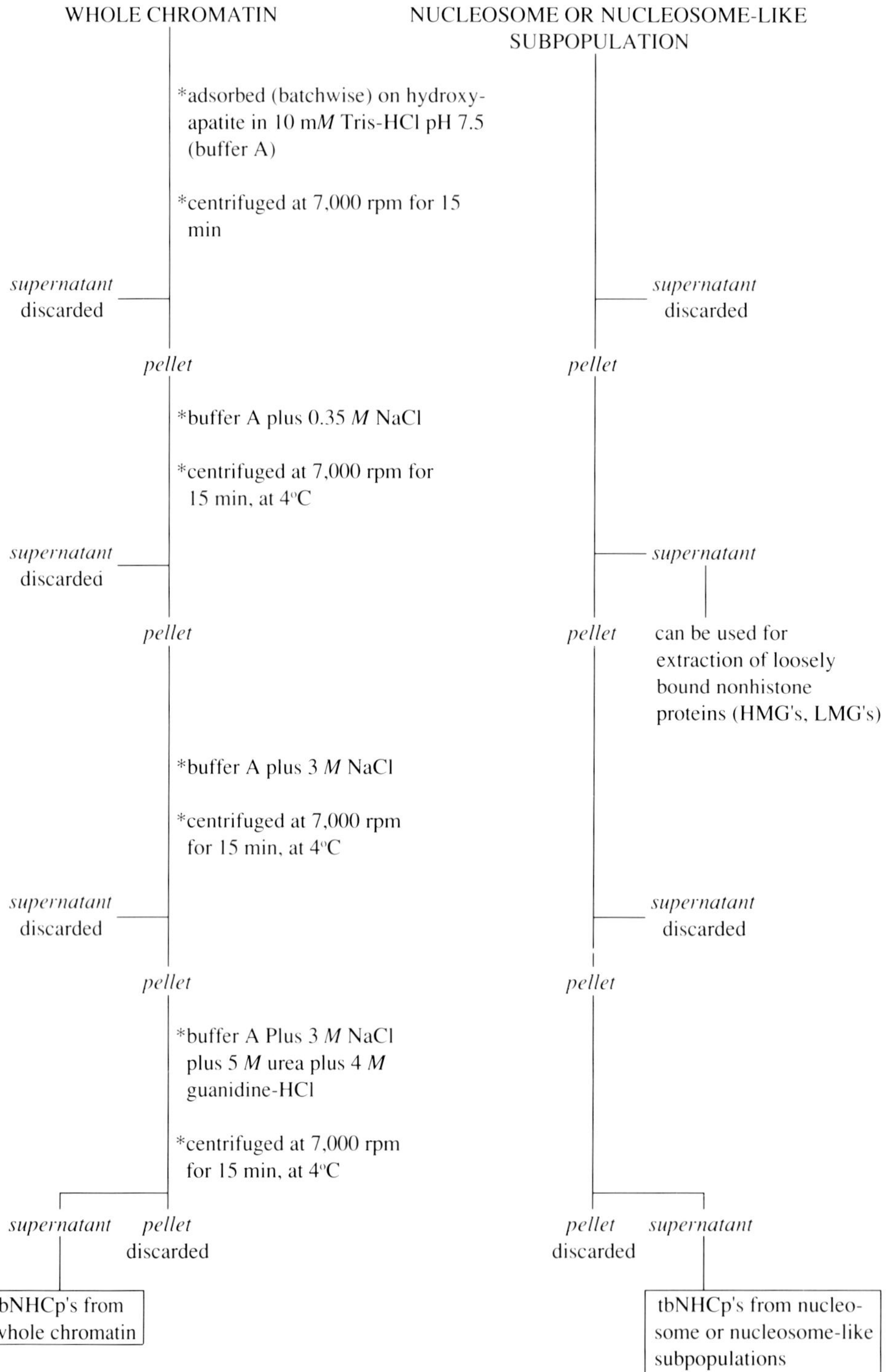

FIGURE 2.   Protocol for the preparation of tbNHCp by hydroxyapatite adsorption-desorption of either whole chromatin, or of nucleosome or nucleosome-like particles.

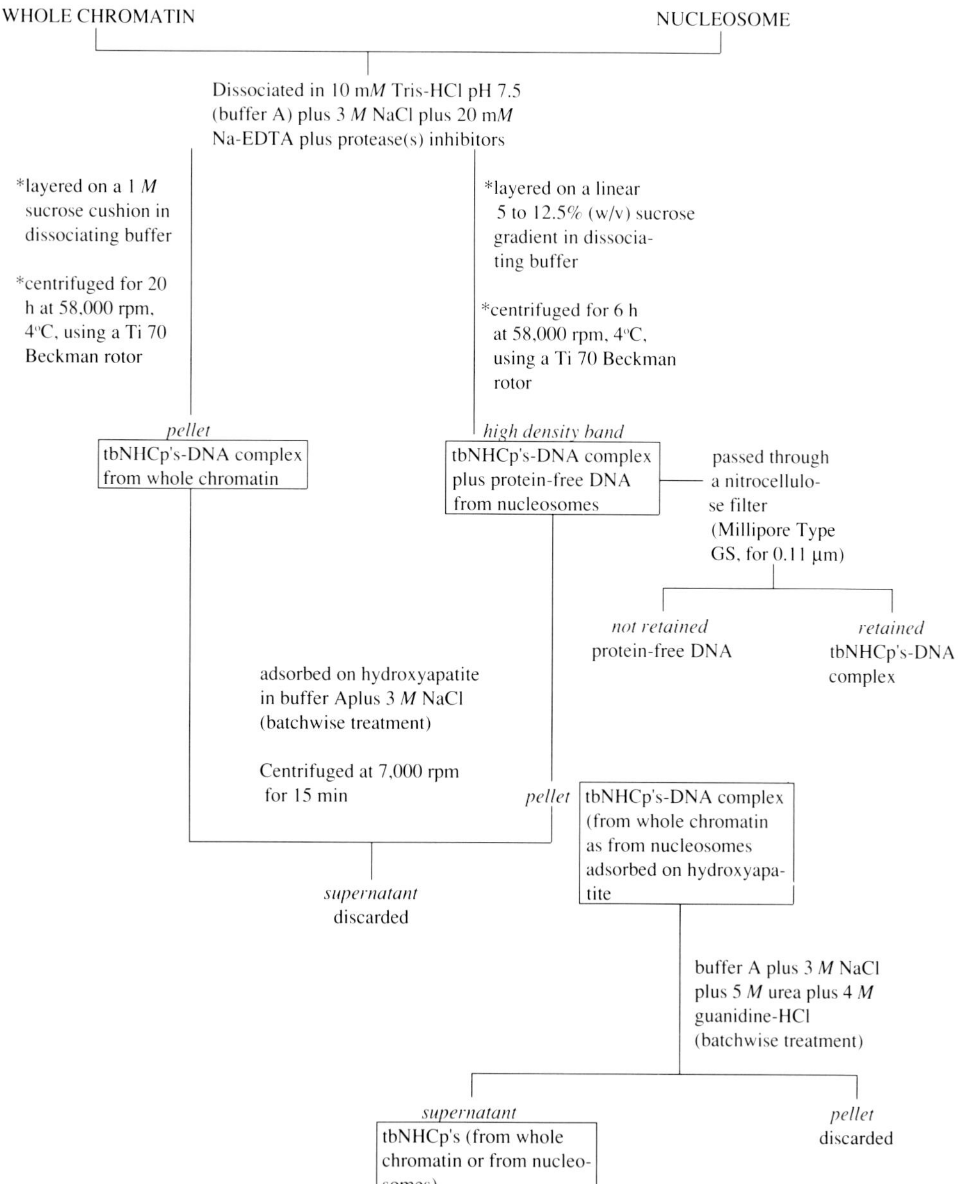

FIGURE 3.   Protocol for the preparation of tbNHCp by preliminary preparation of tbNHCp-DNA complexes and subsequent hydroxyapatite adsorption-desorption.

present irrespectively of the purification procedure used; the method involving adsorbtion of chromatin or nucleosomes to hydroxyapatite in conditions of low ionic strength and followed by washing with different buffer, described in Figure 2, was therefore preferred as being more practical and giving equally satisfactory results.

The alternative method, described in Figure 3, was however necessarily performed in order to obtain the tbNHCp-core DNA complex, so as to allow the determination, by filtration on nitrocellulose,[26] of the relative amounts of protein-free and of tbNHCp-bound core particle DNA.

Tightly-bound nonhistone proteins were characterized by two-dimensional gel electrophoresis performed essentially according to O'Farrel.[28]

Isoelectric focusing was performed in the 8 to 4 pH range (protein bands out of this range failing, therefore, to be separated), the samples having been solubilized according to Peters and Comings.[29]

For the run in the second dimension, 10% acrylamide slab gels, calibrated for mean molecular mass by comparison with known proteins (bovine serum albumin, ovalbumin, carbonic anhydrase, and bovine trypsin inhibitor) were used, the slab gels being thereafter subjected to silver stain.[30]

Quantitative determination of proteins was carried out according to Lowry et al.[31] and Bradford[32] and occasionally verified with an amino acid analyser LKB 4400 after hydrolysis of the proteins in 6 M HCl at 110°C for 24 h.

G. Characterization of Nucleosome Subpopulation "N_1"

The nucleosome subpopulation "N_1", obtained by staphylococcal nuclease digestion from pig kidney chromatin, was purified by a linear sucrose gradient at a density ranging between 1.058 and 1.066 (Figure 1). The relative yield[12] of this nucleosome subpopulation was 30%, with respect to the DNA from whole chromatin taken as 100% (Table 1).

Its DNA, analyzed by agarose slab gel electrophoresis, showed an approximate size of 145 bp, typical of DNA core particle (more precisely between 125 and 171 bp), with some contamination by dimeric particles (Figure 4).

When examined for its 5-methylcytosine content[12] by gas chromatographic-mass spectrometric analysis of the silylated formic acid hydrolysate, the nucleosome subpopulation "N_1" was found to have, as compared to whole chromatin DNA, a higher relative abundance of 5-methylcytosine and a lower amount of cytosine *plus* methylcytosine (Table 1, column 3).

As for its histone protein content (Figure 5), all core histones as well as histone H_1 were present, with a histones/DNA (w/w) ratio of 1.08 (Table 2).

There were only small amounts of loosely-bound nonhistone proteins, with very low values (0.003 and 0.008, respectively) of both the HMGp/DNA and LMGp/DNA ratios (Table 2).

The HMG proteins could not, therefore, be characterized in terms of electrophoretic pattern (Figure 6) while LMG proteins showed, in SDS-PAGE, three main bands having mean molecular masses ranging between 58.8 and 52.5 kDa (Figure 7).

The tightly-bound nonhistone proteins are, by weight, about 5% (w/w) relative to the core DNA content (Table 2). In two-dimensional slab gel electrophoresis, these tightly-bound nonhistone proteins were characterized by having molecular masses ranging from 70 to 23 kDa and isoelectric points from pH 7.6 to pH 5.4 (Figure 8).[26]

When the tbNHCp-DNA complex from nucleosomes (dissociated in 3 M NaCl and obtained by linear 5 to 12.5 sucrose gradient, Figure 3) were passed through a nitrocellulose filter,[26] about 10% of the total nucleosomal DNA was retained, showing its association to proteins, the remaining 90% being protein-free. Only 10% of core particles appeared, therefore, to be associated to nonhistone proteins tightly-bound to DNA.

If we consider that the tbNHCp/DNA ratio is 0.05 in the total subnucleosomal fraction "N_2", and assuming a molecular mass of 100 kDa for the core DNA and an average molecular mass of 50 kDa for this protein class (Figure 9a), we can conclude that this 10% of nucleosomes associated with tbNHCp contained, on the average, one protein molecule for each core DNA.[26]

We also investigated the presence of endogenous RNA polymerase, finding that the nucleosome subpopulation "N_1" is eight times less rich in endogenous RNA polymerase activity as compared to the nucleosome-like subpopulation "N_2"[12] (Table 3).

H. Characterization of Nucleosome-Like Subpopulation "N_2"

The nucleosome-like subpopulation "N_2" was obtained, after staphylococcal nuclease digestion, at a floating density ranging beteen 1.024 and 1.029 in a linear sucrose gradient

Table 1
DISTRIBUTION OF 5-METHYLCYTOSINE IN THE FIRST LEVEL OF CHROMATIN ORGANIZATION AND STUDENT'S t ANALYSIS

DNA from:	Relative DNA recovery (%)	Relative 5-methyl cytosine abundance (vs. cytosine *plus* 5-methylcytosine) $\dfrac{m^5C \times 100}{C + m^5C}$	Relative thymine abundance (vs. all pyrimidine bases) $\dfrac{T \times 100}{Py}$	Significance, by Student's t analysis, of differences in 5-methylcytosine abundance		
				vs. "N$_1$"	vs. "N$_2$"	vs. "N$_3$"
Whole chromatin	100	4.6 ± 0.3	56.5	$p < 1\%$	$p < 1\%$	$1\% < p < 5\%$
Nucleosome "N$_1$"	30	9.9 ± 0.8	70.6	—	$p < 1\%$	$1\% < p < 5\%$
Nucleosome "N$_2$"	2	7.2 ± 0.5	61.5	$p < 1\%$	—	$p < 1\%$
Nucleosome "N$_3$"	24	13.4 ± 1.0	68.8	$1\% < p < 5\%$	$p < 1\%$	—

Note: The amounts of cytosine, 5-methylcytosine, and thymine were estimated by gas chromatography of the BSTFA derivatives of the formic hydrolysates of chromatin and nucleosomal DNAs.

From Caiafa, P., Attina, M., Cacace, F., Tomassetti, A., and Strom, R., *Biochim. Biophys. Acta,* 867, 195, 1986. With permission.

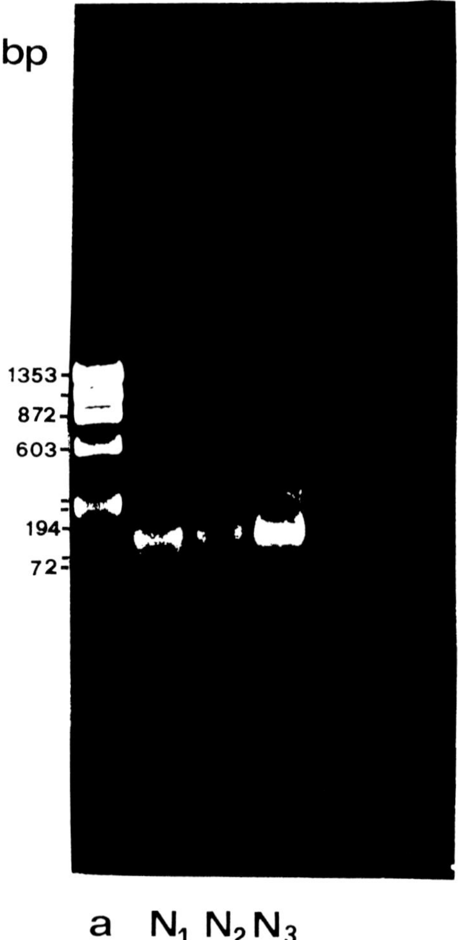

FIGURE 4.   Agarose slab gel electrophoresis (2%) of DNA. (a) α X 174 DNA Hae III restriction fragments; (N$_1$, N$_2$, N$_3$) DNA obtained from nucleosome or nucleosome-like subpopulations. (From Caiafa, P., Attina, M., Cacace, F., Tomassetti, A., and Strom, R., *Biochim. Biophys. Acta*, 867, 195, 1986. With permission.)

(Figure 1). The relative yield of this nucleosome subpopulation was 2% with respect to the DNA from whole chromatin taken as 100% (Table 2).

Its DNA, analyzed by agarose slab gel electrophoresis, showed an approximate size of 145 bp typical of DNA core particle (more precisely between 125 and 171 bp), with some contamination by dimeric particles, (Figure 4).

When examined for its 5-methylcytosine content,[12] the nucleosome subpopulation "N$_2$" was found to have, as compared to whole chromatin DNA, a higher relative abundance of 5-methylcytosine and a lower amount of cytosine *plus* methylcytosine (Table 1, column 3). Its DNA is therefore less hypermethylated than the DNA of nucleosome subpopulation "N$_1$" and even less if compared to the DNA obtained from the nucleosome-like subpopulation "N$_3$".

When examined for its histone proteins content, striking differences were found, both qualitatively and quantitatively (Figure 5 and Table 2), with respect to the nucleosome subpopulation "N$_1$", with a histones/DNA (w/w) ratio of 0.83 (as compared to 1.08). Histone H$_1$ was notably absent from this "N$_2$" subpopulation, and there were lower amounts of H$_{2b}$.

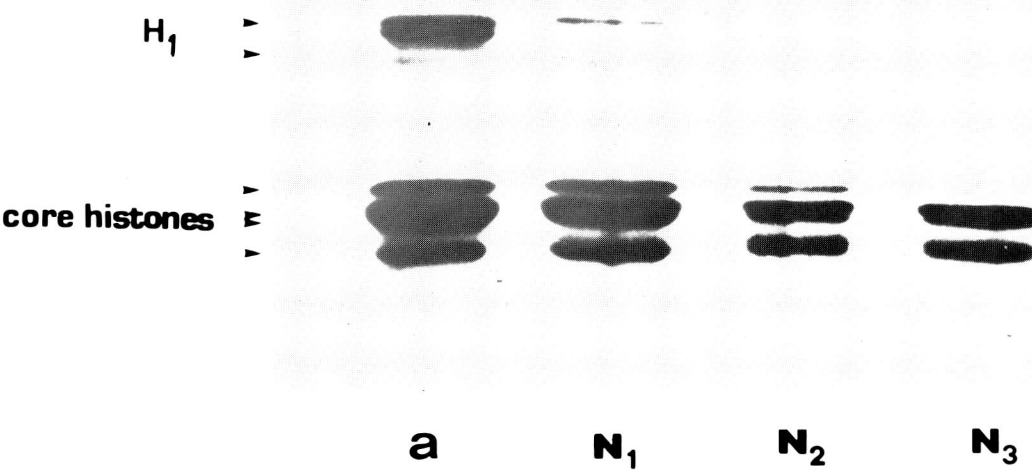

FIGURE 5.   Polyacrylamide slab gel (18%) electrophoretic analysis of histone proteins obtained from 50 μg (expressed as DNA) of each nucleosome subpopulation. (a) Histone protein standards; (N_1, N_2, N_3) histone protein samples obtained from each nucleosome or nucleosome-like subpopulations. (From Caiafa, P., Attina, M., Cacace, F., Tomassetti, A., and Strom, R., *Biochim. Biophys. Acta*, 867, 195, 1986. With permission.)

Table 2
CHARACTERIZATION OF PROTEINS FROM THE DIFFERENT SUBPOPULATIONS OF NUCLEOSOME OR NUCLEOSOME-LIKE PARTICLES

	"N_1"	"N_2"	"N_3"
Histones/core DNA (w/w) ratio	1.08	0.83	0.75
Histones			
H_1	*Present* (but less intense than other histones)	Absent	Absent *(a priori)*
H_3	Present	Present	Present (less intense)
H_{2a}	Present	Present	Present
H_{2b}	Present	Present (but somewhat less intense)	Present (but definitively less intense)
H_4	Present	Present	Present
HMGs/core DNA (w/w) ratio	0.003	0.020	Absent *(a priori)*
HMGps			
HMG_{14}	Practically absent	Present	Absent *(a priori)*
HMG_{17}	Practically absent	Present	Absent *(a priori)*
LMGps/core DNA (w/w) ratio	0.008	0.018	Absent *(a priori)*
Range of molecular masses	kDa 58.8 to 52.5	kDa 61.7 to 44.7	Absent *(a priori)*
LMGps			
Heterogeneity (number of bands)	3	7	Absent *(a priori)*
tbNHCps/core DNA (w/w) ratio	0.05	0.04	0.35
Range of molecular masses	kDa 70 to 23	kDa 69 to 45	kDa 69 to 60
tbNHCps			
Range of isoelectric points	pH 7.6—5.4	pH 7.5—5.4	pH 7.5—5.5
Heterogeneity (number of spots)	40	17	11

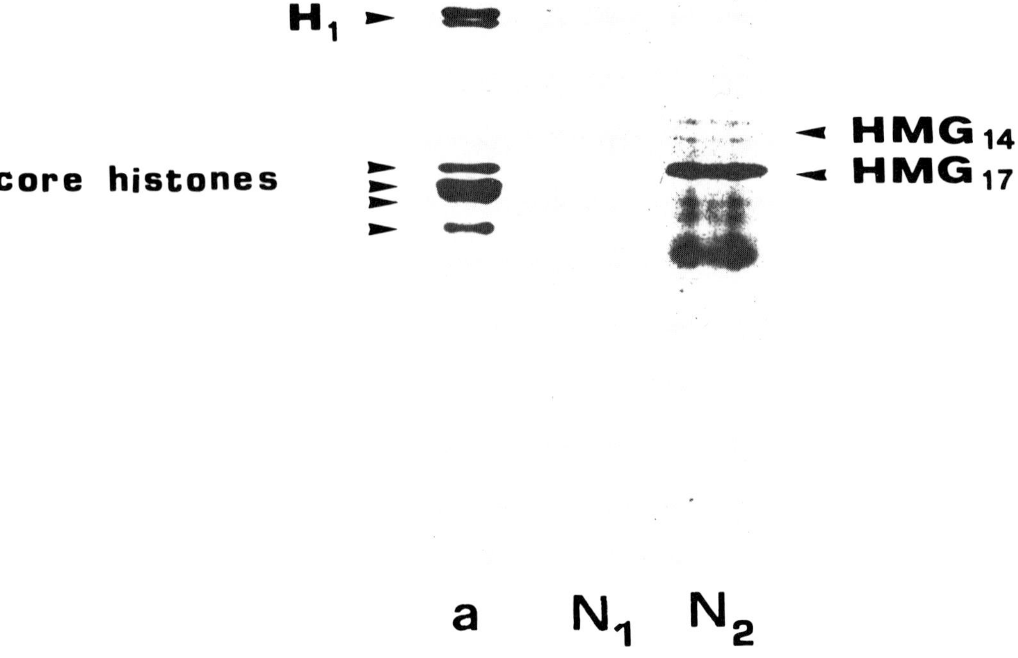

FIGURE 6.   Polyacrylamide slab gel (15%) electrophoretic analysis of HMG proteins obtained from 1.4 mg (expressed as DNA) of nucleosome "N₁" or nucleosome-like "N₂". (a) Histone protein standards; (N₁, N₂) HMG protein samples from nucleosome "N₁" or nucleosome-like "N₂". (From Caiafa, P., Attina, M., Cacace, F., Tomassetti, A., and Strom, R., *Biochim. Biophys. Acta,* 867, 195, 1986. With permission.)

This nucleosome subpopulation contained a significant amount of HMG_{14} and relatively large amounts of HMG_{17} *plus* some other components of lower molecular masses (Figure 6).

The quantitative HMGp/DNA (w/w) ratio was 0.02, while the LMGp/DNA (w/w) ratio was 0.018 (Table 2). The electrophoretic pattern of LMG proteins showed seven protein bonds with molecular masses ranging between 61.7 and 44.7 kDa (Figure 7).

When examined for the presence of tightly-bound nonhistone chromatin proteins, this nucleosome-like subpopulation had, for the tbNHCp/DNA (w/w) ratio, a value of 0.04 (Table 2). Figure 9b shows the two-dimensional slab gel electrophoresis pattern of these tbNHCp: molecular masses ranged from 69 to 45 kDa and isoelectric points from pH 7.5 to pH 5.5.

We investigated the presence of endogenous RNA polymerases, and found that the nucleosome-like subpopulation "N₂" was eight times richer in endogenous RNA polymerase activity than "N₁" (Table 3). This activity was not inhibited by 100 µg/ml of α-amanitin. suggesting it to be RNA polymerase I.[12]

I. Characterization of Nucleosome-Like Subpopulation "N₃"

The nucleosome-like subpopulation "N₃" was released from staphylococcal nuclease digested chromatin only when the ionic strength of the medium had been increased by addition of 0.6 *M* NaCl.[13] It was then purified by a linear sucrose gradient at a floating density ranging from 1.063 to 1.086 (Figure 1). the relative yield of this nucleosome subpopulation was 24%, with respect to the DNA from whole chromatin taken as 100% (Table 2). Its DNA, analyzed by agarose slab gel electrophoresis, showed an approximate size of 145 bp typical of DNA core particle (more precisely between 125 and 171 bp), with some contamination by dimeric particles (Figure 4).

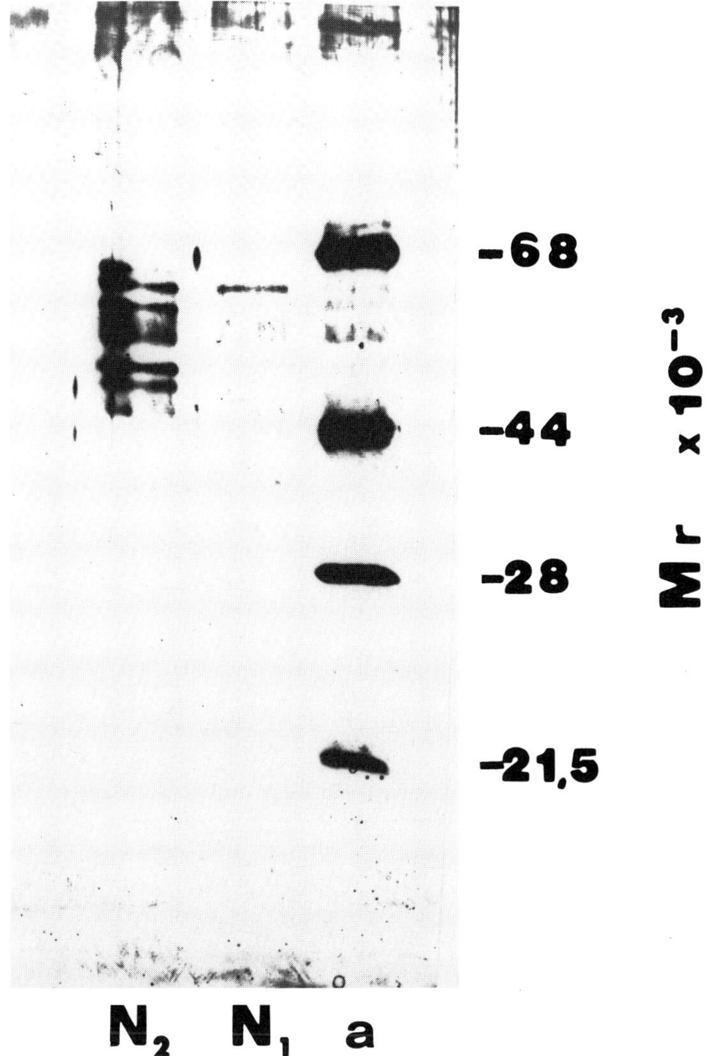

FIGURE 7. Polyacrylamide slab gel (10%) electrophoretic analysis of LMG proteins obtained from 1.4 μg (expressed as DNA) of nucleosome "N_1" or nucleosome-like "N_2". (a) Protein standards; (N_1, N_2) LMG protein samples from nucleosome "N_1" or nucleosome-like "N_2".

When examined for its 5-methylcytosine contents,[12] it was found to have, as compared to whole chromatin DNA, a higher relative abundance of 5-methylcytosine and a lower amount of cytosine *plus* methylcytosine (Table 1, column 3). Its DNA is hypermethylated in comparison to DNA from nucleosome subpopulation "N_1" and from nucleosome-like subpopulation 'N_2".

When examined for its histone proteins content (Table 2 and Figure 5), a histones/DNA (w/w) ratio of 0.75 was found, the electrophoretic bands of histone H_3 and H_{2b} were less intense, and (almost by definition, since this nucleosome has been obtained at a relatively high ionic strength) there was no histone H_1.

The isolation at a relatively high ionic strength resulted also in the absence, in this nucleosome-like subpopulation, of HMG and LMG proteins.

When the presence of tightly-bound nonhistone proteins in this nucleosome subpopulation

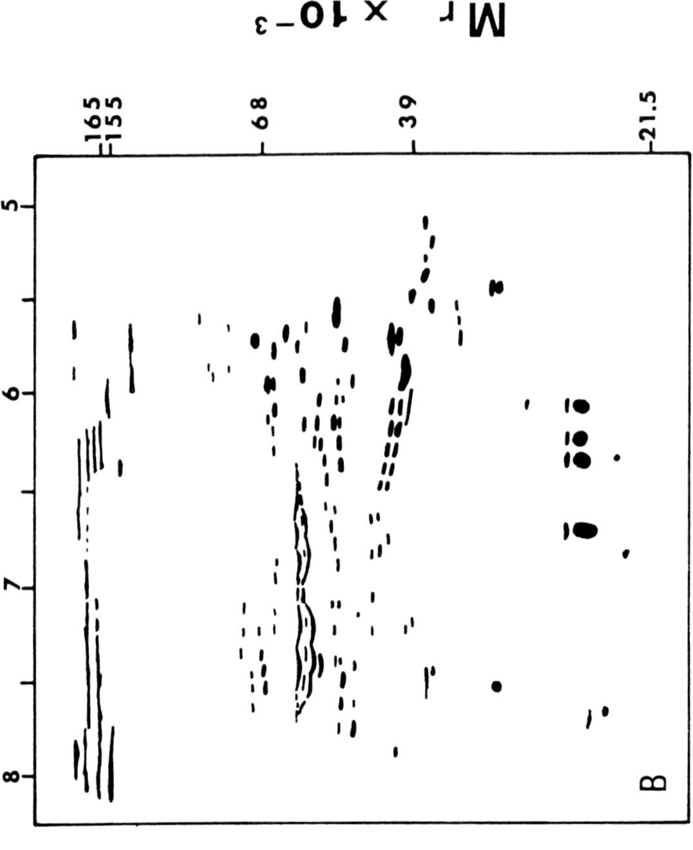

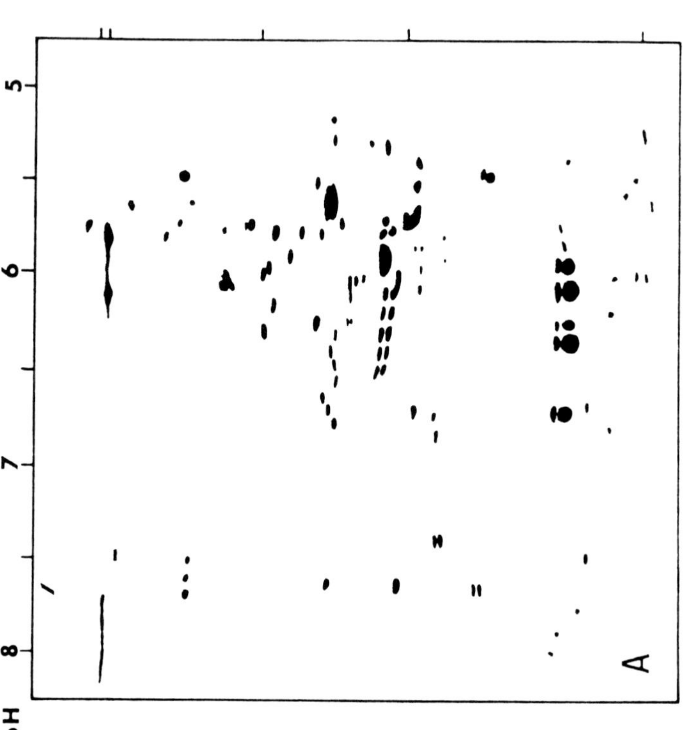

FIGURE 8. Two-dimensional slab gel electrophoresis of tightly-bound nonhistone proteins from (A) whole liver chromatin; (B) whole kidney chromatin. (From Caifa, P., Tomassetti, A., Mastrantonio, S., Reale, A., Spinelli, M., and Strom, R., *Cell Biochem. Function*, 6, 39, 1988. Reprinted by permission of John Wiley & Sons, Ltd.)

was examined, we found that these proteins are very abundant, the tbNHCp/DNA (w/w) ratio reaching a value as high as 0.35 (Table 2). In two-dimensional slab gel electrophoresis, the tbNHCp are characterized by molecular masses ranging from 69 to 60 kDa and isoelectric points from pH 7.5 to pH 5.5 (Figure 9c).

As for the presence of endogenous RNA polymerase activity, "N_3" showed an activity higher than "N_1", though significantly lower than "N_2" (Table 3).

III. DISCUSSION

This paper deals with several currently debated questions concerning the chromatin components and their organization, notably: (1) the existence of different nucleosome or nucleosome-like subpopulations; (2) the distribution of 5-methylcytosine between nucleosomal or internucleosomal DNA and the methylation of DNA from chromatin fractions differently involved in the gene expression; (3) the structural and/or functional role of tightly-bound nonhistone proteins in chromatin organization.

Results obtained by our research group on the characterization of nucleosome subpopulations confirm that by sucrose density gradient centrifugation in the presence or absence of Na-EDTA and at different ionic strengths, it is possible to obtain well-defined nucleosome subpopulations. A DNA size of about 145 bp is the only common characteristic the qualitative and quantitative analysis of the histone proteins from different subpopulations induce to consider the particles of subfractions "N_2" and "N_3" as nucleosome-like.

The various subpopulations show also other distinct and significant differences: they differ not only in their ability to sediment under different conditions and in the 5-methylcytosine contents of their DNA, but also in their protein contents (HMG, LMG, tbNHCp) and in the endogenous RNA polymerase activity associated with them.

The 5-methylcytosine level in DNA can be considered as a chemical probe indicating the derivation of a given nucleosome or nucleosome-like subpopulation from chromatin fractions differently involved in gene expression.

This distribution of 5-methylcytosine between nucleosomal and internucleosomal DNA has been the object of conflicting reports.[33-35] Our results on the distribution of 5-methylcytosine confirm essentially the results obtained by Razin and Cedar[33] and Solage and Cedar,[36] i.e., that nucleosomal DNA is indeed hypermethylated with respect to DNA from whole chromatin, the 5-methylcytosine levels in DNA nucleosome subpopulations ranging between 7.2 to 13.0%, as opposed to 4.6% in DNA from whole chromatin.[13] (Table 1). Nucleosomal DNA could, however, also be shown to have lower amounts of cytosine *plus* methylcytosine, Table 1, column 3. The 5-methylcytosine level of internucleosomal DNA can therefore be estimated to be low, around 0.3%, but still nonnegative.

It is moreover worth noting that the three nucleosome and nucleosome-like subpopulations are differently methylated[12] (Table 1). The reason for such differences in the methylation level of these nucleosome and nucleosome-like subpopulations can presumably be traced to their peculiar characteristics (Figure 1 to 7, 9, and Tables 1 to 3). The methylation of DNA from chromatin fractions differently involved in the gene expression has been extensively investigated, and many data show an undermethylation of DNA from chromatin fraction involved in gene expression.[37-42] Ball et al.,[43] by use of specific antibodies against 5-methylcytosine, have concluded that this methylated base is localized primarily in nucleosomes which contain histone H_1, a lower level being present in nucleosomes lacking H_1 or possessing HMG proteins. Weisbrod[10] has found, by nearest-neighbor analysis, that active nucleosomes are undermethylated as compared to bulk nucleosome DNA.

The characteristics of our nucleosome and nucleosome-like subpopulations are consisting with the following hypotheses: (1) nucleosome "N_1" is the typical nucleosome originating from chromatin fraction(s) not engaged in template activity; (2) nucleosome-like "N_2"

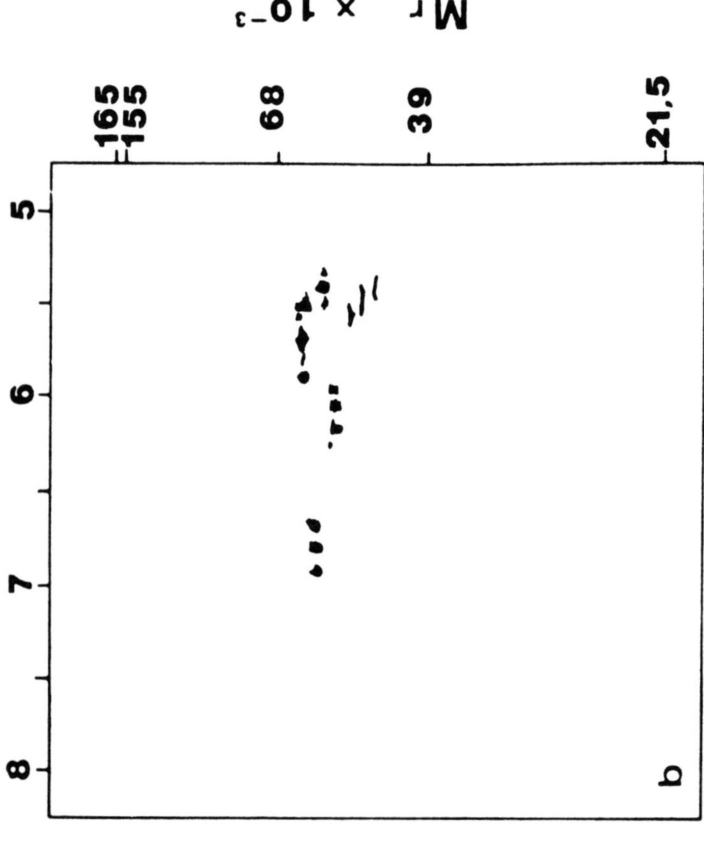
Mr x 10⁻³
165
155
68
39
21.5
5
6
7
8
b

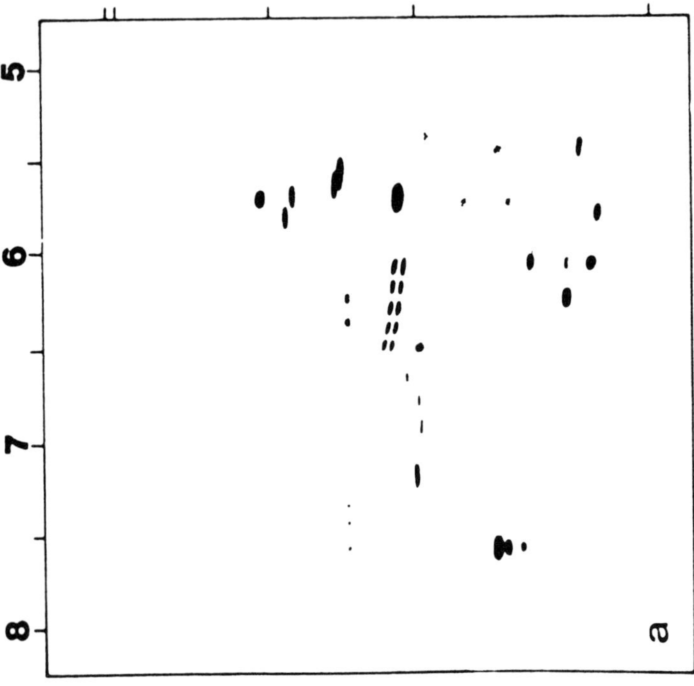
5
6
7
8
a

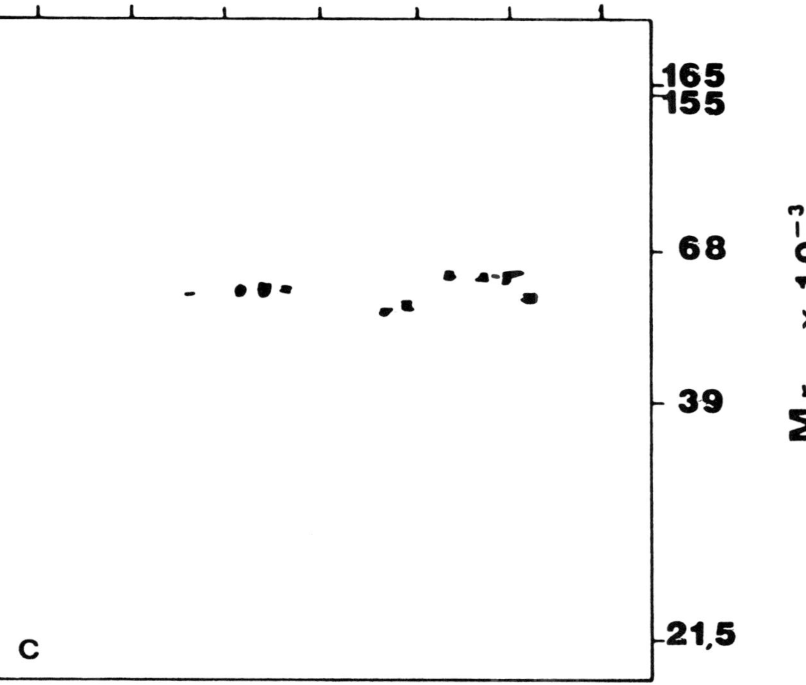

FIGURE 9. Two-dimensional slab gel electrophoresis of tightly bound non-histone proteins from (a) nucleosome "N$_1$"; (b) nucleosome-like "N$_2$"; (c) nucleosome-like "N$_3$". (From Caiafa, P., Tomassetti, A., Mastrantonio, S., Reale, A., Spinelli, M., and Strom, R., *Cell Biochim. Funct.*, 6, 30, 1988. With permission.)

Table 3
ENDOGENOUS RNA POLYMERASE ACTIVITY ASSAYED
IN NUCLEOSOME SUBPOPULATIONS
"N_1", "N_2", AND "N_3"[a]

	"N_1"	"N_2"	"N_3"
dpm In ^{3}H - UTP incorporated as RNA after 13 min at 28°C	2977 ± 1059	24,574 ± 3547	8679 ± 1275

Note: Endogenous RNA polymerase activity was assayed by incubating at 28°C for 13 min each nucleosome subpopulation (60 μg DNA/ml) with ^{3}H-UTP (6.7 μCi/ml), specific activity 8.4 mCi/m mol) in 50 mM Tris/HCl buffer (pH 8) additioned with 10% glycerol (v/v), 5 mM MgCl$_2$, 0.01 M Na-EDTA, 0.5 mM each of GTP, CTP, ATP, and 0.08 mM UTP.

[a]   Data express mean ± S.E. from ten experiments.

presumably originates from a chromatin fraction involved in template activity, in accordance to the criteria defined by Ball et al.,[43] Weisbrod,[10] and Baer and Rhodes;[7] (3) nucleosome-like "N_3" is likely to originate from a particular compact and inactive chromatin fraction, i.e., presumably from condensed heterochromatin.[44,45] This last conclusion is supported by the high level of methylation in this nucleosome-like subpopulation and by the condition of high ionic strength needed to obtain it.

The nucleosome-like N_2 subpopulation, which possess HMG proteins, appears therefore from our results (which are in accordance with data from other authors[10,43]) to have a relatively low methylation level, while DNA from the "N_1" nucleosome particles, possessing H_1 histone protein, is hypermethylated.[43] Such results raise obviously the question of the validity and significance of the relationship between the presence of these proteins and the level of DNA methylation. It may be worth recalling that Felsenfeld et al.,[46] working with synthetic polynucleotides, have found no significant evidence for a preferential binding of histones or HMG proteins to methylated DNA. It would probably be interesting to consider the possibility that other nonhistone proteins play an important role in the regulation of nucleosome structure and/or functions. Such an investigation is difficult by their low abundance with respect to DNA in the nucleosomes.

Controversial reports have indeed appeared in the presence of other, less characterized nonhistone proteins in the nucleosome.[24,26,47-51] We have focused our attention on the tightly-bound nonhistone proteins so called in consideration of their need of urea and guanidine-HCl to be released from DNA,[24-26] since there is experimental indication that they might play an important role in the modulation of gene expression. These tightly-bound nonhistone proteins are tissue-specific,[26,52-56] sequence specific, and randomly distributed on chromatin DNA, but associated with DNA sequences enriched in tissue-specific active gene sequences,[27,57-61] nonrandomly distributed on chromatin organization,[26,50,62,63] responsible for maintenance of the tightly ordered metaphase chromosome structure,[64,65] nonrandomly distributed in chromatin fractions obtained by DNAaseII digestion.[23,56]

In pig kidney chromatin, these proteins are, by weight, about 10% with respect to the whole chromatin DNA and, as shown in Figure 8A, B, are highly heterogeneous. Comparing the two-dimensional slab gel electrophoretic patterns of this protein class obtained from whole pig kidney or liver chromatin, it appears that in both tissues these proteins are heterogeneous and that some components are tissue-specific.[26,56]

The characterization of this protein class shows quantitative and qualitative differences among nucleosome "N_1",[26] and nucleosome-like subpopulations "N_2" and "N_3".[68] They are 5% by weight in nucleosome "N_1", 4% in nucleosome-like "N_2", and 35% in nucleosome-like "N_3", Table 2. Two-dimensional slab gel electrophoreses show that the protein components from the three subpopulations are different (Figure 9 a-c).

Burdon et al.[66] and Drahovsky et al.[67] have described a form of DNA methyltransferase associated to the matrix structure. In preliminary experiments, concerning the distribution in the third level of chromatin organization of tightly-bound nonhistone proteins from human placenta (high levels of DNA methyltransferase activity occurring in this tissue), we have found this enzymatic activity associated to loop DNA-tbNHCp complex and also to chromatin matrix but absent from chromatin matrix digested with DNAase I.[68]

Whether tbNHCp play a role in the enzymatic process leading to DNA methylation cannot however be yet ascertained, as it is so far impossible to identify their role in nucleosome structure. It can, however, be suggested that some DNA methyltransferase be associated in chromatin structure, with the tightly-bound nonhistone proteins.

ACKNOWLEDGMENTS

Fruitful discussion with Professor Roberto Strom is gratefully acknowledged.

Financial support was provided by Italian Ministry of Education through the University of Rome "La Sapienza" (Progetto di Ateneo) and by the Commission of the European Communities (contract number B16-0196-I).

REFERENCES

1. **Tsanev, R.,** The substructure of nucleosomes, in *The Cell Nucleus,* vol. 4, Busch, H., Ed., Academic Press, New York, 1978, 107.
2. **McGhee, D. J. and Felsenfeld, G.,** Nucleosome structure, *Annu. Rev. Biochem.,* 49, 1115, 1980.
3. **Klug, A., Rhodes, D., Smith, J., Finch, J. T., and Thomas, J. O.,** A low resolution structure for the histone core of the nucleosome, *Nature,* 287, 509, 1980.
4. **Bloom, K. S. and Anderson, J. N.,** Fractionation of hen oviduct chromatin into transcriptionally active and inactive regions after selective micrococcal nuclease digestion, *Cell,* 15, 141, 1978.
5. **Beatriz, L. W., and Dixon, G. H.,** Partial purification of transcriptionally active nucleosomes from trout testis cells, *Nucleic Acids Res.,* 5, 4155, 1978.
6. **Prior, C. P., Cantor, C. R., Johnson, E. M., Littau, V. C., and Allfrey, V. G.,** Reversible changes in nucleosome structure and histone H_3 accessibility in transcriptionally active and inactive states of rDNA chromatin, *Cell,* 34, 1033, 1983.
7. **Baer, B. W. and Rhodes, D.,** Eukaryotic RNA polymerase II binds to nucleosome cores from transcribed genes, *Nature,* 301, 482, 1983.
8. **Matis, D., Dudet, P., and Chambon, P.,** Structure of transcribing chromatin, *Prog. Nucl. Acids Res.,* 24, 1, 1980.
9. **Weisbrod, S., Groudine, M., and Weintraub, H.,** Interaction of HMG 14 and 17 with actively transcribed genes, *Cell,* 19, 269, 1980.
10. **Weisbrod, S. T.,** Properties of active nucleosomes as revealed by HMG_{14} and HMG_{17} chromatography, *Nucleic Acids Res.,* 10, 2017, 1982.
11. **Czupryn, M., Solnica, L., and Toczko, K.,** An altered conformation of nucleosomal core particle in the active chromatin of Physarum polycephalum, *Biochim. Biophys. Acta,* 866, 252, 1986.
12. **Caiafa, P., Attinà, M., Cacace, F., Tomassetti, A., and Strom, R.,** 5-methylcytosine levels in nucleosome subpopulations differently involved in gene expression, *Biochim. Biophys. Acta,* 867, 195, 1986.
13. **Sorgen, D. R. and Butterworth, H. W.,** Eukaryotic ternary transcription complexes: transcription complexes of RNA polymerase II are associated with histone-containing, nucleosome-like particles in vivo, *Nucleic Acids Res.,* 13, 3805, 1985.
14. **Wu, R. S., Pamusz, H. T., Hatch, C. L., and Bonner, W. M.,** Histones and their modifications, *CRC Crit. Rev. Biochem.,* 20, 201, 1986.
15. **Goodwin, G. H. and Mathew, G. P.,** Role in gene structure and function, in *The HMG Chromosomal Proteins,* Johns, E. W., Ed., Academic Press, London, 1982, chap. 9.

16. **Richwood, D. and Birnie, G. D.**, Preparation, characterization and fractionation of chromatin, in *Subnuclear Components*, Birnie, G. D., Ed., Butterworths, London, 1976, chap. 4.

17. **Rill, R. L., Shaw, R. B., and Van Holde, K. E.**, Isolation and characterization of chromatin subunits, in *Methods in Cell Biology*, Vol. 18, Stein, G., Stein, J., and Kleinsmith, L. J., Eds., Academic Press, New York, 1978, chap. 6.

18. **Burton, K.**, Determination of DNA concentration with diphenylamine, in *Methods in Enzymology*, Vol. 12 (part B), Colowick, S. P. and Kaplan, N. O., Eds., Academic Press, New York, 1968, 163.

19. **Gehrke, C. W. and Lakings, D. B.**, Gas-liquid chromatography of purine and pyrimidine bases, *J. Chromatogr.*, 61, 45, 1971.

20. **Monahan, J. J. and Hall, R. H.**, Fractionation of chromatin components, *Can. J. Biochem.*, 51, 709, 1973.

21. **Nicolas, R. H. and Goodwin, G. H.**, Isolation an analysis, in *The HMG Chromosomal Proteins*, Johns, E. W., Ed., Academic Press, London, 1982, chap. 3.

22. **Laemmli, U. K.**, Cleavage of structural proteins during the assembly of the bacteriophage T4, *Nature*, 227, 680, 1970.

23. **Lonigro, I. R., Altieri, F., Allegra, P., and Caiafa, P.**, Distribution of tightly bound non-histone proteins in chromatin fractions obtained by DNAase II digestion, *Cell Biochem. Funct.*, 3, 223, 1985.

24. **Marzluff, W. F. and Huang, R. C. C.**, Transcription of RNA in isolated nuclei, in *Transcription and Translation*, Hames, B. D. and Higgins, S. J., Eds., IRL Press, Oxford, 1984, chap. 4.

25. **Bloom, K. S. and Anderson, J. M.**, Fractionation and characterization of chromosomal proteins by the hydroxyapatite dissociation method, *J. Biol. Chem.*, 235, 4446, 1978.

26. **Altieri, F., Allegra, P., Lonigro, I. R., Scarpa, S., and Caiafa, P.**, Distribution of tissue-specific tightly bound non-histone proteins in the first level of repeating chromatin structures, *Eur. J. Biochem.*, 154, 147, 1986.

27. **Bekhor, I. and Mirell, G. J.**, Simple isolation of DNA hydrophobically complexed with presumed gene regulatory proteins (M$_3$), *Biochemistry*, 18, 609, 1979.

28. **O'Farrel, P. H.**, High resolution two-dimensional electrophoresis of proteins, *J. Biol. Chem.*, 250, 4007, 1975.

29. **Peters, K. E. and Comings, D. E.**, Two-dimensional gel electrophoresis of rat liver nuclear washes, nuclear matrix and HnRNA proteins, *J. Cell. Biol.*, 16, 135, 1980.

30. **Morrisey, J. H.**, Silver stain for proteins in polyacrylamide gels: a modified procedure with enhanced uniform sensitivity, *Anal. Biochem.*, 117, 307, 1981.

31. **Lowry, O. K., Rosebrough, M. J., Far, A. L., and Randall, R. J.**, Protein measurement with the Folin phenol reagent, *J. Biol. Chem.*, 193, 265, 1951.

32. **Bradford, M. M.**, Rapid and sensitive method for the quantitation of microgram quantities of protein utilizing the principle of protein-dye binding, *Anal.Biochem.*, 72, 248, 1976.

33. **Razin, A. and Cedar, H.**, Distribution of 5-methylcytosine in chromatin, *Proc. Natl. Acad. Sci. U.S.A.*, 74, 2725, 1977.

34. **Adams, R. L. P., McKay, E. L., Douglas, J. T., and Burdon, R. H.**, Methylation of nucleosomal and nuclease sensitive DNA, *Nucleic Acids Res.*, 4, 3097, 1977.

35. **Hatayama, T., Nakamura, T., and Yukioka, M.**, Undermethylation of DNA in mononucleosomes solubilized by micrococcal nuclease digestion of HeLa cell nuclei, *Biochem. Int.*, 9, 251, 1984.

36. **Solage, A. and Cedar, H.**, Organisation of 5-methylcytosine in chromosomal DNA, *Biochemistry*, 17, 2934, 1978.

37. **Razin, A. and Riggs, A. D.**, DNA methylation and gene function, *Science*, 210, 604, 1980.

38. **Razin, A. and Friedman, J.**, DNA methylation and its possible biological roles, *Prog. Nucl. Acid Res. Mol. Biol.*, 25, 33, 1981.

39. **Adams, R. L. P. and Burdon, R. H.**, DNA methylation in eukaryotes, *Crit. Rev. Biochem.*, 13, 349, 1982.

40. **Kuo, M. T., Mandel, J. L., and Chambon, P.**, DNA methylation: correlation with DNase I sensitivity of chicken ovalbumin and conalbumin chromatin, *Nucleic Acids Res.*, 7, 2105, 1979.

41. **Groudine, M., Eisenman, R., and Weintraub, H.**, Chromatin structure of endogenous retroviral genes and activation by an inhibitor of DNA methylation, *Nature (London)*, 292, 311, 1981.

42. **Keshet, I., Lieman-Hurwitz, J., and Cedar, H.**, DNA methylation affects the formation of active chromatin, *Cell*, 44, 535, 1986.

43. **Ball, D. J., Gross, D. S., and Garrard, W. T.**, 5-methylcytosine is localized in nucleosomes that contain histone H$_1$, *Proc. Natl. Acad. Sci. U.S.A.*, 80, 5490, 1983.

44. **Doerfler, W.**, DNA methylation and gene activity, *Annu. Rev. Biochem.*, 52, 93, 1983.

45. **Cartwright, I. L., Abmayr, S. M., Fleischmann, G., Lowenhaupt, K., Elgin, S. C. R., Keene, M. A., and Howard, G. C.**, Chromatin structure and gene activity: the role of nonhistone chromosomal proteins, *Crit. Rev. Biochem.*, 13, 1, 1982.

46. **Felsenfeld, G., Nickol, J., Behe, M., McGhee, J., and Jackson, D.,** Methylation and chromatin structure, *Cold Spring Harbor Symp. Quant. Biol.,* 47, 577, 1983.
47. **Bakayev, V. V., Bakayeva, T. G., Schmatchenko, V. V., and Georgiev, G. P.,** Non-histone proteins in mononucleosomes and subnucleosomes, *Eur. J. Biochem.,* 91, 291, 1978.
48. **Chan, P. K. and Leiw, C. C.,** Identification of nonhistone chromatin proteins in chromatin subunits (or mononucleosomes) devoid of histone H₁, *Can. J. Biochem.,* 57, 666, 1979.
49. **Schlaeger, E. J., Van Telgen, H. J., Klempnaver, K. H., and Knippers, R.,** Association of DNA polymerase with nucleosomes from mammalian cell chromatin, *Eur. J. Biochem.,* 84, 95, 1978.
50. **Caiafa, P., Scarpati-Cioffari, M. R., Altieri, F., Allegra, P., and Turano, C.,** Tightly bound non-histone proteins in nucleosomes from pig liver chromatin, *Eur. J. Biochem.,* 121, 15, 1981.
51. **Matsumoto, H., Tohno, Y., Tohno, S., and Takakuso, A.,** Non-histone chromatin proteins associated with nucleosome cores from rat ascites hepatoma cells, *Cell. Mol. Biol.,* 29, 549, 1983.
52. **Wakabayashi, K., Wang, S., and Hnilica, L. S.,** Immunospecificity of non-histone proteins in chromatin, *Biochemistry,* 13, 1027, 1974.
53. **Chiu, J. F., Wang, S., Fujitani, H., and Hnilica, L. S.,** DNA binding chromosomal non-histone proteins. Isolation, characterization and tissue specificity, *Biochemistry,* 14, 4552, 1975.
54. **Yaneva, M., Beltchev, B., and Tsanev, R.,** Tissue specificity of tightly bound non-histone chromosomal proteins, *Cell Differ.,* 9, 351, 1980.
55. **Pumo, D. E., Wierzbicki, R., and Chiu, J. F.,** Chicken reticulocyte nuclear antigen: its identification and relation in transcriptive activity in erythropoietic cells, *Biochemistry,* 19, 2362, 1980.
56. **Lonigro, I. R., Allegra, P., Altieri, F., Tomassetti, A., and Caiafa, P.,** Distribution of tissue-specific tightly bound non-histone proteins in chromatin fractions obtained by DNAase II digestion, *Physiol. Chem. Phys. Med. NMR,* 17, 219, 1985.
57. **Gates, D. N. and Bekhor, I.,** Distribution of active gene sequences: a subset associated with tightly bound chromosomal proteins, *Science,* 207, 661, 1979.
58. **Norman, G. and Bekhor, I.,** Enrichment of selected active human gene sequences in the placental deoxyribonucleic acid fraction associated with tightly bound non-histone chromosomal proteins, *Biochemistry,* 20, 3568, 1981.
59. **Robinson, S. I., Small, D., Idzerda, R., McKnight, G. S., and Vogelstein, B.,** The association of transcriptionally active genes with the nuclear matrix of chicken oviduct, *Nucleic Acids Res.,* 16, 5113, 1983.
60. **Robinson, S. I., Nelkin, B. D., and Vogelstein, B.,** The ovoalbumin gene is associated with the nuclear matrix of chicken oviduct cells, *Cell,* 28, 99, 1982.
61. **Kuo, M. T.,** Distribution of tightly bound proteins in the chicken ovalbumin gene region, *Biochemistry,* 21, 321, 1982.
62. **Lonigro, I. R., Tomassetti, A., and Caiafa, P.,** Non-histone proteins tightly bound to loops and matrix from pig kidney chromatin, *Cell. Mol. Biol.,* 32, 319, 1986.
63. **Allegra, P., Lonigro, R. I., Altieri, F., Tomassetti, A., Reale, A., and Caiafa, P.,** Tightly bound non-histone proteins: distribution of tissue-specific components in the chromatin organization, *Basic Appl. Histochem.,* 31, 247, 1987.
64. **Paulson, J. R. and Laemmli, U. K.,** The structure of histone-depleted metaphase chromosomes, *Cell,* 12, 817, 1977.
65. **Razin, S. V., Mantieva, V. L., and Georgiev, G. P.,** The similarity of DNA sequences remaining bound to scaffold upon nuclease treatment of interphase nuclei and methaphase chromosomes, *Nucleic Acids Res.,* 7, 1713, 1979.
66. **Burdon, R. H., Qureshi, M., and Adams, R. L. P.,** Nuclear matrix-associated DNA methylase, *Biochim. Biophys. Acta,* 825, 70, 1985.
67. **Drahovsky, D., Pfeifer, G. P., Grünwald, S., Hirth, H. P., Vogel, M., Brzoska, M. J., and Palitti, F.,** Characterization of DNA methylating enzymes by use of monoclonal antibodies, in *Progress in Clinical and Biological Research,* Cantoni G. L. and Razin, A., Eds., Alan R. Liss, New York, 1985, 67.
68. **Caifa, P., Mastrantonio, S., Cacace, F., Attina, M., Rispoli, M., and Strom, R.,** Localization in human placenta of tightly bound form of DNA methylase in the higher order of chromatin organization, *Biochim. Biophys. Acta,* 351, 182, 1988.

Chapter 6

DISCONTINUITIES OF PEPTIDE NATURE IN DNA

Dieter Werner and Beatrice Neuer-Nitsche

TABLE OF CONTENTS

I. INTRODUCTION

Covalently bound peptides or polypeptides in DNA have a long history and have been discussed most controversially. The earliest conclusions were based on the finding that a small but persistent amount of peptide material is inevitably copurified with DNA.[1-6] Together with the shear-induced loss of viscosity of DNA solutions, this has led to models in which relatively short double-stranded DNA segments are tandemly joined[7] or cross-linked[8] by peptides. However, about 10 years later such non-DNA linkers joining short double-stranded DNA pieces could be ruled out unequivocally. At that time methods became available which allowed isolating and investigating completely unfolded and extremely long DNA molecules. Sedimentation analyses showed that DNA strands much longer in size than the previously anticipated double-stranded DNA subunits could be released from cells if shear forces and secondary enzymatic changes of DNA were reduced as far as possible.[9-11] In addition, electron microscopical techniques allowed visualizing unfolded DNA strands longer than 75μm (230 kbp).[9,12] Finally, Kavenoff and Zimm[13] showed by viscoelastic measurements that double-stranded DNA from *Drosophila* cells is continuous throughout the length of the chromosome. This led to the general view that there exists no subunit structure of double-stranded DNA and that DNA is most probably of chromosome length.

However, viscoelastic measurments, sedimentation analyses, and electron microscopical inspections could not rule out single-stranded nicks, nondeoxynucleotide linkers, nor covalently bound peptides/polypeptides at internal DNA ends; native DNA strands containing such discontinuities could still behave as long and continuous molecules as long as these discontinuities were not arranged opposite or close to each other in the antiparallel strands. Although this possibility was explicitly discussed by Kavenoff and Zimm[13] and although this had been experimentally proven by Hays and Zimm,[14] this fact was widely ignored.

In the context of anchorage or attachment of DNA at subnuclear structures (nuclear cage;[15] nuclear matrix[16,17]), peptides or polypeptides tightly associated with DNA again became of interest. Parallel to this development, enzymes have become known that control the super-helical conformation of DNA and form at least transient covalent complexes with DNA.[18,19] In light of these new results, it was of interest to reinvestigate tight DNA/peptide complexes with respect to their possible involvement in genome organization.

II. ANALYSIS OF THE MOST TIGHTLY BOUND PROTEINS IN DNA

A. Physico-Chemical Characteristics and Detection

DNA isolated by procedures generally considered to be most efficient for purifying DNA still contains detectable peptide components. These residual peptides are not dissociated from DNA by high salt, SDS, proteases, and phenol.[20-24] A portion of the peptides associated with isolated native DNA can be released by alkali treatment. Their characteristics and biological sigificances are described in other sections of this volume[25] or elsewhere.[26] However, even alkali-treated DNA is still associated with residual material of peptide nature. The latter material can only be released by methods which degrade DNA.[22,27] Even prolonged treatment of denatured DNA with various proteases in the presence of SDS cannot remove all peptides from DNA.[27] This points to an unusually tight interaction of proteinaceous material with DNA. Moreover, the peptides involved in such complexes must be considered to be somehow protected by the DNA strands, because after digestion of DNA by DNase I the residual peptides become sensitive to proteolytic degradation.[22,27] Even trace amounts of proteases occasionally present in DNase I preparations are sufficient to digest the residual peptide materials.

The detection of residual peptides in DNA has been performed by various biochemical techniques. The most conventional technique involves acid hydrolysis of highly purified

DNA followed by amino acid analyses.[6,20] However, this method allows only quantitation of proteinaceous material in DNA solutions and it gives no information about the characteristics of the peptide material and its interaction with DNA. A large scale procedure for the isolation of the residual polypeptides in DNA and their characterization by Coomassie-stained SDS polyacrylamide gels will be described in more detail below.[22] However, the latter method needs relatively large amounts of highly purified DNA. Other methods were designed which detect more directly the DNA/protein association and the stability of such complexes. For example, radiolabeling of rapidly growing cells with [35S]methionine/[3H]thymidine over several cell cycles results in significant 35S-label cobanding with [3H]DNA on alkaline caesium sulfate gradients or cosedimenting with DNA through alkaline sucrose gradients.[20] Moreover, DNA fragments associated with the 35S-label can be enriched by filtration through glass fiber filters under high salt conditions.[28] According to Bodnar et al. DNA/protein complexes of similar characteristics as those described by us can be detected and isolated by gel exclusion chromatography.[23] A sedimentation method has been applied by Bekhor and Mirell[24] which allows separation of DNA strands tightly bound to hydrophobic proteins from those DNA strands free of protein. The latter protein fraction comprises apparently the most tightly bound proteins in DNA. Another convenient method for the detection of protein complexes in DNA involves treatment of highly purified DNA with 125I, heat degradation of the complexes, SDS polyacrylamide gel electrophoresis, and finally autoradiography.[20,21,28-30] According to such autoradiographs, most of the radiolabel remains at the top of the gels, which is due to irreversible denaturation of the proteinaceous components. However, a portion of the radiolabeled material becomes solubilized by ionic detergents and migrates with an apparent molecular weight of 54/68 kDa. In addition, variable amounts of material smaller than 30 kDa are also seen on autoradiographs. Although this method is most sensitive and although it needs only 1 OD_{260} unit or even less of DNA, it should be noted that this method has been criticized in so far as unspecific contaminants of the same size as the polypeptides involved in tight DNA complexes may be detected.[31] Especially β-mercapto ethanol previously used to reduce the excess of iodine may contain skin proteins which become radiolabeled under the iodination conditions and which may migrate on gels like the polypeptides bound to DNA. Although correct handling of β-mercapto ethanol and proper controls eliminate such problems, the detection by radioiodination alone appears insufficient to prove the presence of specific proteinaceous material in DNA (see also Note Added in Proof).

In the following sections we shall describe in more detail additional techniques which clearly indicate that the polypeptides copurifying with DNA are indeed of nuclear origin and intrinsically associated with DNA.

B. Visualization

The proteins most tightly attached to DNA cannot be visualized directly in the electron microscope. As will be shown in sections below, their average frequencies in unfractioned DNA are in the order of 5 to 10 kbp which is equivalent to several μm of spread DNA. A pattern of such low frequency in spread DNA cannot be reliably identified. Moreover, the size of these polypeptides is too small to be clearly resolved from the background. However, by a combination of two techniques it was possible to localize proteinaceous material associated with purified DNA. (1) DNA fragments several hundred base pairs in size (Alu I digests) which are associated with the residual proteinaceous material can be enriched by a glass fiber binding technique originally developed for the isolation of the adenovirus terminal protein complex.[32] Although the protein-associated DNA fragments comprised in Alu I digests of chromosomal DNA are not quantitatively retained on glass fiber filters,[28] their enrichment is sufficiently high to localize the DNA-associated proteins. (2) The protein molecules associated with DNA fragments in the enriched fraction can be indirectly visualized

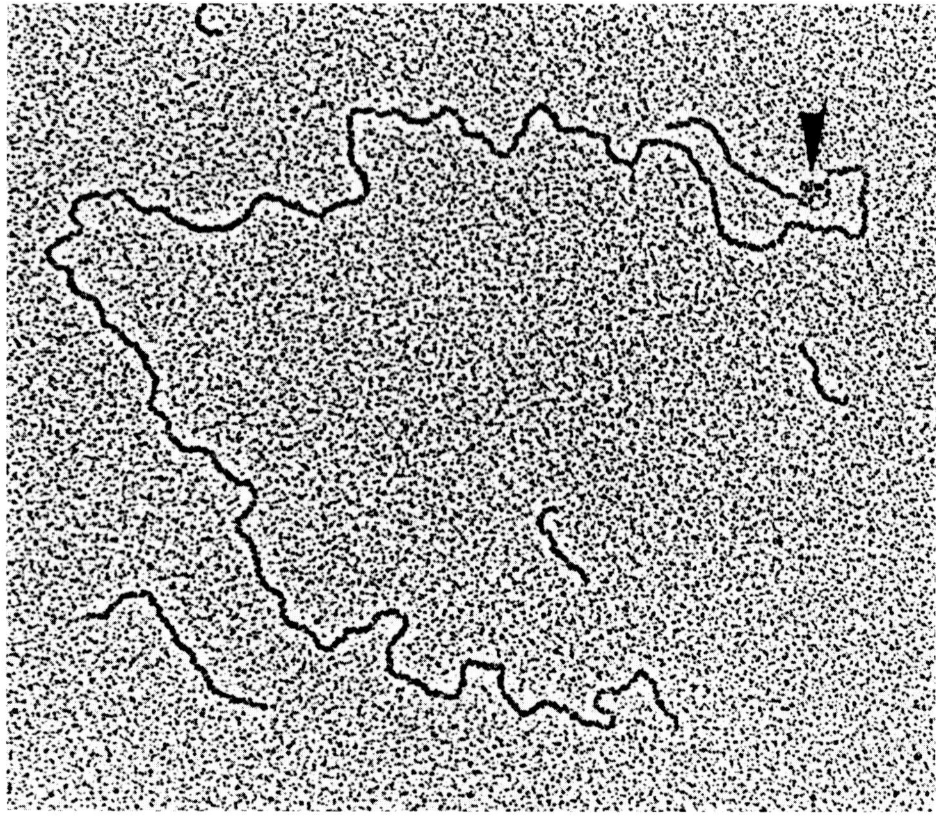

FIGURE 1. Visualization of DNA/protein complexes in DNA. Alu I digested DNA fragments retained on glass fiber filters were incubated with fluro dinitro oenzene under conditions designed to modify proteins with dinitro phenyl groups. This was followed by decoration with antibodies to dinitro phenyl residues. The antibody marker is indicated by an arrowhead. (From Spiess, E., Neuer, B., and Werner, D., *Biochem. Biophys. Res. Commun.*, 104, 548, 1982. With permission.)

after their chemical modification by dinitro fluoro benzene (DNFB) followed by decoration with monoclonal antibodies to dinitro phenyl residues.[28] Figure 1 exhibits spread DNA from the filter-fraction treated with DNFB and antibodies. Identically prepared DNA from the filtrate fraction is decorated by the antibodies with much lower frequency.[28]

C. Immunolocalization

According to our experience, denatured DNA of high purity is released from cells by prolonged alkaline cell lysis in the presence of SDS followed by vigorous phenol extraction. Such DNA preparations can be considered to be free of proteins except for the most tightly and alkali stably bound material.[22] After degradation of this DNA with protease-free DNase I, the residual polypeptides are insoluble in buffers which do not contain ionic detergents. Only freshly prepared material can be partially dissolved in SDS buffer and analysed on SDS polyacrylamide gels. Two main bands of polypeptides are resolved with some faint bands in between (Figure 2). The upper main band (UMB) shows an apparent molecular weight of 68 kDa. The lower main band (LMB) is in the order of 54 kDa. The position of this set of proteins on SDS polyacrylamide gels is characteristic and identical when isolated from cells of different sources.[22,23] However, the relative intensities of the bands and the amount of material retained on the stacking gel can differ in repeated experiments. Occasionally most of the material is retained on the stacking gel which is indicative for its self-

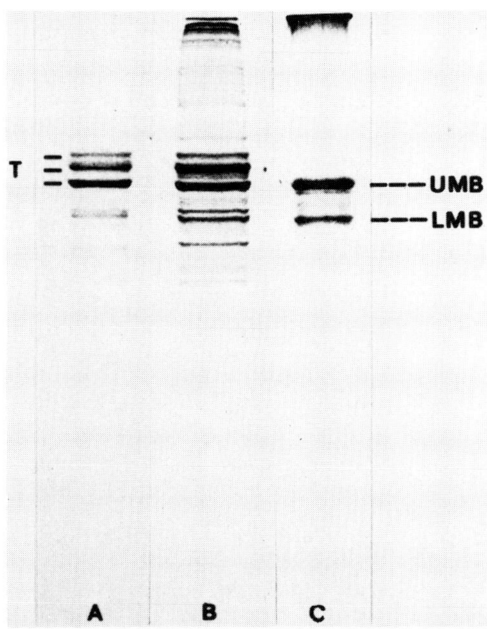

FIGURE 2.   Coomassie stained SDS polyacrylamide gel indicating the size of the polypeptides co-purifying with DNA in alkali and released by digestion of the DNA (C). The upper main band (UMB) reflects a polypeptide with an apparent molecular weight of 68 kDa, the lower main band (LMB) is in the order of 54 kDa. On parallel slots the major nuclear matrix proteins are shown in different intensities. (A) The lamin triplet and (B) the next prominent nuclear matrix associated proteins. (From Werner, D., Zimmerman, H. P., Rauterberg, E., and Spalinger, J., *Exp. Cell Res.*, 133, 149, 1981. With permission.)

aggregation. Once formed, aggregates cannot be dissolved even in buffers containing SDS, urea, and β-mercapto ethanol. It should be noted that this characteristic has also been described in the above section for the radioiodinated complexes. Similar characteristics have been reported by Bodnar et al.[23]

Antibodies were raised against the polypeptides contained in the two main bands by inoculation of minced gel slices into the lymph nodes of rabbits.[22] The immunostain of nuclei by means of such antibodies reveals that the polypeptides copurifying with the DNA and migrating on SDS polyacrylamide gels with an apparent molecular weight of 54/68 kDa reflect nuclear antigens. It should be noted that the antibodies stain a nuclear substructure (Figure 3). The fibrogranular *in situ* networks immunostained in nuclei by these antibodies may reflect the nuclear matrix association of these antigens. This is supported by the finding that protein blots of nuclear matrix proteins are also immunostained at the 54/68 kDa position.[33] Moreover, partial peptide maps of the proteins released from DNA and of nuclear matrix polypeptides reveal homologies in peptide sequences.[23] This indicates that a significant fraction of the 54/68 kDa polypeptides is retained during nuclear matrix preparations.

Unexpectedly, condensed chromosomes were not immunostained by the antibodies (Figure 4). It is not yet clear whether this reflects inaccessibility of DNA/protein complexes in condensed chromosomes, a release of these polypeptides from DNA during chromosome condensation, or whether the immunostain of nuclei is due to a large fraction of these polypeptides not associated with DNA while a smaller fraction of these polypeptides associated with DNA escapes serological detections.[22]

Alternatively, it could be argued that the antibodies raised against the isolated polypeptides could eventually be directed against epitopes which are not accessible in the complex.

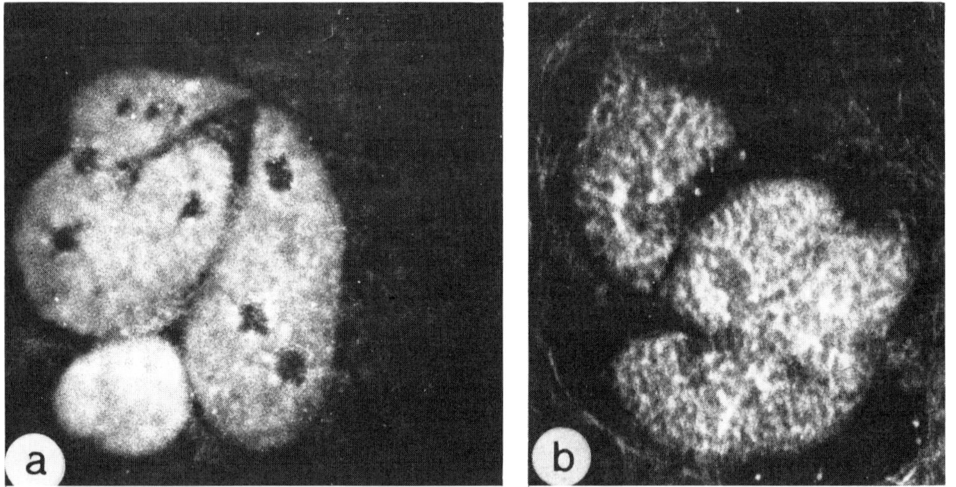

FIGURE 3.   Immunolocalization of the 54/68 kd proteins co-isolating with DNA in *in situ* nuclei of poly-nuclear giant rat-kangaroo (PtKl) cells. (a) Antibody to the UMB and (b) antibody to the LMB polypeptide. (From Werner, D., Zimmermann, H. P., Rauterberg, E., and Spalinger, J., *Exp. Cell Res.*, 133, 149, 1981. With permission.)

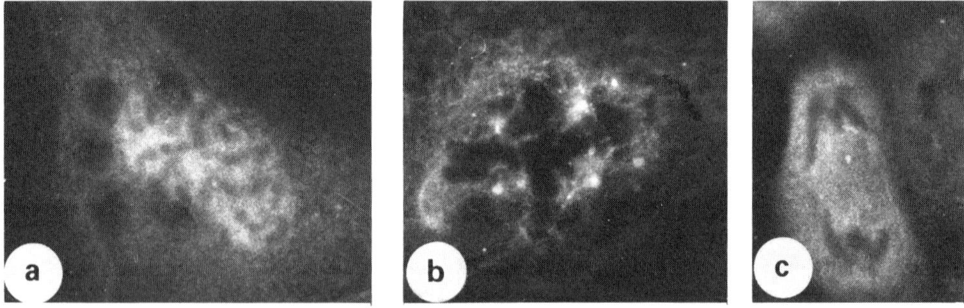

FIGURE 4.   Mitotic figures of rat-kangaroo (PtKl) cells in indirect immunofluorescence microscopy after treatment with antibodies to the LMB (54 kDa antigen). (From Werner, D., Zimmerman, H. P., Rauterberg, E., and Spalinger, J., *Exp. Cell Res.*, 133, 149, 1981. With permission.)

However, it could be shown that protein-associated DNA fragments retained on nitrocellulose filters are decorated on the filter by the antibodies.[33]

D. Phosphodiester Bonds between Peptides and Chromosomal DNA

Although the results described in the above sections point to covalent bonds between peptides and isolated chromosomal DNA, it was of interest to demonstrate such bonds more directly. This could be achieved by the sequence of experimental steps designed in Scheme 1. For such experiments, DNA was released by alkali and purified by phenol extraction. This highly purified DNA absorbed [125]Iodine under conditions designed to radioiodinate almost exclusively tyrosine and only negligible amounts of histidine[27] in peptides. Following this radioiodination, the material which had absorbed the radiolabel could be precipitated repeatedly by ethanol from solutions containing SDS, urea, β-mercapto ethanol, or alkali. Since the radiolabeled material remained also associated with DNA during repeated treatments with phenol, this showed that it behaved like DNA. Since the DNA itself is not radiolabeled (Figure 6a) there remain two interpretations: (1) the radioiodinated material is

DNA   PROTEIN

Step 1: .....AGCTAGCTN – – Seq1 – – Seq2 – – Seq3 – – Seq4 – – Seq5 – – Seq6....

Radioiodination *

Step 2: .....AGCTAGCTN – – Seq1 – – Seq2* – Seq3* – Seq4 – – Seq5* – Seq6....

Protease digestion

Step 3: .....AGCTAGCTN – – Seq1 – – Seq2* Seq3* Seq4 Seq5* Seq6....

A portion of the radiolabel remains

at DNA

Step 4: .....AGCTAGCTN – – Seq1 – – Seq2*

Nuclease digestion

Step 5:   N – – Seq1 – – Seq2*

Phosphodiesterase digestion

Step 6:   N   Seq1 – – Seq2*

The radiolabel is released and migrates

on TLC sheets

SCHEME 1.   Experimental steps designed to demonstrate covalent bonds between peptides and internal ends of DNA.

chemically bound to DNA or, alternatively, (2) the radioiodinated material has the same physico-chemical characteristics as DNA with respect to its solubility in buffers and its insolubility in ethanol and phenol.

To distinguish between these possibilities, the radiolabeled material was submitted to prolonged proteolysis under extreme conditions (SDS, proteinase K 6 mg $\times$ ml^{-1}, 37°C, 16 h). From this treatment, it can be expected that all nonspecific proteinaceous components eventually associated with the DNA should be digested. Moreover, all DNA-bound peptides susceptible to proteolytic digestion should be released from DNA. Expectedly, most of the radiolabel appeared in the supernatants of further ethanol precipitations. This radioiodinated material released from DNA migrated on thin-layer chromatography sheets in solvent systems designed to separate amino acids and peptides (Figure 5a). However, even the harsh treatment with protease could not release all radiolabeled material from DNA. In repeated experiments, we found that about 12% of the radiolabel originally absorbed by the DNA preparation still remained associated with the DNA during precipitations with ethanol. Additional treatment with pronase could not release more radiolabel from the DNA.

The radioiodinated material which could not be removed from DNA by proteolysis followed by ethanol precipitations did not migrate when submitted to thin-layer chromatography under conditions designed to separate amino acids and peptides (Figure 5b). Experiments performed to cleave this residual complex with chemical methods revealed its alkali stability and its stability against mild acids. Only harsh treatments which are expected to cleave chemical bonds in peptide-chains and thus releasing the radiolabel from the terminal amino acid involved in the binding site at the DNA resulted in radiolabeled peptides migrating on

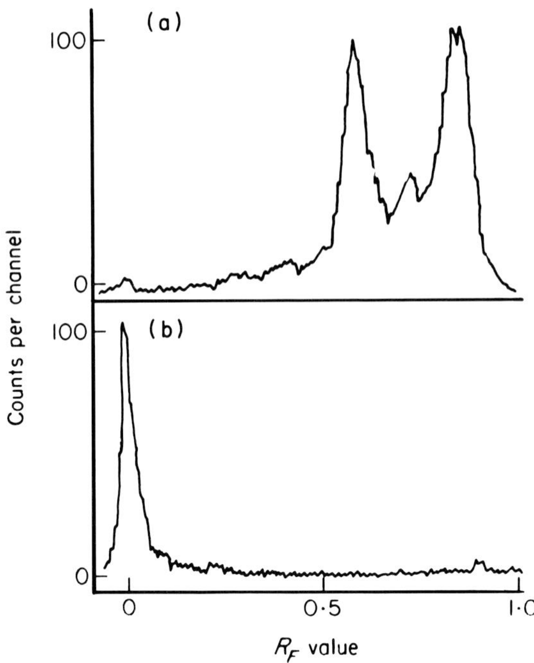

FIGURE 5.   Radiochromatography of the radioiodinated proteinaceous materials co-purifying with DNA in alkali. (a) Radioiodinated peptides that are released from DNA by prolonged digestion with proteinase K (see Scheme 1 Step 3). (b) Radioiodinated peptides that remain bound to DNA after prolonged proteolysis (see Scheme I, Step 3). (From Neuer, B., Plagens, U., and Werner, D., *J. Mol. Biol.*, 164, 213, 1983. With permission.)

thin-layer chromatography sheets.[27] This unusual chemical stability of the complex between DNA and the residual peptides points to O^4-tyrosyl-phosphodiester linkages between the residual peptides and DNA.[34,35] Since tyrosine in O^4-linkage cannot react with iodine, it follows that there remain peptides bound to DNA which have to comprise additional tyrosine(s) in peptide linkages.

In order to isolate the linking group, the DNA was digested by mixtures of DNase I and Sl nuclease. Following this digestion, no radioiodinated deoxynucleotides migrated on thin-layer chromatography sheets developed with solvent systems which allow nucleotides and peptides to migrate (Figure 6a). This indicates that the radiolabel was definitely not incorporated into intrinsic DNA components, e.g, pyrimidines.

According to the sequence of the experimental steps, the linking group material could be expected to consist of one (or a few) deoxynucleotides bound to a peptide comprising at least one tyrosine not involved directly in the binding site at the DNA. The immobility of this linking group material on thin-layer chromatography sheets and the expected mobility of the radiolabeled peptides after their release from DNA was used to characterize it further by treatment with phosphodiesterases. The phosphodiesterase from calf spleen cleaves preferentially phosphodiesters between DNA and non-DNA materials in 3′-position.[36,37] In contrast, snake venom phosphodiesterase cleaves such phosphodiesters in 5′-position.[38,39] Consequently, the linking group material was treated with these enzymes resulting in the expected mobility of the radiolabeled components (Figures 6b and 6c). This confirmed the results described above and can be taken as indication for phosphodiester bonds between the residual peptides in DNA and DNA. Moreover, the results are consistent with two different linking groups at 3′ and 5′ ends of DNA. The individual phosphodiesterases could not release all radiolabeled material. A fast moving peptide was preferentially released from

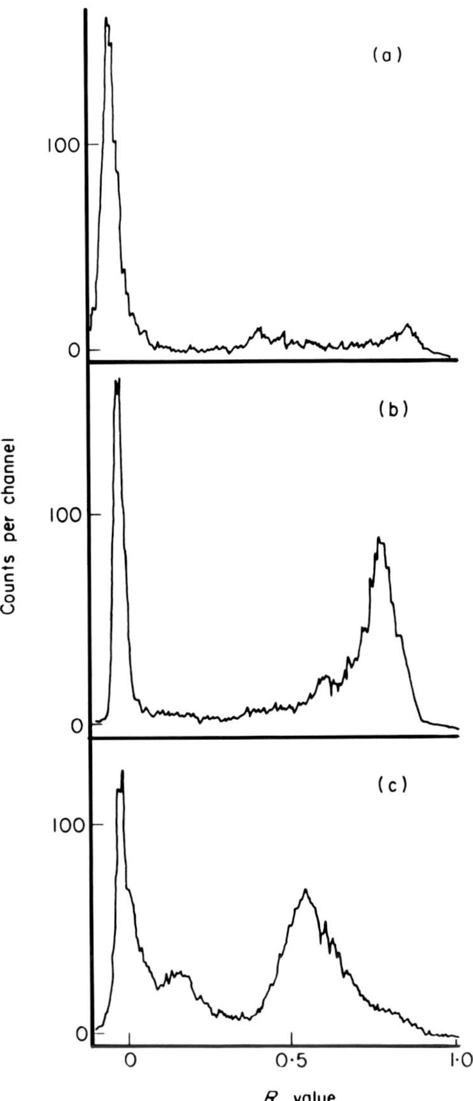

FIGURE 6.   Radiochromatography of the linking-groups (a) before treatment with phosphodiesterases (Scheme 1, Step 5), (b) after treatment with phosphodiesterase from calf spleen, and (c) after treatment with phosphodiesterase from snake venom (Scheme I, Step 6). (From Neuer, B., Plagens, U., and Werner, D., *J. Mol. Biol.*, 164, 213, 1983. With permission.)

3′ ends, and two slower moving components were typical for peptides bound at 5′ ends of DNA. However, incubation of the linking group material with the two phosphodiesterases released the radiolabel almost completely.

Thus, the sequence of steps designed in Scheme 1 resulted in clear evidence that alkali-denatured and protease-treated chromosomal DNA contains covalently bound peptide material at 3′ and 5′ ends, respectively. We consider these results at present as the strongest arguments for covalently bound peptides in chromosomal DNA. It appears that this method is more reliable than others because it demonstrates directly chemical bonds between the proteinaceous material and DNA. It should be noted that by means of this method essentially identical results have been obtained with DNA from different sources by other authors.[40] However, the peptides found at 3′ and 5′ ends of denatured and protease-digested DNA

Table 1
DISTANCES OF DISCONTINUITIES OF PROTEIN
NATURE IN DNA ESTIMATED BY VARIOUS
METHODS

	Distances		
Method	**Kilo Bp**	**Daltons**	**Ref.**
Amino acid quantitation	5—10	3.3—6.6×10^6	20
Sedimentation	12.8	8.5×10^6	10, 11, 41
Electron microscopy	10.5	6.9×10^6	12
Affinity filtration	10.6—15	7—10×10^6	45

reflect only remnants of larger polypeptides associated with native DNA in a not yet clearly identified fashion.

III. PROTEASE INDUCIBLE ALKALI LABILITY OF EUKARYOTIC DNA

The amount of peptide material most tightly associated with DNA is difficult to quantitate. According to our experience, reasonable data are obtained by estimations of amino acids present in acid hydrolysates of isolated DNA purified by sedimentation through alkaline sucrose gradients. Based on such estimations,[20] we calculate that the amount of residual peptides in DNA which cannot be removed except by degradation of DNA is in the order of 1 to 2% by weight. Under the assumption that the molecular weight of the most tightly bound polypeptides is in the order of 54/68 kDa, one would expect 3 to 6×10^5 of such molecules per average genome (3×10^6 kbp, 1.98×10^{12} Da). This is equivalent to one polypeptide molecule per 5 to 10 kbp of dsDNA (3.3 to 6.6×10^6 Da) (Table 1). If these protein molecules are bound to internal ssDNA ends it follows that the antiparallel strands should be interrupted at an average distance of 10 to 20 kb. Consequently, after alkali denaturation, single-stranded DNA should be as short as 10 to 20 kb. However, since single-stranded DNA released from cells by direct alkaline cell lysis is more than ten times larger[9-12] it was suggested that single-stranded DNA subunits 10 to 20-kb long may be linked by the tightly bound polypeptides in an alkali stable fashion.[10,11]

Although there are no hard facts which could rule out that alkali stably bound proteins in DNA may join ssDNA subunits, there is at present no experimental design available which could prove this unequivocally. However, some experimental evidence is seen in the protease inducible alkali lability of eukaryotic DNA.[10,11,41] Most of the DNA strands released by alkaline cell lysis are sedimenting with a molecular weight larger than 10^8 Da. However, if the cells are lysed initially in buffers containing SDS and high concentrations of proteases followed by alkali-denaturation and sedimentation through alkaline sucrose gradients, the single-stranded DNA is much shorter and the average molecular weight of ssDNA is in the order of 8.5×10^6 Da. These results are compatible with the view that single strands in DNA are eventually tandemly joined by proteins. It should be noted that the DNA released from cells by SDS/protease digestion followed by alkali denaturation is in the same order as the calculated distances of binding sites of the most tightly bound polypeptides in DNA (Table 1). Numerous control experiments have been performed which ruled out simple explanations for the protease inducible change in size of ssDNA, such as nucleases eventually present in the protease preparations. Most of these results have been published and explicitly discussed.[10,11,41]

Here we wish to summarize only some arguments which support the view that a change in ssDNA size may be somehow induced by a protease-catalyzed reaction. (1) The change in size of single-stranded DNA is dependent on the protease concentrations.[11] (2) There is a limit size of single-stranded DNA in the order of 8.5×10^6 Da when the DNA is released with increasing concentrations of proteases.[11] (3) The protease inducible change in size of single-stranded DNA is inhibited by protease inhibitors.[11] (4) The change in size of a small amount of DNA from [³H]thymidine prelabeled cells is not competed by a 2400-fold amount of unlabeled DNA.[11] (5) Phage DNA is not changed by the protease treatment.[10,12] (6) DNA released in the presence of high concentrations of proteases contains nicks which are S1 nuclease sensitive,[12] however, these nicks cannot serve as starting signals for *Escherichia coli* polymerase I.[11]

From this it appears that alkali stably bound proteins in DNA are correlated with the phenomenon of the protease inducible alkali lability of DNA. This correlation may be based on protein linkers joining ssDNA subunits or on other complex structures (e.g., alkali-stable phosphotriester bonds between polypeptides and DNA could also be rendered alkali sensitive after proteolysis).

In order to prevent confusion, we wish to stress that the changes in sedimentation rates observed after pretreatment of cell lysates with proteases reflect true changes in size of single-stranded DNA and not conformational changes. This could be clearly confirmed by electron microscopical inspections of native and denatured DNA released from cells by different protease concentrations.[12] The results of these investigations with respect to the lengths of single-stranded DNA released from cells by various concentrations of proteases were almost identical with those obtained from sedimentation analyses (Table 1). This is of interest because other authors[42,43] have interpreted discrete changes in sedimentation rates induced by prolonged alkali treatment and/or irradiation as indicative of single-stranded DNA subunits. However, in the light of more recent results on the unfolding and stepwise release of DNA from cell nuclei,[15,44] the earlier results[42,43] are now best explained by alkali or irradiation induced conformational changes such as unfolding of supercoiled DNA loops present in partially lysed or histone-depleted chromatin.[15,44]

IV. ISOLATED DNA/PEPTIDE COMPLEXES

A. Isolation Procedure

In order to characterize the DNA sequences involved in tight DNA/peptide complexes, we worked out a method that allows the separation of protein-associated DNA fragments from the bulk of protein-free fragments in isolated DNA.[45] As mentioned in an above section, filtration through glass fibre filters results only in a partial enrichment of the population of DNA fragments that are associated with the proteins or peptides co-isolating with chromosomal DNA. However, filtration through nitrocellulose filters appears to retain the DNA/protein complexes almost quantitatively. This method is based on the following observations: (1) only denatured DNA fragments show a salt-dependent affinity to nitrocellulose,[46] (2) native DNA fragments, that are not associated with any non-DNA material can be considered to pass nitrocellulose at any salt concentration,[47-49] and (3) native DNA fragments associated with DNA binding proteins are completely retained on nitrocellulose if submitted at salt concentrations stabilizing DNA/protein complexes and inducing interaction between the protein residue of the complex and the filter material.[47-49] Since the naturally occurring covalent DNA/peptide complexes could be expected to be stable at any salt concentration, it was anticipated that the DNA/protein complexes contained in highly purified genomic DNA should be retained on nitrocellulose filters at salt conditions inducing affinity of protein to nitrocellulose.

For such experiments we used DNA purified by repeated treatments with proteinase K, ribonucleases, and phenol, essentially the procedure described by Gross-Bellard et al.[9] In order to avoid mechanical interaction of long DNA strands with filters, the DNA preparations were fragmented by digestion with Alu I which results in an average fragment length of 0.6

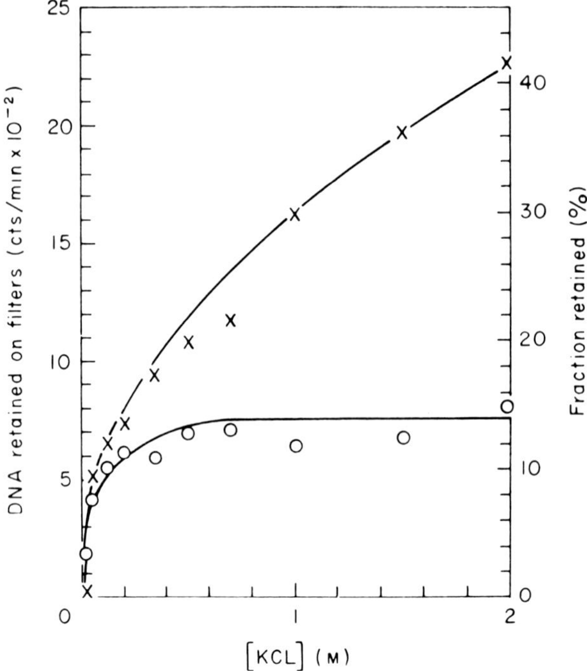

FIGURE 7.   Isolation of a distinct fraction of native DNA fragments by nitrocellulose affinity filtration. Uniformly radiolabeled, Alu I digested, native DNA fragments were filtered at increasing salt concentrations through nitrocellulose filters under water stream vacuum (○). DNA that is heat-denatured (X) before filtration shows a different binding characteristic from that of native DNA. (From Neuer, B. and Werner, D., *J. Mol. Biol.*, 181, 15, 1985. With permission.)

to 0.9 kbp.[28] As shown in Figure 7, there exists a population of native DNA fragments that are retained on nitrocellulose filters with binding characteristics different from that of denatured DNA. At salt concentrations higher than 500 mM a constant fraction of 13 to 15% of Alu I-fragmented native DNA is retained on filters. In contrast, the degree of binding of denatured DNA increases almost linearly with the salt concentration (Figure 7). From this it can be concluded that only a minor portion of the filter-bound DNA is due to the binding of ssDNA that is present in the DNA preparations: (1) the binding of ssDNA at 500 mM salt is incomplete, (2) there are no parallel binding characteristics between native and denatured DNA fragments, and (3) the fraction of DNA fragments bound to nitrocellulose after treatment with S1 nuclease is only slightly reduced and exhibits the same binding characteristics as shown in Figure 7 for DNA not treated with S1 nuclease.

Various control experiments published in detail[45] indicated that the fraction of DNA fragments retained on nitrocellulose filters consists of a DNA subpopulation in which most of the DNA fragments must be associated with a non-DNA factor that shows high affinity to nitrocellulose. Unspecific trapping or ssDNA binding play negligible roles in the nitrocellulose affinity of this fraction: (1) the 13 to 15% fraction of DNA is retained in one filtration step; refiltration of the DNA that passes the filter in the first filtration step results only in unspecific trapping of about 1 to 2% of DNA (Figure 8); (2) differently labeled DNAs prefiltered and not prefiltered were mixed and submitted to filtrations; only DNA fragments that were not prefiltered showed significant binding rates to the filters.[45]

In order to study the factor that induces the binding of the fraction of native DNA fragments to nitrocellulose, recovery experiments were performed.[45] The highest rate of recovery of native DNA fragments was obtained with buffers containing zero salt. In this case the recovery of filter-bound DNA was in the order of 50%. The efficiency was higher when

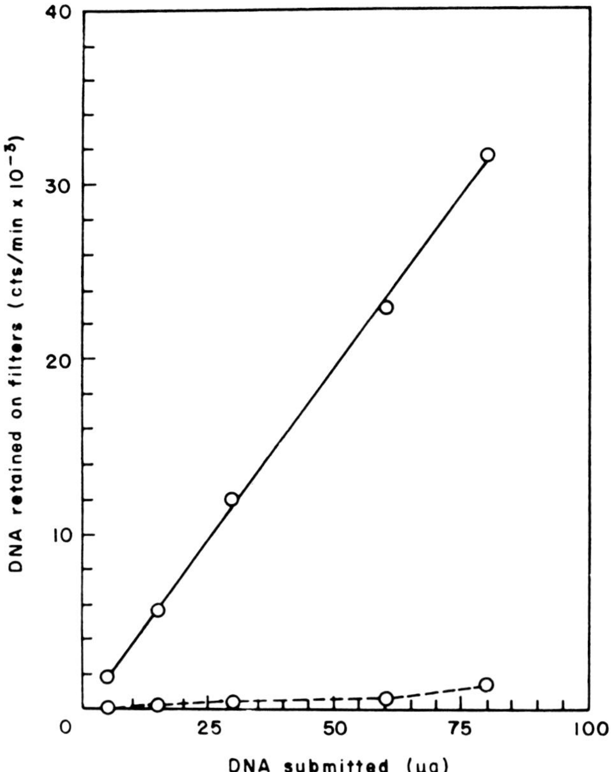

FIGURE 8. Native DNA fragments with high affinity to nitrocellulose filters are separated from the bulk of DNA fragments with low affinity by one single filtration step. Increasing amounts of native DNA specified in Figure 7 were filtered through nitrocellulose filters at 0.5 M KCl. A constant portion of the DNA submitted to filtration (13 to 15%) is retained on filters (O — O). The DNA passing the filters during the first filtration (85 to 87%) was submitted to refiltration through new filters (O - - O). No significant amounts of DNA are retained in the refiltration step. (From Neuer, B. and Werner, D., *J. Mol. Biol.*, 181, 15, 1985. With permission.)

the DNA was recovered under alkaline conditions. Up to 90% could be released from the filters with 0.05 N sodium hydroxide. The denatured DNA recovered from nitrocellulose filters was used to investigate its enrichment in associated proteinaceous material. It was found that the DNA eluted from filters absorbed 17.5 times more [125]Iodine than the fraction passing the filters. This indicates that about 94% of the material which absorbed [125]Iodine was retained on filters together with 13 to 15% of the DNA. The radioiodinated material recovered from nitrocellulose filters and submitted to the experimental steps listed in Scheme 1 showed the same characteristics as the materials analysed in Figures 5 and 6. This indicates that the DNA fragments retained on nitrocellulose during filtration are associated with the peptides covalently bound to DNA. Thus it is highly likely that the factors inducing the binding of a population of native DNA fragments to nitrocellulose are identical with the protein discontinuities that are found in unfractioned DNA. Consequently, it is justified to conclude that the fraction of DNA retained on nitrocellulose filters reflects the population of DNA strands associated with the tightly bound proteins co-isolating with DNA.

The amount of native DNA strands retained on nitrocellulose filters allows a rough estimation of the distances of the protein discontinuities in DNA. Based on experimental evidences[45] it can be assumed that the 13 to 15% fraction of DNA fragments retained on nitrocellulose filters may contain small amounts of unspecifically trapped DNA and some DNA which is retained on the filters due to single strands. However, at least 8 to 9% of

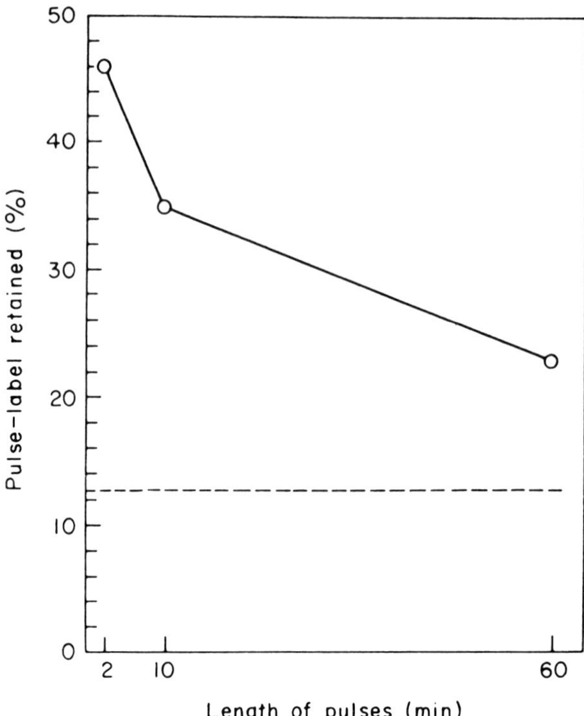

FIGURE 9.   High degree of adsorption of pulse-labeled DNA on nitrocellulose filters during affinity filtration. Identical amounts of DNA pulse-labeled for different pulse lengths were submitted to nitrocellulose affinity filtration under conditions described in the legend of Figure 8. The portion of pulse-labeled DNA retained on filters decreases with increasing pulse lengths. Pulse-label (-O-), amount of DNA retained (---). (From Neuer, B. and Werner, D., *J. Mol. Biol.*, 181, 15, 1985. With permission.)

the DNA fragments can be considered to be specifically retained on the filters due to their protein association. Since the average molecular weight of the Alu I fragments of mouse DNA is in the order of 0.6 to 0.9 $\times$ 10^6 Da,[28] and since one out of 11 to 12 fragments is retained, it can be calculated that the average distances of the proteinaceous discontinuities inducing the affinity to nitrocellulose are in the order of 7 to 10 $\times$ 10^6 Da. This is in agreement with the results obtained by other methods (Table 1).

B. DNA Sequence Characteristics in Isolated Complexes

The method which allows separating the DNA fragments associated with residual peptides from peptide-free DNA fragments[45] was used to investigate whether the distribution of the DNA/peptide complexes found in isolated DNA is random or restricted to a subset of DNA sequences. Restriction to a subset of DNA sequences would clearly point to their physiological significance. Moreover, in the latter case, information can be expected about the class of DNA sequences most tightly interacting with proteinaceous nuclear materials. Such information is of interest with respect to the attachment sites of chromosomal DNA at nuclear skeletons and their potential involvement in genome functions.

It has been shown that the DNA sequences in the vicinity of attachment points of genomic DNA in nuclear matrix are enriched in newly replicated DNA.[17] In order to test the relation between the fraction of DNA fragments retained on nitrocellulose filters and the DNA sequences attached to the nuclear matrix, pulse-labeled Alu I fragmented DNA was submitted to nitrocellulose filtration. In Figure 9 it is shown that after short pulses, 46% of the pulse-label is retained on filters, while only a total of 13 to 15% of the DNA is bound. This high enrichment was almost identical if the DNA submitted to filtration was pretreated with S1 nuclease. This indicates that the enrichment of the pulse-label in the filter fraction is not due to binding of highly radiolabeled nascent ssDNA pieces (Okazaki-type pieces). In longer pulses, the specific radioactivity in the retained fraction becomes smaller. These labeling kinetics of DNA fragments retained on filters are in agreement with the labeling kinetics of residual DNA strands attached to the nuclear matrix.[17] From this it is tempting to conclude that the factors inducing the binding of DNA fragments to nitrocellulose filters and those which are involved in anchorage of replicatively active DNA in the nuclear matrix may be identical.

Further experiments were performed which resulted in additional evidence that DNA sequences retained on nitrocellulose filters reflect a subset of DNA sequences. For example the hybridization characteristics of the DNA fraction recovered from nitrocellulose filters with radiolabeled poly A$^+$ RNA populations were different from those of the DNA fraction passing the filters.[45] Moreover, Southern analysis of the various DNA fractions reveal that the sequence complexity in the DNA fraction retained on nitrocellulose filters is different from that of unfractioned DNA or DNA passing the filters. For this, Southern blots of the various DNA fractions were hybridized with radiolabeled DNA recovered from nitrocellulose filters. As is shown in Figure 10, the self-hybridization of the DNA recovered from filters is about ten times more intense than the hybridization with unfractioned DNA. Consistently, the DNA fragments passing the filters are poor in sequences retained on filters. These results indicate that in most of the genomes of a distinct cell population the residual peptides inducing the binding of DNA fragments to nitrocellulose filters have to be associated with a nonrandom subset of DNA sequences. If radiolabeled nuclear matrix DNA was used as a probe in experiments shown in Figure 10, identical results were obtained (see Note Added in Proof): the nuclear matrix DNA probe hybridized much more intensely with the DNA sequences retained on nitrocellulose filters than with the fraction passing the filters. This indicates that the fraction of DNA with high affinity to nitrocellulose filters and the fraction of DNA which is not released during nuclear matrix extraction are overlapping populations of DNA sequences.

V. BIOLOGICAL IMPLICATIONS

While covalent DNA/protein complexes have been accepted for a long time as a fact in various prokaryotic and eukaryotic viral systems,[35] such complexes in chromosomal DNA have been put into question during the last decade although the experimental evidences for their existence appear to be rather convincing. Since we are now about to demonstrate that these complexes are located on a subset of DNA sequences, we are expecting a new dimension of this field of research. DNA strands which can be separated from the bulk of DNA by nitrocellulose filtration comprise, in addition to their letter code, further information: they derive from sites where the genome is most tightly interacting with non-DNA components of the cell nucleus. Our preliminary results already point to two ambivalent characters of such interactions: one type of interaction appears to be dynamic in so far as all DNA sequences of a proliferating cell have to go through this interaction during the S phase at least once which is revealed by the high enrichment of replicatively active DNA on the filters (Figure 9). On the other hand, it appears that a subset of DNA sequences is permanently in this tight interaction with proteins (Figure 10). Work is going on to select DNA probes which are typical for the specific interactions between distinct DNA sequences and the proteins (see Note Added in Proof).

The proteinaceous discontinuities in isolated DNA have to be discussed also in relation to other lines of research concerned with DNA/non-DNA interactions, with special emphasis on (1) topoisomerases and (2) anchorage of DNA in the nuclear ground substance which is

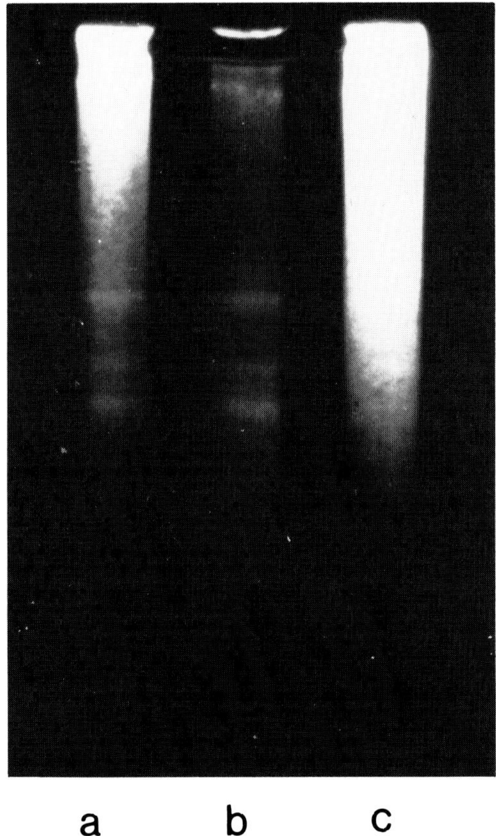

FIGURE 10. A subset of DNA sequences is retained during nitrocellulose affinity filtration. Equal amounts (1 μg) of unfractioned DNA (a), DNA passing nitrocellulose filters (b), and DNA recovered from nitrocellulose filters with 0.05% sodium hydroxide (c) were submitted to alkaline agarose (2%) gel electrophoresis followed by Southern-transfer to a nitrocellulose membrane. The blot was probed with radioactively labeled (^{32}P, nick-translation) DNA recovered from the filter under neutral conditions. The autoradiograph shows that the DNA retained on filters hybridizes more intensely than unfractioned DNA or DNA passing the filters. This indicates a much lower sequence complexity of the DNA fraction retained on filters compared to that of total genomic DNA. Identical results were obtained when the blots were probed with radioactively labeled residual nuclear matrix DNA.

generally termed nuclear matrix. In the light of recent results the latter fields are most probably connected.[50]

Topoisomerases are known to form transient covalent complexes with DNA. Cell nuclei contain up to 10^6 soluble topoisomerase I molecules and about 10^5 soluble topoisomerase II molecules.[19] The fraction of topoisomerase molecules which is not extracted by high salt buffers and which remains bound to the nuclear matrix is unknown at present. Since topoisomerases are considered to be involved in various nuclear processes like control of supercoiling, DNA replication, transcription, recombination, transposition, and repair,[19] it is conceivable to conclude that at any time a fair number of topoisomerase molecules are in an active state. Consequently, if cells are lysed and DNA is isolated there *must* be found a considerable amount of proteinaceous material covalently linked to DNA. From this one would suggest that the covalently bound polypeptides found in isolated chromosomal DNA reflect topiosomerase molecules "in action". However, the characteristics of the alkali stably bound polypeptides in DNA are divergent from the known physico-chemical characteristics of topiosomerases. The latter are readily soluble proteins with molecular weights larger than 100 kDa.[19] Moreover, eukaryotic topoisomerase I is only covalently linked to 3′ ends of

DNA and topoisomerase II is only covalently linked to 5′ ends of DNA. Due to these different characteristics of topoisomerases and the alkali stably bound proteins in DNA, we suggested previously that these classes of polypeptides may be different.[51] However, more recently it has been shown that fragments of topoisomerase I in the order of 54/68 kDa exhibit full enzymatic activities.[19] Moreover, solubilities of topoisomerases could be changed by covalent binding to the terminal nucleotide(s) which remain(s) bound to the protein after degradation of DNA by DNase I. Finally, topoisomerases exhibit *in vitro* reactions (knotting, catenation) which do not seem biologically meaningful. Conversely, these enzymes could catalyse reactions *in vivo* which are not assayed under *in vitro* conditions. The latter reactions could be induced by chromatin proteins which show intimate and not well understood interactions with topoisomerases and which are usually not present in the *in vitro* standard assays.[19] Based on such considerations, it seems to be open whether the covalently bound polypeptides in isolated DNA reflect proteins identical to or different from topoisomerases (see also Note Added in Proof).

Protein discontinuities in DNA have to be discussed also in relation to higher order structures in chromatin. In high salt extracted nuclei, DNA is histone depleted and tremendously extended. However, superhelical DNA loops remain still anchored and restrained in a nuclear ground substance which has been termed nuclear matrix. There is evidence indicating that RNP is involved in the stability of this subnuclear structure and in at least one type of DNA anchorage. A summary of these arguments has been published elsewhere.[52] A visualization of this interaction between residual nuclear matrix DNA strands and RNP is shown in Figure 11.[53] On the other hand, at least a portion of the proteins most tightly interacting with DNA is nuclear matrix associated[22,23,33] and DNA sequences involved in these complexes exhibit characteristics of nuclear matrix DNA, e.g., similarities in sequence complexity (Figure 10) and enrichment in replicatively active DNA.[45] This suggests that the most stable interaction between DNA and protein is somehow involved in anchorage sites of DNA strands in high salt extracted nuclei. Since the frequencies of these anchorage sites have been determined by various authors[17] to be in the order of 0.5×10^5 to 1×10^5 which is about 6 to 12 times lower than the frequencies of the protein discontinuities in isolated DNA (Table 1), it could be suggested that not all of the protein discontinuities found in isolated DNA originate from anchorage structures of residual DNA in DNase I and high salt extracted nuclei. However, the DNA sequences anchored in the nuclear matrix and those associated with the most tightly bound proteins are clearly overlapping.

It should be noted also that the quality and possibly the number of anchorage sites of DNA seem to be different if high salt extraction of nuclei is replaced by other methods designed to destabilize nucleosomes and to extract histones. In this case, it appears that topoisomerase II plays a major role in the attachment structure.[50] Since the different procedures result in other DNA sequences attached to insoluble nuclear material, it has been argued that the high salt extraction procedure may induce nonphysiological attachment. However, it is likely as well that additional or other attachment sites are formed during other extractions, e.g., by reaction of nuclear matrix associated topoisomerase II with DNA strands comprising sequences preferentially attacked by this enzyme.[50]

Since all procedures used so far to detect the *in situ* anchroage sites of chromosomal DNA apply conditions far from being physiological, it cannot be decided at present which sites of DNA are physiologically anchored in higher order structures. The RNP nuclear matrix, the topoisomerases, and the discontinuities of protein nature in DNA may represent pieces of a puzzle which is at present still incomplete.

NOTE ADDED IN PROOF

A number of new relevant results have been published during the preparation of this volume. For example, conditions have been described that allow ^{32}P-radiolabeling and

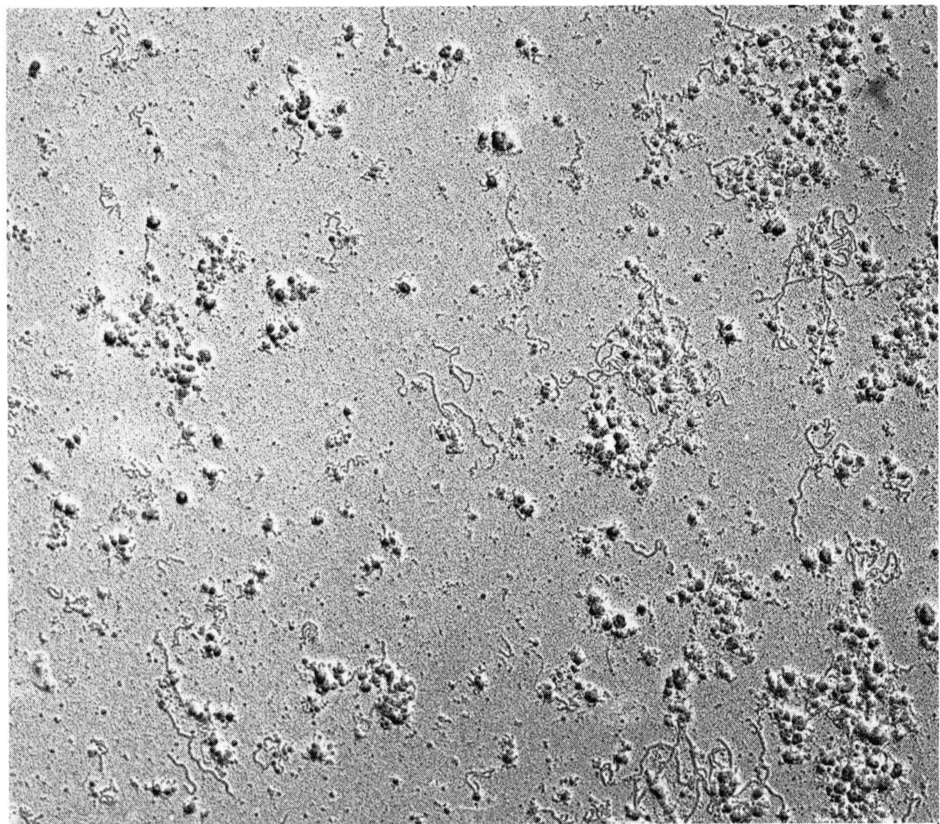

FIGURE 11.   Tight interaction between residual nuclear matrix DNA and RNP. Isolated nuclear matrix was disrupted by aerodynamic shear forces (particle fragmentation[53]) and visualized by electron microscopy. The picture shows that the entire nuclear matrix complex consists of RNP particles of different size classes. There are no DNA strands that are not associated with RNP.

detection of the tight complexes between DNA and polypeptides by nick-translation in purified bulk DNA and in residual nuclear matrix DNA.[54] The physico-chemical characteristics of the complexes radiolabeled by ^{32}P are identical with those identified previously by treatment of DNA with ^{125}Iodine.[20,21,29] Moreover, a high degree of homology has been found between residual nuclear matrix DNA sequences and the DNA sequences associated with the tightly bound polypeptides which indicates that the complexes originate from sites where the genome is salt-stably anchored in the nuclear matrix.[55] Finally, it should be noted that the covalent nature of the DNA/polypeptide complexes first shown in DNA of Ehrlich ascites cells[27] has been confirmed in another system.[56] The latter work presents also evidence for different peptide patterns of topoisomerases and the polypeptides co-purifying with DNA. A first DNA sequence decisively involved in the site-specific location of covalent DNA/peptide complexes in the mouse genome has been published recently.[57]

REFERENCES

1. **Kirby, K. S.,** A new method for the isolation of deoxyribonucleic acids: evidence on the nature of bonds between deoxyribonucleic acid and protein, *Biochem. J.,* 66, 495, 1957.
2. **Kirby, K. S.,** Preparation of some deoxyribonucleic protein complexes from rat liver homogenates, *Biochem. J.,* 70, 260, 1958.

3. **Kirby, K. S.,** The preparation of deoxyribonucleic acids by the p-amino-salycylate-phenol method, *Biochim. Biophys. Acta,* 36, 117, 1959.
4. **Borenfreund, E., Fitt, E., and Bendich, A.,** Isolation and properties of deoxyribonucleic acid from mammalian sperm, *Nature,* 191, 1375, 1961.
5. **Hilgartner, C. A.,** The binding of DNA to residual protein in mammalian nuclei, *Exp. Cell Res.,* 49, 520, 1968.
6. **Salser, J. S. and Balis, M. E.,** Investigations of amino acids bound to DNA, *Biochim. Biophys. Acta,* 149, 220, 1967.
7. **Bendich, A., Borenfreund, E., Korngold, G. C., Krim, M., and Balis, M. E.,** Amino acids or small peptides as punctuation in the genetic code of DNA, In *Symp. Nucleic Acids and Their Biological Roles,* Tipografia Successori Fusi, Pavia, Italy, 1964, 214.
8. **Dounce, A. L. and Hilgartner, C. A.,** A study of DNA nucleoprotein gels and the residual protein of isolated cell nuclei, *Exp. Cell Res.,* 36, 228, 1964.
9. **Gross-Bellard, M., Oudet, P., and Chambon, P.,** Isolation of high molecular weight DNA from mammalian cells, *Eur. J. Biochem.,* 36, 32, 1973.
10. **Hershey, H. V. and Werner, D.,** Evidence for non-deoxynucleotide linkers in Ehrlich ascites tumour cell DNA, *Nature,* 262, 148, 1976.
11. **Werner, D., Krauth, W., and Hershey, H. V.,** Internucleotide protein linkers in Ehrlich ascites cell DNA, *Biochim. Biophys. Acta,* 608, 243, 1980.
12. **Schroeter, D., Meinzer, P., and Werner, D.,** Size of native and denatured DNA of Ehrlich ascites tumour cells isolated in the presence of different protease concentrations, *Eur. J. Cell Biol.,* 24, 131, 1981.
13. **Kavenoff, R., Klotz, L. C., and Zimm, B. H.,** On the nature of chromosome-sized DNA molecules, *Cold Spring Harbor Symp. Quant. Biol.,* 38, 1, 1973.
14. **Hays, J. B. and Zimm, B. H.,** Flexibility and stiffness in nicked DNA, *J. Mol. Biol.,* 48, 297, 1970.
15. **Cook, P. R. and Brazell, I. A.,** Supercoils in human DNA, *J. Cell Sci.,* 19, 261, 1975.
16. **Berezney, R. and Coffey, D. S.,** The nuclear protein matrix: isolation, structure and function, *Adv. Enzyme Regul.,* 14, 63, 1976.
17. **Berezney, R.,** Organisation and functions of the nuclear matrix, In *Chromosomal Non-Histone Proteins,* Vol. 4, Hnilca, L. S., Ed., CRC Press, Boca Raton, FL, 1984, 119.
18. **Geider, K. and Hoffmann-Berling, H.,** Proteins controlling the helical structure of DNA, *Annu. Rev. Biochem.,* 50, 233, 1981.
19. **Vosberg, H. P.,** DNA topoisomerases: enzymes that control DNA conformation, *Curr. Top. Microbiol. Immunol.,* 114, 19, 1985.
20. **Krauth, W. and Werner, D.,** Analysis of the most tightly bound proteins in eukaryotic DNA, *Biochim. Biophys. Acta,* 564, 390, 1979.
21. **Capesius, I., Krauth, W., and Werner, D.,** Proteinase K resistant and alkali-stably bound proteins in higher plant DNA, *FEBS Lett.,* 110, 184, 1980.
22. **Werner, D., Zimmermann, H. P., Rauterberg, E., and Spalinger, J.,** Antibodies to the most tightly bound proteins in eukaryotic DNA, *Exp. Cell Res.,* 133, 149, 1981.
23. **Bodnar, J. W., Jones, C. J., Coombs, D. H., Pearson, G. D., and Ward, D. C.,** Proteins tightly bound to HeLa cell DNA at nuclear matrix attachment sites, *Mol. Cell. Biol.,* 3, 1567, 1983.
24. **Bekhor, I. and Mirell, C. J.,** Simple isolation of DNA hydrophobically complexed with presumed gene regulatory proteins, *Biochemistry,* 18, 609, 1979.
25. **Gianfranceschi, G. I., Coderoni, S., Miano, A., Bramucci, M., Felici, F., Paperelli, M., and Amici, D.,** chapter 7, this volume.
26. **Hillar, M.,** personal communication, 1985.
27. **Neuer, B., Plagens, U., and Werner, D.,** Phosphodiester bonds between polypeptides and chromsomal DNA, *J. Mol. Biol.,* 164, 213, 1983.
28. **Spiess, E., Neuer, B., and Werner, D.,** Isolation and visualisation of alkali stable protein/DNA complexes, *Biochem. Biophys. Res. Commun.,* 104, 548, 1982.
29. **Plagens, U.,** Effect of salt-treatment on manually isolated polytene chromosomes, *Chromosoma (Berl.),* 68, 1, 1978.
30. **Nehls, P., Rajewsky, M. F., Spiess, E., and Werner, D.,** Highly sensitive sites for guanine-O^6 ethylation in rat brain DNA exposed to N-ethyl-N-nitrosourea in vivo, *EMBO J.,* 3, 327, 1984.
31. **Ochs, D.,** Protein contaminants of SDS-polyacrylamide gels, *Anal. Biochem.,* 135, 470, 1983.
32. **Coombs, D. H. and Pearson, G. D.,** Filter-binding assay for covalent DNA-protein complexes: adenovirus DNA-terminal protein complex, *Proc. Natl. Acad. Sci. U.S.A.,* 75, 5291, 1978.
33. **Werner, D., Shi Chanpu, Mueller, M., Spiess, E., and Plagens, U.,** Antibodies to the most tightly bound proteins in eukaryotic DNA: formation of immuno complexes with nuclear matrix components, *Exp. Cell Res.,* 151, 384, 1984.
34. **Shapiro, B. M. and Stadtman, E. R.,** 5′-Adenylyl-O-tyrosine, *J. Biol. Chem.,* 243, 3769, 1968.

35. **Tse, Y.Ch., Kirkegaard, K., and Wang, J. C.,** Covalent bonds between protein and DNA, *J. Biol. Chem.,* 255, 5560, 1980.
36. **Razzell, W. E. and Khorana, H. G.,** Studies on polynucleotides. X. Enzymatic degradation. Some properties and mode of action of spleen phosphodiesterase, *J. Biol. Chem.,* 236, 1144, 1961.
37. **Bernardi, A. and Bernardi, G.,** Studies on acid hydrolases. V. Isolation and characterization of spleen nucleoside polyphosphatase, *Biochim. Biophys. Acta,* 155, 371, 1968.
38. **Razzell, W. E. and Khorana, H. G.,** Studies on polynucleotides III. Enzymatic degradation. Substrate specificity and properties of snake venom phosphodiesterase, *J. Biol. Chem.,* 234, 2105, 1959.
39. **Bernardi, A. and Bernardi, G.,** Spleen acid exonuclease, In *The Enzymes,* Vol. 4, Boyer, P. D., Ed., Academic Press, New York, 1971, 329.
40. **Tsanev, R., Avramova, Z., and Tasheva, B.,** chapter 3, this volume.
41. **Werner, D., Hadjiolov, D., and Neuer, B.,** Protease inducible alkali lability of DNA from proliferating and non-proliferating cells, *Biochem. Biophys. Res. Commun.,* 100, 1047, 1981.
42. **Lett, J. T., Klucis, E. S., and Sun, C.,** On the size of DNA in the mammalian chromosome. Structural subunits, *Biophys. J.,* 10, 277, 1970.
43. **Elkind, M. M.,** Sedimentation of DNA released from chinese hamster cells, *Biophys. J.,* 11, 502, 1971.
44. **Nakane, M., Ide, T., Anzai, K., Ohara, S., and Andoh, T.,** Supercoiled DNA folded by non-histone proteins in cultured mouse carcinoma cells, *J. Biochem.,* 84, 145, 1978.
45. **Neuer, B. and Werner, D.,** Screening of isolated DNA for sequences released from anchorage sites in nuclear matrix, *J. Mol. Biol.,* 181, 15, 1985.
46. **Southern, E. M.,** Detection of specific sequences among DNA fragments separated by gel electrophoresis, *J. Mol. Biol.,* 98, 503, 1975.
47. **Riggs, A. D. and Bourgeois, S.,** On the assay, isolation and characterization of the lac repressor, *J. Mol. Biol.,* 34, 361, 1968.
48. **Riggs, A. D., Bourgeois, S., Newby, R. F., and Cohn, M.,** DNA binding of the lac repressor, *J. Mol. Biol.,* 34, 365, 1968.
49. **Jack, R. S., Gehring, W. J., and Brack, C.,** Protein component from drosophila larval nuclei showing sequence specificity for a short region near a major heat-shock protein gene, *Cell,* 24, 321, 1981.
50. **Cockerill, P. N. and Garrard, W. T.,** Chromosomal loop anchorage of the kappa immunoglobulin gene occurs next to the enhancer in a region containing topoisomerase II sites, *Cell,* 44, 273, 1986.
51. **Werner, D. and Petzelt, Ch.,** Alkali-stably bound proteins in eukaryotic and prokaryotic DNAs show common characteristics, *J. Mol. Biol.,* 150, 297, 1981.
52. **Rest, R., Mueller, M., and Werner, D.,** Fragmentation of nucleoskeletal elements by metrizamide/2M salt isopycnic centrifugation, *Exp. Cell Res.,* 167, 144, 1986.
53. **Mueller, M., Spiess, E., and Werner, D.,** Fragmentation of nuclear matrix on a mica target, *Eur. J. Cell Biol.,* 31, 158, 1983.
54. **Werner, D. and Rest, R.,** Radiolabelling of DNA/polypeptide complexes in isolated bulk DNA and in residual nuclear matrix DNA by nick-translation, *Biochem. Biophys. Res. Commun.,* 147, 340, 1987.
55. **Neuer-Nitsche, B. and Werner, D.,** Sub-set characteristics of DNA sequences involved in tight DNA/polypeptide complexes and their homology to nuclear matrix DNA, *Biochem. Biophys. Res. Commun.,* 147, 335, 1987.
56. **Avramova, Z. and Tsaner, R.,** Stable DNA-protein complexes in eukaryotic chromatic, *J. Mol. Biol.,* 196, 437, 1987.
57. **Neuer-Nitsche, B., Lu, X., and Werner, D.,** Functional role of a highly repetitive DNA sequence in anchorage of the mouse genome, *Nucleic Acids Res.,* 16, 8351, 1988.

Chapter 7

SMALL PEPTIDES BOUND TO CHROMATIN DNA ARE INVOLVED IN THE REGULATION OF TRANSCRIPTION

G. L. Gianfranceschi, S. Coderoni, A. Miano, M. Bramucci, F. Felici, M. Paparelli, and D. Amici

TABLE OF CONTENTS

I. INTRODUCTION

We have previously reported the isolation from prokaryotic and eukaryotic chromatin of low molecular weight peptides with high specific activity in inhibiting transcription in cell and cell-free systems.[1-9] Control of chromatin and DNA transcription *in vitro* is apparently mediated by a specific interaction between peptides and DNA which appears pH-dependent and, over the pH range 6 to 9.5, it decreases rapidly after pH 7.5. Consequently, we perform the isolation and purification of DNA at slightly acidic pH (pH 6) and then the extraction of peptides from deproteinized DNA by using an acetic acid/ammonia alkaline buffer (pH 9.5). The purified peptides contain prevailingly glutamic acid, serine, aspartic acid, glycine, and alanine.

The extraction of peptides from DNA causes a relevant amplification of DNA template capacity for prokaryotic and eukaryotic RNA polymerase.[9] This could imply a potential novel control mechanism of gene expression. Recently, the presence of small peptides bound to the nucleic acids has been demonstrated in several laboratories.[10-12] The isolated peptides have common amino acid composition, molecular weight (about 1000), and display regulatory effects on gene expression *in vitro*. Our preliminary studies on the purified peptides structure by N- and C-terminal analysis were unsuccessful suggesting the presence of blocked terminal groups. A remarkable difficulty in the chemical characterization of the controlling transcription peptides extracted from purified DNA is also represented by the small amount of purified material available. This takes into account the low ratio (w/w) between peptide and DNA which ranges from 0.001 to 0.03 in the DNA-peptide complexes from various sources.

Moreover, the complete purification of low molecular weight active compounds from other substances which may be present in the crude extract (other peptides, amino acids, nucleotides, sugars, salts) is quite difficult. In this paper we report on the biochemical characterization of the chromatin peptides and on the elucidation of the activity regulating the gene expression in cell and cell-free systems. The possible origin and function of these peptides are discussed.

II. ISOLATION AND BIOCHEMICAL CHARACTERIZATION OF THE CHROMATIN PEPTIDES

The DNA-binding peptides are isolated from the nuclei of several mammalian tissues (thymus, testicle, liver, spleen) or from prokaryotic nucleoid following the procedure previously reported.[9] The isolated nuclei were suspended in 0.15 M NaCl/0.015 M trisodium citrate (SSC) adjusted to pH 6 with HCl. Lysis of the nuclei was effected with 2.5% (final concentration) sodium dodecyl sulfate (SDS) and NaCl added to a final concentration of 1.5 M.

The isolation of DNA-binding peptides from prokaryotic nucleoid *Escherichia coli* B cells were carried out in 8 volumes SSC (acidified at pH 6.5 with HCl)/1.5 M sucrose, and lysis was done in 2.5% SDS (final concentration) at 60°C for 20 min, and NaCl was added to a final concentration of 1.5 M. The nucleic acid fractions were extensively deproteinized by cold chloroform/isoamyl alcohol (24:1 by volume) (two deproteinization steps) and cold phenol (two deproteinization steps). The DNA was precipitated from the final aqueous phase by adding 2 volumes ethanol (acidified to pH 5 with HCl) and immediately redissolved in 0.1 × SSC (pH 6). The redissolved DNA was then incubated at 37°C with pancreatic ribonuclease (50 µg/ml, Calbiochem) and ribonuclease T1 (50 µg/ml Calbiochem) for 60 min. The enzymes were removed by deproteinization with cold phenol, the DNA was precipitated with acidified ethanol, and redissolved in 0.1 × SSC (pH 6).

The DNA preparations from various sources were then subjected to alkali extraction by

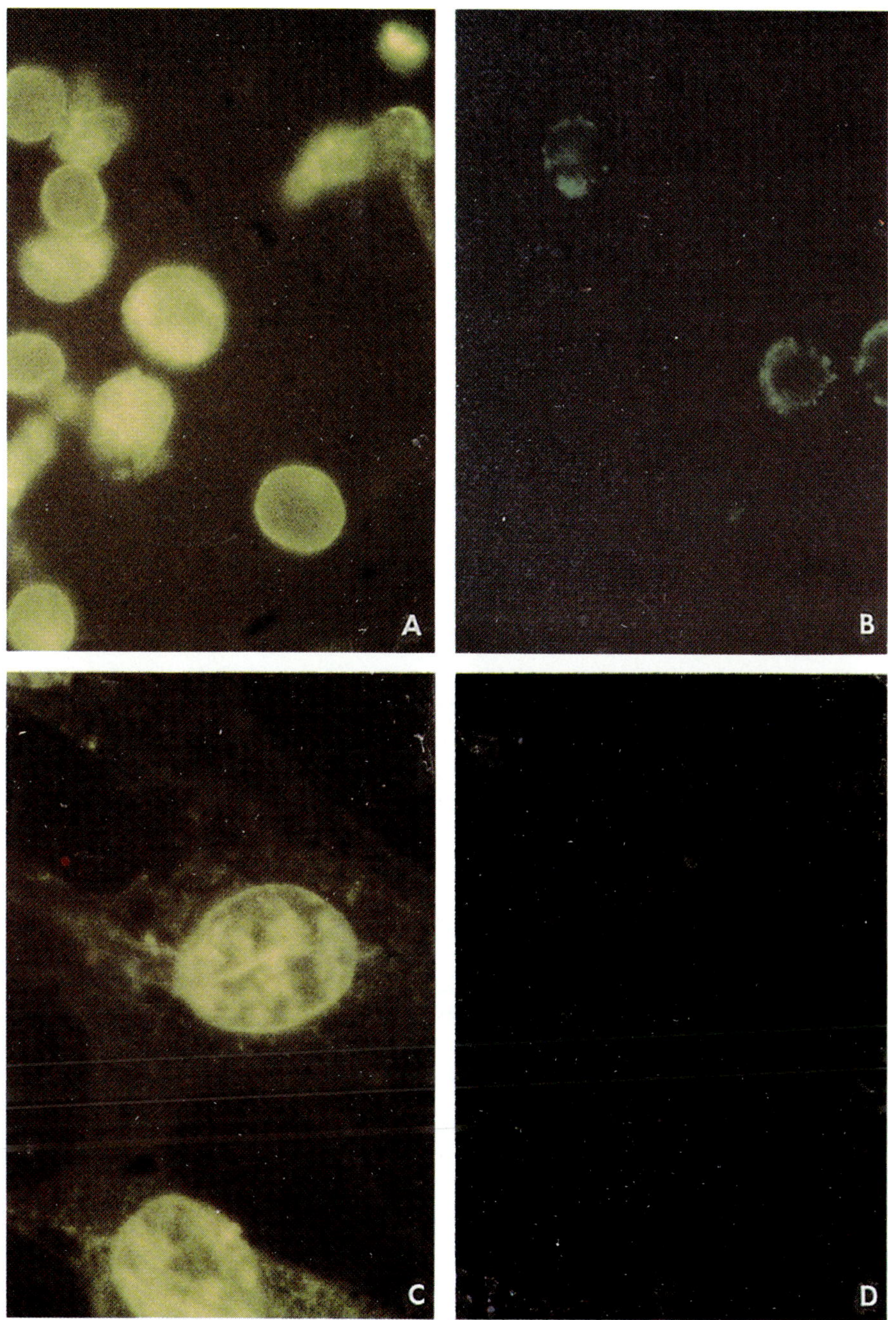

PLATE 1. Chapter 1.   Indirect immunofluorescence verifying nuclear location of B_2 phosphoprotein. Rat liver nuclei (10^5) were first fixed with 4% paraformaldehyde and then permeabilized with 0.2% Triton in phosphate buffered saline (PBS). The nuclei were then probed with B_2 protein IgM monoclonal antibody (diluted 1:100) (A) or control mouse serum (1:100) (B), and after extensive washing with PBS, binding was detected by fluorescein isothiocyanate conjugated goat anti-mouse IgG + IgM (diluted 1:50). (Magnification × 600.) Similarly, mouse BC3H1 muscle glioma cells were grown on cover slips and then sequentially extracted with nucleases and high salt (to obtain an *in situ* nuclear matrix) before being fixed and probed with B_2 protein IgM monoclonal antibody (1/100) (C) or control mouse serum (1/100) (D). Fluorescein specific binding was detected as described above. (Magnification × 1250.)

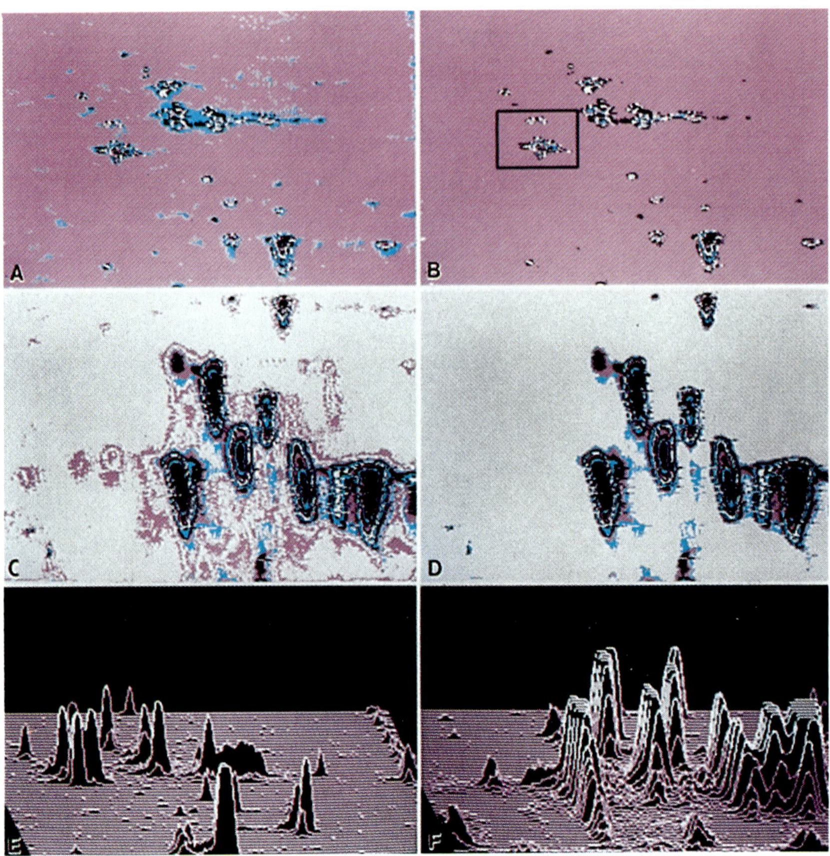

PLATE 1. Chapter 9.   Computer generated two-D gel scans of the results shown in Figure 2, as computed from a 9.3-cm laser scan using the PCI (Biomed Instruments, Inc.) program. (A) Contour analysis of the boxed area shown in Figure 2A with a low baseline setting to visualize both the low and high intensity radiolabeled protein products; (B) same as in (A) except the contour analysis was done at a high baseline setting; (C) contour analysis of the crystallin mRNA products (alpha, beta, gamma) as shown in the lower half of Figure 2A with a low baseline setting; (D) same as in (C) except the analysis was done at a high baseline setting to accentuate the major protein products; (E) pictorial mapping of the data shown in (B); (F) pictorial mapping of the contours shown in (D). Boxed area in (B) was amplified as shown in Plate 4A. Color changes from the periphery of the contours to the center signify increase in intentsity within that spot.

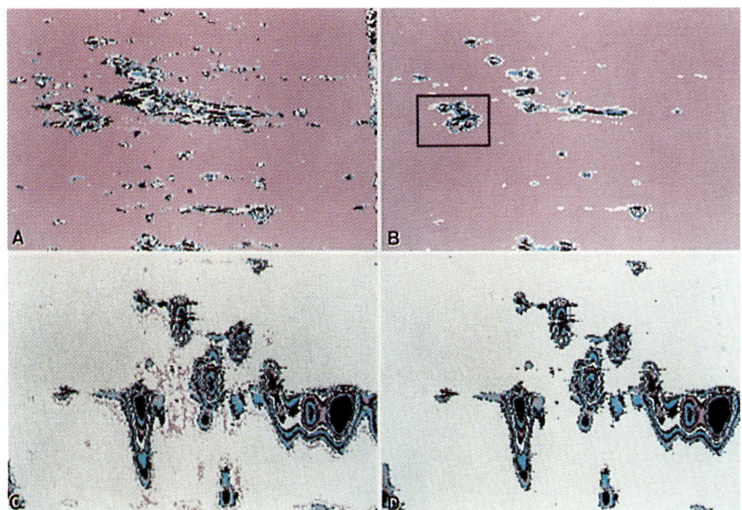

PLATE 2. Chapter 9.   (A) Contour analysis at a low baseline setting of the boxed area in Figure 2B, showing the noncrystallin radiolabeled protein products from mRNA of lens undergoing reversal of cataracts; (B) same as (A) at a high baseline setting; (C) contour analysis of the lower half of the fluorogram shown in Figure 2B at a low baseline setting; (D) same as (C) at a high baseline setting. The scan was done at 9.3-cm setting. Boxed area is that amplified in Plate 4B.

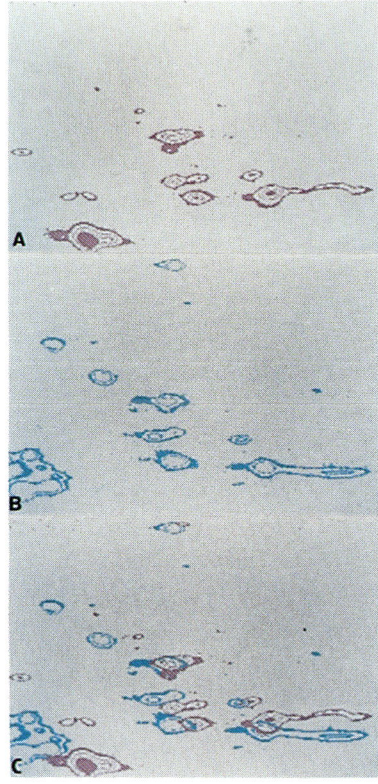

PLATE 3. Chapter 9.   (A) Amplification of a portion of the contours shown in upper half of Plate 1B, 32-d cataractous lens; (B) amplification of the same portion of the contours as shown in Plate 2B, 12-d reversal of the 32-d cataractous lens; (C) computer superimposition of the data shown in (A) and (B) above. Amplification is done by carrrying out a second scan at a setting of 5 cm.

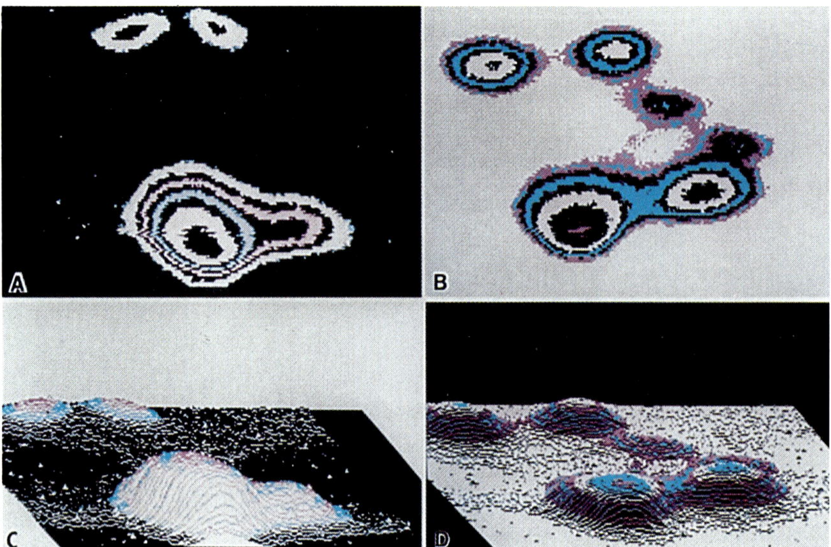

PLATE 4. Chapter 9.   Amplification of a small section, shown in box in Plate 1B (A) and in Plate 2B (B), and pictorial mapping of the same data (C for A above; and D for B above). The data clearly suggest presence of a new species of mRNA in the reversed cataractous lens. The scan was done at a setting of 1.2 cm.

Table 1
SPECIFIC ACTIVITY IN INHIBITING
TRANSCRIPTION *IN VITRO* EXERTED BY
THE ACTIVE PEPTIDE FRACTIONS
FOLLOWING THE PURIFICATION
PROCEDURE

	Transcription inhibition (%)	
Peptide purification step	**60—70**	**90—100**
Crude extract	0.6	1.2
Sephadex G25 fraction	0.4	0.8
Trisacryl GF05 fraction (Figure 1)	0.4	0.8
Sephadex G10 fraction I (Figure 2)	0.05	0.1
Sephadex G10 fraction II (Figure 2)	0.5	1.0
HPLC fraction 3a (Figure 4a)	0.01	0.02
HPLC fraction 2b (Figure 4b)	0.2	0.4

Note: Incubation mixture and other conditions for DNA transcription *in vitro* are given in Reference 9; incorporation of ^{3}H-UMP after 10 min of incubation. Values are ratios peptide: DNA.

adding 1 volume 0.4 *M* ammonium acetate/ammonia buffer (pH 9.5). The solution was allowed to stand for 1 h with occasional shaking and the DNA precipitated again with ethanol. The ethanol supernatant from the last precipitation (ethanol extract of DNA) was used to purify the chromatin peptide component. The gel filtration of the crude peptide fraction on Sephadex G25 or Trisacryl GFO.5 or Sephadex G15 shows that the peptide(s) is active in inhibiting transcription (Table 1) and it migrates as a major peak with an eluted volume/void volume (V_e/V_o) ratio equal to 2, 2.5, and 1.9, respectively (Figure 1). The other minor fractions eluting at different volumes are almost completely inactive. Gel filtration of chromatin peptides on Sephadex G25 column calibrated with dextran blue, bacitracin, oxytocin, oxidized glutathione, and potassium dichromate demonstrates that the elution volume of the active fraction corresponds to a molecular weight of about 1000. The chromatin peptides obtained from the various gel filtration chromatographies show very similar specific activity in the control of calf thymus DNA transcription *in vitro* (60 to 70% inhibition at a peptide/DNA ratio of 0.4, w/w). The amino acid analysis demonstrates the predominant presence of aspartic acid, serine, glutamic acid, glycine, and alanine. The chromatin peptides from Sephadex G25 or Trisacryl GF05 gel filtration are subsequently chromatographed on Sephadex G10 column equilibrated and eluted by water (Figure 2). The determination of the distribution of the inhibiting-transcription activity shows that the two major peaks are active (Fraction I and II). The data in Table 1 demonstrate that the first peak shows higher specific activity.

Figure 3 shows the absorption spectra of the active fractions. The patterns of analytical separation by high performance liquid chromatography (HPLC) of the active fractions I and II of chromatin peptides (from trout testis) following Sephadex G10 chromatography are shown in Figure 4 a and b, respectively. The fractionation of chromatin peptides by HPLC is not completely satisfactory because the active compounds are quickly eluted by reverse phase columns indicating that they are strongly hydrophilic.

The fractions 3a (Figure 4a) and 2b (Figure 4b) show a sharp activity in inhibiting transcription *in vitro* (70% block of transcription with protein/DNA ratio 0.01 for the fraction 3a and 0.2 for the fraction 2b). The results in Table 1 show the increase in the specific activity of the chromatin peptides following the purification procedure.

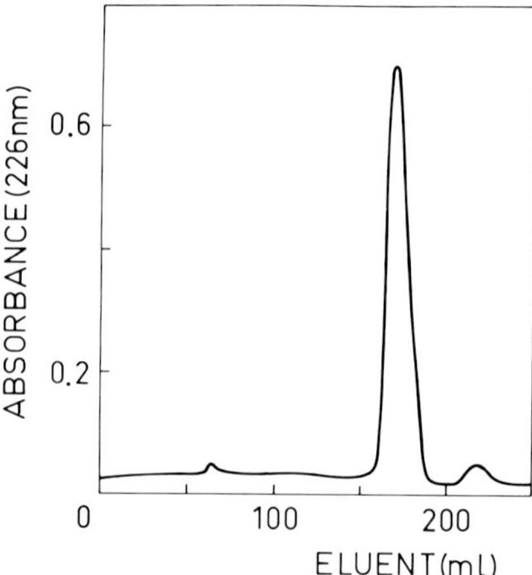

FIGURE 1.   Chromatography in Trisacryl GF05 (L.K.B.) of active peptides extracted by alkaline buffer from trout testis chromatin DNA. 1 mg chromatin peptides were chromatographed; the column (2 × 100 cm) was equilibrated and eluted with 50 mM ammonium acetate/acetic acid (pH 4.7). Flow rate, 28 ml/h. The profile represents continuous monitoring.

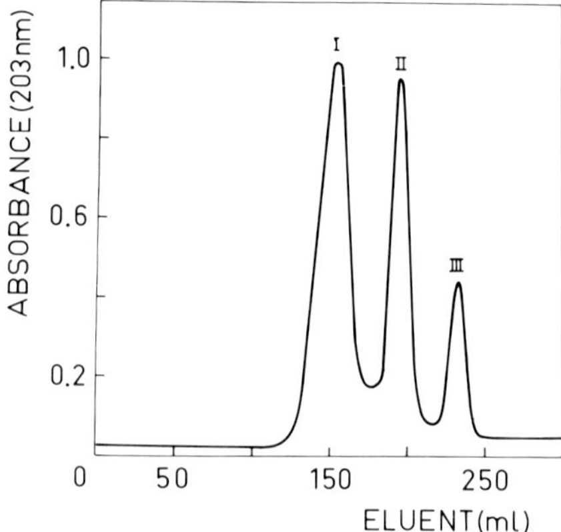

FIGURE 2.   Chromatography on Sephadex G10 (Pharmacia) of the active peptide fraction from trout testis DNA following Trisacryl GF05 chromatography. 500 μg peptides were chromatographed; the column (2.5 × 180 cm) was equilibrated and eluted with bidistilled water. Flow rate, 40 ml/h. The absorbance profile at 203 nm represents continuous monitoring.

The amino acid composition of the most purified peptide fractions obtained by gel filtrations and HPLC from calf thymus, calf testicle, trout testicle, and *E. coli* are reported in Table 2.

Studies of the chromatin peptides structure by N-terminal analysis using the dansyl chloride procedure was unsuccessful, suggesting the presence of a blocked NH_2 group. According to the results obtained with the dansyl chloride procedure, the activity of the peptide fraction is not destroyed after digestion by aminopeptidase (leucine aminopeptidase, Sigma, 100 μg/ml, 3 h).

On the other hand, digestion of the chromatin peptides by pyroglutamate aminopeptidase for different times (Sigma; enzyme/peptide ratio = 0.02), results in loss of transcriptional activity (Figure 5), thus indicating that pyroglutamic acid could be the N-terminal.

At the same time the digestion of chromatin peptides extracted from trout testis DNA by

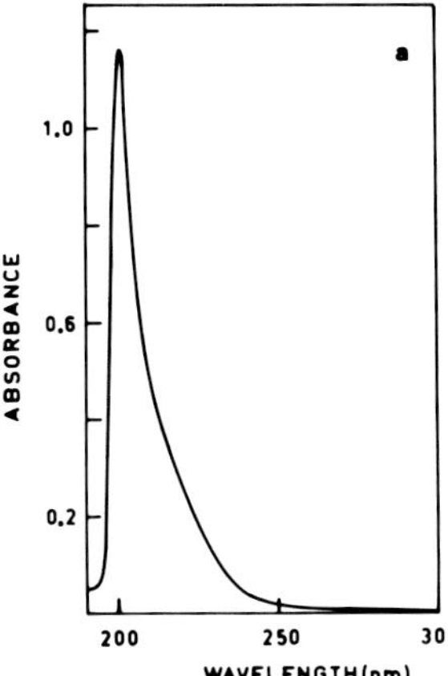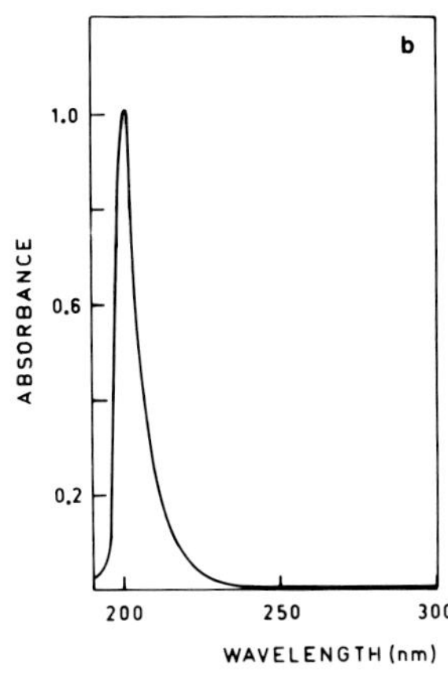

FIGURE 3.   Absorption spectrum of the peptide fractions I (Figure 3a) and II (Figure 3b) from Sephadex G10 chromatography (Figure 2). Solvent is 1 mM NaCl. The profile represents continuous monitoring.

carboxypeptidases A and Y (Sigma; 100 µg/ml, 3 h) does not cause the release of any amino-acid.

The active peptide fraction from Sephadex (G25 or G15) or Trisacryl GFO 5 chromatographies, has been analyzed by paper electrophoresis. The electrophoresis performed at pH 3 shows that the chromatin peptides are separated in two fractions (Figure 6). One of them (fraction A) migrates towards the positive electrode, the other one (fraction B) practically does not migrate. The electrophoretic conditions have been chosen following a series of experiments performed at different pH and ionic strength. The results of these experiments demonstrate that a component of the peptide fraction is strongly acidic and migrates towards the positive pole until pH 1.9 indicating the probable presence of phosphoric radicals. This is confirmed by the positivity to molybdate (ammonium)-perchloric acid reagent (Hanes reagent). To determine the activity distribution of chromatin peptides following electrophoresis, unstained electrophoretic strips were cut every 1 cm and the subfractions reextracted by water and analyzed for the controlling transcription activity. The results demonstrate that both the fractions A and B are active, but the fraction A shows higher specific activity (about 10 to 20 times) in the RNA polymerase reaction *in vitro* while the fraction B is much more active in the control of RNA synthesis in cell systems. It is possible that the phosphorylated fraction A cannot enter into the cell, whereas the fraction B may be incorporated and processed by the cell. Determinations of the phosphate content in fraction A in solution, following the procedure described by Buss and Stull,[13] show that in the acid active fraction the ratio (mol/mol) of bound phosphate/peptide ranges from 1 to 2, assuming an average molecular weight for the peptide about 1000. The electrophoretic pattern of the fraction I and II from Sephadex G10 chromatography in the same conditions described in Figure 6 shows that the fraction I is rich in the component which migrates towards the positive pole, while the fraction II is almost completely composed by the component which does not migrate.

In order to obtain further information on the presence of phosphates in the active fractions, peptides were extracted from DNA of *E. coli* cells grown in the presence of ^{32}P-ortho-phosphate. This labeled peptide fraction was subjected to electrophoresis as described in Figure 6 and the electrophoretic strips analyzed by autoradiography. The results, in Figure 7 (a and b), clearly confirm the presence of phosphates in the peptide structure; in fact, autoradiographies performed with peptides purified from DNA of *E. coli* cells grown in the presence of ^{14}C-glutamic acid show a peptide fraction with the same electrophoretic mobility

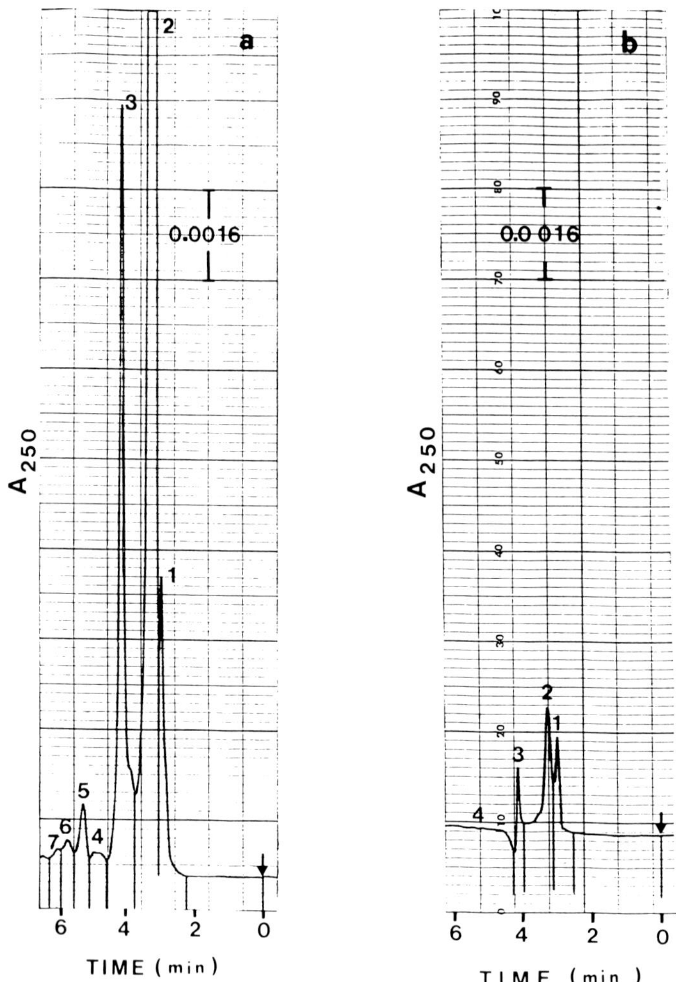

FIGURE 4.   Pattern of analytical fractionation by HPLC of trout testis active fraction I (Figure 4a) and II (Figure 4b) from Sephadex G10 chromatography (see Figure 2). The HPLC was performed on a Perkin Elmer chromatograph (Series 2/2) using a μBondapak C18 column (4.6 mm × 25 cm). 5 to 10 μg peptides were injected and elution was performed with 5% formic acid; flow rate 1 ml/min. The absorbance pattern was measured at 250 nm. The arrow indicates the injection. The subfractions from 20 chromatographies were pooled and lyophilized.

of the phosphorylated compound. Moreover, the peptide fraction labeled with ^{32}P-orthophosphate has been subjected to acid hydrolysis in 2 M HCl for 4 h at 110°C which, according to Krajewski et al.[14] released, at least partially, phosphoserine from the peptide structure. The phosphorylated compound after this hydrolysis, subjected to electrophoresis, shows a slight increase in the migration (Figure 7c). On the other hand strong acid hydrolysis in 6 M HCl for 18 h at 100°C causes the release from the acidic peptide fraction of 60 to 70% free phosphate while the 30 to 40% appears still bound to active compound (Figure 7d; the sample e represents the electrophoretic migration of ^{32}P-orthophosphate). These results do not allow definite conclusions on the presence of phosphoserine in the chromatin peptides, however Juodka[15] reported that the stability of phosphoserine esters is related to the amino acids sequence and peptide structure. We have also investigated if the phosphoric groups can be associated to the chromatin peptides through nucleoside or sugar component, but these experiments did not give positive results.

Table 2
**AMINO ACID COMPOSITION, AFTER ACID
HYDROLYSIS, OF ACTIVE PEPTIDE FRACTIONS
ISOLATED FROM DNA OF CALF THYMUS, CALF
TESTICLE, TROUT TESTICLE, AND *E. COLI***

Amino acid	Chromatin peptides			
	Calf thymus	**Calf testicle**	**Trout testicle**	***E. coli***
Aspartic acid	1.6	2.9	2.1	2.2
Threonine	—	—	—	1.4
Serine	1.8	2.5	1.2	3.0
Glutamic acid	5.0	8.4	4.8	5.8
Glycine	1.9	3.7	2.0	4.1
Alanine	1.1	3.6	1.2	2.6
Valine	1.2	4.1	0.8	2.0

Note: The chromatin peptides were isolated from DNA of various sources and purified by gel filtrations and HPLC following the procedures shown in Figures 1, 2, and 4. The amino acid analysis and the quantitative determination of chromatin peptides were performed with an LKB 4400 amino acid analyzer after hydrolysis of the samples with 6 M HCl in sealed evacuated tubes. Values are expressed as nmol of amino acid per fraction. Values for amino acids present in traces are not given.

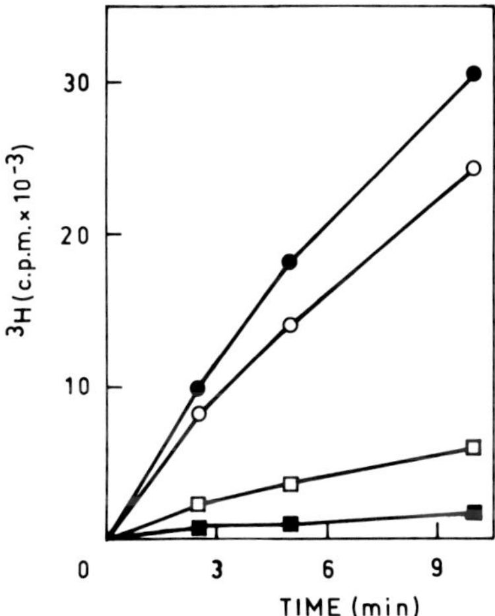

FIGURE 5.   Time-course of RNA synthesis directed by calf thymus DNA: incorporation of ^{3}H-CMP. Inhibition of cell-free transcription by trout testis peptides from Trisacryl GF05 chromatography (peptide/ DNA ratio = 0.5). Incubation mixture for DNA transcription *in vitro* by *E. coli* K-12 RNA polymerase and determination of radioactivity incorporated into the acid-insoluble fraction were performed as reported in Reference 9. Control, no peptide, (●); chromatin peptides, (■); peptides subjected to digestion by pyroglutamate aminopeptidase (Sigma, enzyme/peptide ratio = 0.02, pH 8.0, 37°C) for 2 h (□) and 12 h (○).

III. CONTROL OF DNA TRANSCRIPTION *IN VITRO*

The chromatin peptides purified by gel filtration and HPLC show very high specific activity in inhibiting the RNA polymerase reaction *in vitro* directed by various templates: calf thymus DNA, *E. coli* DNA, λ phage DNA, and poly[d(A-T).d(A-T)] (Table 3). The

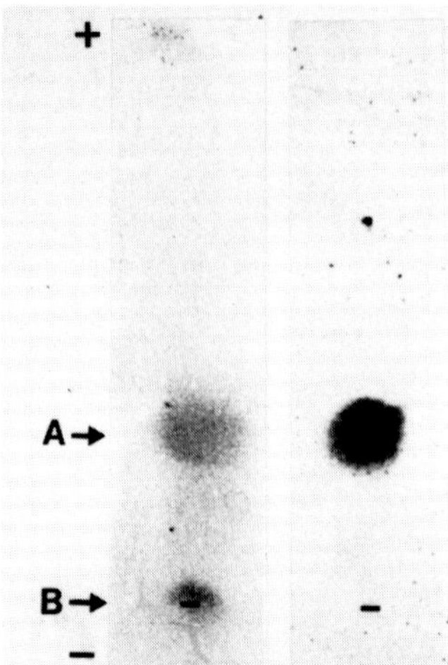

FIGURE 6.   Electrophoretic pattern of trout testis peptides from Trysacryl GF05 chromatography on paper strips (Whatman 3MM) stained with ninhydrin (left) and ammonium molybdate (right). The electrophoresis (300 V; 20 V/cm); 8 mA/strip) was performed for 30 min in 30 m*M* ammonium acetate/6% acetic acid, pH 3.

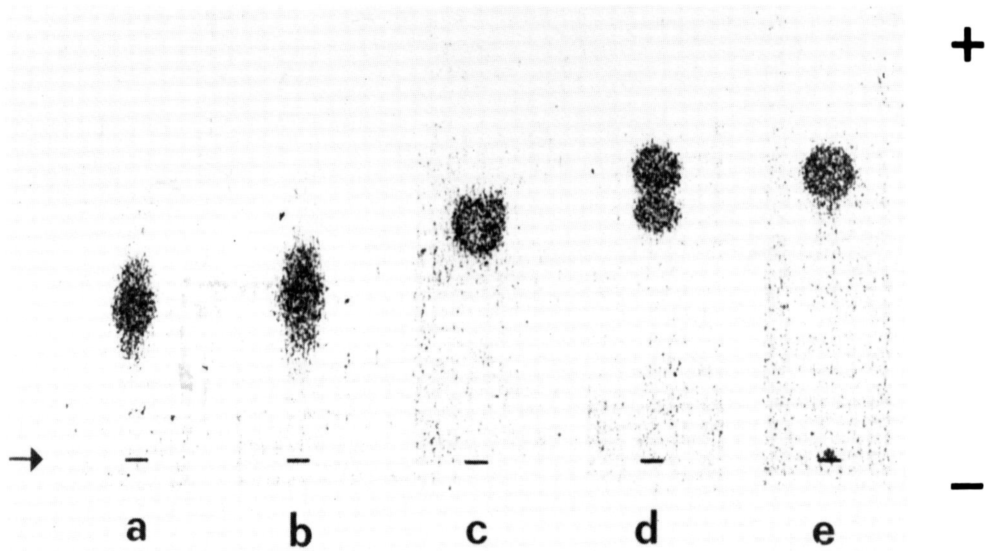

FIGURE 7.   Autoradiography of chromatin peptides labeled with ^{32}P and subjected to electrophoresis in the same conditions described in Figure 6. The peptides were isolated from deproteinized DNA of *E. coli* cells grown in the presence of ^{32}P-orthophosphate and purified by gel filtration as described in Figures 1 and 2. (a) 1 μg peptide; (b) 2 μg peptide; (c) 2 μg peptide after hydrolysis with HCl 2 *M* for 4 h at 110°C; (d) 2 μg peptide after hydrolysis with HCl 6 *M* for 18 h at 110°C; (e) ^{32}P-orthophosphate.

percentage inhibitions obtained with the less concentration of chromatin peptides demonstrate that the controlling activity is more evident in λ phage and poly[d(A-T).d(A-T)] transcription. Transcription *in vitro* of calf thymus DNA by RNA polymerase II isolated from rat liver nuclei following the method of Stein and Hausen[16] shows very similar controlling activity by chromatin peptides. We have reported[2] that the active peptides are also able to control

Table 3
EFFECT OF CHROMATIN PEPTIDES ON THE
TEMPLATE CAPACITY OF CALF THYMUS DNA,
***E. COLI* DNA, PHAGE λ DNA, AND POLY[d(A-T).d(A-T)]**

Template DNA (2.5 μg)	Peptides (μg)	^{3}H- UMP incorporation (cpm) 3 min	8 min
Calf thymus DNA	0	15884 ± 1912	31093 ± 2831
	0.04	7530 ± 936 (53)	15632 ± 1911 (50)
	0.10	2836 ± 648 (82)	4918 ± 831 (84)
E. coli DNA	0	9841 ± 1218	21238 ± 2418
	0.04	4613 ± 734 (53)	9569 ± 1060 (55)
	0.10	1325 ± 381 (86)	2536 ± 518 (88)
λ Phage DNA	0	20438 ± 2316	54594 ± 4001
	0.04	4910 ± 980 (76)	13562 ± 1288 (75)
	0.10	2034 ± 411 (90)	2818 ± 349 (95)
Poly[d(A-T).d(A-T)]	0	74981 ± 4935	141858 ± 7312
	0.04	8585 ± 1125 (88)	36845 ± 3236 (74)
	0.10	1485 ± 289 (98)	2634 ± 431 (98)

Note: Transcription by *E. coli* RNA polymerase. 2.5 μg of DNA in 50 μl incubation mixture; aliquots of 10 μl were used to determine the radioactivity incorporated into the acid-insoluble fraction of RNA. Other ingredients of incubation mixture are as given in Reference 9. The chromatin peptides were isolated from calf testicle DNA and purified by gel filtration and HPLC.[9] Values are expressed as means of triplicate determinations ± S.E. Numbers in parentheses represent percentages of transcription inhibition.

the RNA polymerase reaction directed by rat liver chromatin. Moreover, we have demonstrated[9] that the DNA-binding peptides control the initiation reaction of transcription while the propagation reaction is almost completely unaffected.

Control of DNA and chromatin transcription is apparently mediated by a bond between peptides and DNA; in fact, the presence of chromatin peptides causes a significant increase of the DNA melting temperature.[4,8] These results have been confirmed also with the poly[d(A-T).d(A-T)]) (Table 4); this experiment shows that the peptides increase also the temperature at which the half of the complete renaturation occurs, thus demonstrating that the polynucleotide-peptide bond facilitates the renaturation process.

We have repeatedly demonstrated[1,9] that the chromatin peptides following purification by means of gel filtration do not contain ribonuclease activity and that in the transcription experiments they do not cause the template precipitation.

It is noteworthy to examine whether the chromatin peptides can bind to DNA fragments or if they are able to recognize at least preferentially some nucleotide sequences. Calf thymus DNA has been incubated with several concentrations of chromatin peptides and the DNA-peptide complexes precipitated with ethanol, resolubilized, and transcribed by *E. coli* RNA polymerase. The results in Table 5 show that in this experimental model the transcription of DNA is only partially inhibited. Consequently, it is possible that the peptides bind preferentially to specific DNA sequences. These results are in agreement with observations reported[5] previously.

IV. REGULATORY ACTIVITY OF RNA SYNTHESIS IN CELLULAR SYSTEMS

The RNA synthesis of human leukemic leukocytes and phytohemagglutinin-stimulated lymphocytes is markedly reduced by administration of low-molecular-weight chromatin

Table 4

EFFECT OF CHROMATIN PEPTIDES ON THE DENATURATION AND RENATURATION OF POLY [d(A-T).d(A-T)]

Poly[d(A-T).d(A-T)]	Peptides	$A_{260\,nm}$ (30°C)	$A_{260\,nm}$ (65°C) denaturation	Tm_1	$A_{260\,nm}$ (30°C) renaturation	Tm_2
10 μg/ml	0 μg/ml	0.19	0.30	53.50 ± 0.50	0.20	43.00 ± 0.50
10 μg/ml	1 μg/ml	0.19	0.30	55.50 ± 0.50[a]	0.20	45.00 ± 0.50[b]
10 μg/ml	2 μg/ml	0.19	0.30	57.50 ± 0.50[a]	0.20	48.50 ± 1.00[a]

Note: The melting profiles of poly [d(A-T).d(A-T)] (Sigma), in the presence or absence of chromatin peptides, were obtained by means of a Beckman spectrophotometer (model DU8). The rates of heating and cooling were 1°C and 2°C/min, respectively. Solvent was $0.1 \times$ SSC. Tm_1 = temperature corresponding to half of maximum denaturation; Tm_2 = temperature corresponding to half of maximum renaturation. The chromatin peptides were isolated from calf testicle DNA and purified by gel filtration and HPLC. Values expressed as mean of experiments made in triplicate ± S.D. [a] $p <0.001$; [b] $p <0.05$: significant differences from the control (Student's t-test).

Table 5
TEMPLATE CAPACITY OF CALF THYMUS DNA
COMPLEXED WITH CHROMATIN PEPTIDES

Template DNA (2.5 µg)	Peptides (µg)	^{3}H-UMP incorporation (cpm)	
		3 min	8 min
Calf thymus DNA	0.00	13420 ± 1541	28416 ± 2512
	0.25	10810 ± 916 (19)	22216 ± 1642 (22)
	0.62	10291 ± 1047 (23)	19840 ± 1593 (30)
	1.25	9216 ± 864 (31)	18318 ± 1961 (35)
	2.50	7518 ± 832 (44)	15240 ± 1418 (46)
	5.00	7240 ± 1024 (46)	14518 ± 1119 (49)

Note: Calf thymus DNA (2.5 µg in 10 µl) has been incubated with several concentrations of chromatin peptides (0—5 µg in 10 µl). The DNA-peptide complexes were precipitated with ethanol, resolubilized with 0.1 × SSC, and transcribed by *E. coli* RNA polymerase in 50 µl incubation mixture. In some experiments 30 µg of dextran (200,000 mol wt Serva) were added as carrier to the mixture containing DNA and peptides in order to facilitate the complete recovery of DNA following precipitation with ethanol. Determinations of radioactivity incorporated into the acid-insoluble fraction of RNA and other ingredients of RNA polymerase incubation mixture are as given in Reference 9. The chromatin peptides were isolated from calf testicle DNA and purified by gel filtration and HPLC following the procedures described in Figures 1, 2, and 4. Values are expressed as means of triplicate determinations ± S.E. Numbers in parentheses represent percentages of transcription inhibition.

peptides.[6] Moreover, low molecular weight chromatin peptides exert a dose-dependent inhibition of Dimethylsulfoxide (DMSO)-induced erythroid differentiation of murine Friend leukemia cells (FLC).[6,7] Analysis of globin mRNA amounts in DMSO-treated FLC given the peptides showed a four-to-fivefold decrease of messenger RNA in the cytoplasm with no nuclear storage of globin transcripts. Spectrin accumulation in "induced" FLC is inhibited as well.[7] The effects of the peptides on erythroid markers are reversible upon removal of the compounds. They also appear to be specific for induced gene expression as (1) no effects are observed on cell growth and RNA synthesis in normal nondifferentiating cell lines; and (2) no changes have been detected with regard to the expression of integrated viral genes coding for continuous shedding of viral particles. Recently we have demonstrated that the DNA-binding peptides cause a dose-dependent inhibition of DNA and RNA synthesis in L1210 mouse leukemia cells line *in vitro* (Figure 8). The analysis of data reported in Figure 8 seems to confirm our hypothesis that the RNA synthesis is the first target for the chromatin peptides activity. It is noteworthy that treatment with DNA-binding peptides does not modify the cell viability. In addition, the effect of peptides in metabolic activities of L1210 cells is dependent on the permanence time in the culture medium (Figure 9). In agreement with data in Figure 8 also the cell proliferation of L1210 mouse leukemia cells is affected by peptide effectors (Figure 10). In particular we have observed that suitable doses of peptides stop the L1210 cell growth for several days (3 to 4) without remarkably decreasing the cell viability. This effect may be reversed by washing and reseeding the cells in fresh medium not containing the peptide effectors (Figure 10).

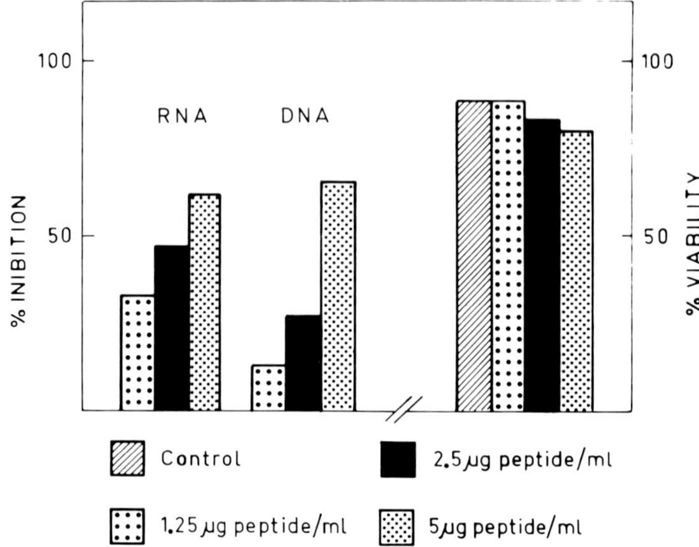

FIGURE 8.   Effect of chromatin peptides on DNA and RNA synthesis in L1210 mouse leukemia cells. The DNA and RNA synthesis was measured by incorporation of ^{14}C-Thymidine and ^{14}C-Uridine, respectively; 2 h of labeling from 4 to 6 h of culture; the peptide fraction isolated from trout testis DNA was added at 0 time.

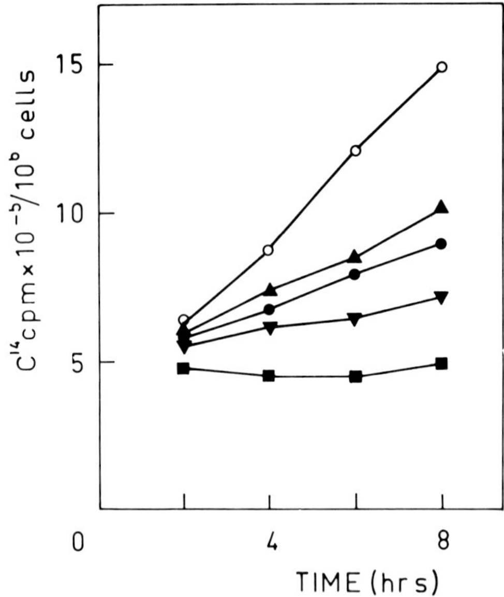

FIGURE 9.   Effect of chromatin peptides on RNA synthesis in L1210 mouse leukemia cells at different times of culture. The RNA synthesis was measured by incorporation of ^{14}C-Uridine; the cells were labeled for 2 h at 0, 2, 4, and 6 h of culture. The peptide fraction isolated from trout testis DNA was added at 0 time. (○) control; (▲) 0.62 µg peptide/ml; (●) 1.25 µg peptide/ml; (▼) 2.5 µg peptide/ml; (■) 5 µg peptide/ml.

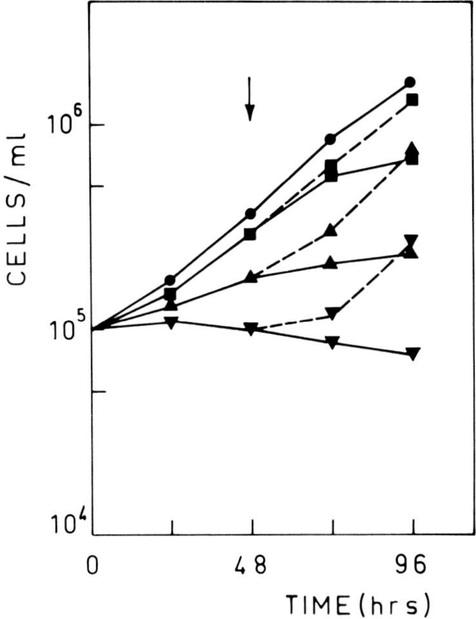

FIGURE 10.   Effect of chromatin peptides on L1210 cells growth expressed as number of cells/ml. Initial inoculum: 1×10^5 cells/ml. The peptides from trout testis were added at 0 time. The cells were washed at the 48th hr of culture (see arrow) and reseeded in fresh medium with (—) and without (---) peptide. (●) control; (■) 0.62 μg peptide/ml; (▲) 1.25 μg peptide/ml; (▼) 2.5 μg peptide/ml.

V. CONCLUDING REMARKS

We have demonstrated that low molecular weight peptides are tightly bound to chromatin DNA of eukaryotic and prokaryotic cells.[17] The chromatin peptides can be extracted from deproteinized DNA by alkali buffer (ammonium acetate/ammonia pH 9.5), and they show high specific activity in the control of transcription in cell-free systems and of RNA synthesis in several cellular systems.

Control of chromatin and DNA transcription *in vitro* is mediated by a specific interaction between peptides and DNA and takes place at the level of the initiation reaction.[9] Accordingly, the chromatin peptides do not inhibit the transcription of endogenous residual RNA polymerase bound to chromatin preparations.[2]

On the other hand, all results showing the regulation of RNA synthesis by peptides in cell systems seem in line with the control of the transcription rate of genes previously activated by independent mechanisms, rather than a selection of genomic portions. The chromatin peptides can constitute a general mechanism regulating gene expression; however, at this stage, several questions remain unanswered: the origin of the peptide, the structure of the peptide, and its physiological function.

The amount of chromatin peptides which can be extracted from DNA by alkali buffer is not modified by performing the DNA purification and peptide extraction in the presence of protease inhibitors as phenylmethylsulfonyl fluoride (Sigma), sodium bisulfite, or bestatin (Sigma)[9]. Such result should exclude the possibility that the chromatin peptides are an artifact due to proteolytic digestion of proteins during the extraction procedure; this with particular reference to protease bound to chromatin structure.[18] However, the hypothesis that these peptides originate physiologically from proteolytic degradation of protein precursor synthesized by the cell should be taken into consideration.

As described in Section II, the structural studies of the chromatin peptides present many difficulties because they probably are distributed in several fractions (a family of acidic low molecular weight peptides) and the N- and C-terminals cannot be easily analyzed by the conventional methods. Experiments by means of mass spectrometry have been recently directed to progress in the chemical elucidation of the controlling transcription peptide effectors. Preliminary results have shown the presence of two fractions with molecular weight 794 and 642. These molecular weights could be consistent, respectively, with the presence of the following amino acids; I° (mol. wt. 794): <glu(1), ala(1), gln(2), gly(1), ser-P(1), asn(1); II° (mol. wt. 642); <glu(1), val(1), gly(2), ser-P(1), asn(1). The possibility that the aspartic acid and glutamic acid residues revealed by amino acid analysis are present in the peptide structure in form of asparagine and glutamine, it appears in agreement with the electrophoretic properties of the chromatin peptides. In fact, the migration towards the positive electrode at various pH (1.9 to 9.0) appears strictly related to the presence of phosphoric radicals and not to other negative charges. Moreover of particular interest is the observation of Rich[21] according to which asparagine and glutamine are potentially able to perform hydrogen bonds with particular sites of double-stranded DNA and, consequently, they can be involved in the recognition of DNA sequences. According to this hypothesis, which would represent the existence of a recognition code for the protein-DNA interactions, the chromatin peptides could be able to recognize the DNA sequences involved in the initiation of transcription and, consequently, to control the rate of gene expression. Accordingly, Cohen[22] suggested that chromatin peptides could originate from processing of precursor proteins, subsequently they could act as regulators of the transcription of the genes involved. The controlling activity by peptides is particularly evident in the transcription of poly [d(A-T).d(A-T)] and λ phage DNA which is also rich in base pairs A to T (50%).

ACKNOWLEDGMENTS

This work was supported by grant from Consiglio Nazionale delle Ricerche (Progetto finalizzato "Oncologia" number 84.00424.44).

REFERENCES

1. **Gianfranceschi, G. L., Amici, D., and Guglielmi, L.,** Evidence for the presence in calf thymus of a peptidic factor controlling DNA transcription in vitro, *Biochim. Biophys. Acta,* 414, 9, 1975.
2. **Gianfranceschi, G. L., Amici, D., and Guglielmi, L.,** Restriction of template capacity of rat liver chromatin by a non-histone peptide from calf thymus, *Nature,* 262, 622, 1976.
3. **Gianfranceschi, G. L., Amici, D., and Guglielmi, L.,** Stabilization of double-stranded DNA molecule by non-histone peptidic effector from calf thymus, *Mol. Biol. Rep.,* 3, 55, 1976.
4. **Guglielmi, L., Gianfranceschi, G. L., Venanzi, F., Polzonetti, A., and Amici, D.,** Specific thymic peptides-DNA interaction. Correlation with the possible stereochemical kinking scheme of DNA, *Mol. Biol. Rep.,* 4, 195, 1978.
5. **Gianfranceschi, G. L., Guglielmi, L., Amici, D., Bossa, F., Barra, D., and Petruzzelli, R.,** Low molecular weight peptides controlling transcription are present in the calf thymus chromatin structure, *Mol. Biol. Rep.,* 3, 429, 1977.
6. **Amici, D., Rossi, G. B., Cioè, L., Matarese, G. P., Dolei, A., Guglielmi, L., and Gianfranceschi, G. L.,** Low-molecular-weight peptide inhibits RNA synthesis in human leukemic and phytohemagglutinin-stimulated leukocytes and globin mRNA transcription in differentiating Friend cells, *Proc. Natl. Acad.Sci. U.S.A.,* 74, 3869, 1977.
7. **Rossi, G. B., Cioè, L., Pulciani, S., Meo, P., Titti, F., Amici D., and Gianfranceschi, G. L.,** Effects of a chromatin low molecular weight peptidic fraction on differentiation markers and virus production in Friend leukemia cells, *Mol. Biol. Rep.,* 5, 241, 1979.
8. **Gianfranceschi, G. L., Amici, D., and Guglielmi, L.,** Amplification of template capacity of mammalian DNA following extraction of low molecular weight peptides, *Physiol. Chem. Phys.,* 11, 507, 1979.

9. **Gianfranceschi, G. L., Barra, D., Bossa, F., Coderoni, S., Paparelli, M., Venanzi, F., Cicconi, F., and Amici, D.,** Small peptides controlling transcription in vitro are bound to chromatin DNA, *Biochim. Biophys. Acta,* 699, 138, 1982.
10. **Hillar, M. and Przyjemski, J.,** Control of transcription and translation by low molecular weight peptides (deprimerones) from chromatin and poly (A)-messenger RNA. Implication in the mechanism of carcinogenesis, *Biochim. Biophys. Acta,* 564, 246, 1979.
11. **Manera, E., Minchiotti, L., and Lugaro, G.,** An improved method for the isolation of deprimerones from spermatozoan chromatin, *IRCS Med. Sci.,* 12, 745, 1984.
12. **Welsh, R. S. and Vyska, K.,** Organization of highly purified calf thymus DNA. I. Cleavage into subunits and release of phosphopeptides, *Biochim. Biophys. Acta,* 655, 291, 1981.
13. **Buss, J. E. and Stull, J. T.,** Measurement of chemical phosphate in proteins, in *Methods in Enzymology,* Vol. 99, Corbin, J. D. and Hardman, J. G., Ed., Academic Press, New York, 1983, chap. 2.
14. **Krajewski, T., Banas, Z., and Cierniewski, C.,** The presence of phosphoserine and phosphothreonine in tryptic digest degradation products of pig fibrinogen, *Biochim. Biophys. Acta,* 322, 95, 1973.
15. **Juodka, B. A.,** Covalent interaction of proteins and nucleic acids. Synthetic and natural nucleotide-peptides, *Nucleos. Nucleot,* 3, 445, 1984.
16. **Stein, H. and Hausen, P.,** A factor from calf thymus stimulating DNA-dependent RNA polymerase isolated from this tissue, *Eur. J. Biochem.,* 14, 270, 1970.
17. **Gianfranceschi, G. L., Coderoni, S., Miano, A., Felici, F., Bramucci, M., and Amici, D.,** Structure and function of small peptides bound to chromatin DNA, presented at Int. Symp. Chromatin Structure and Function, Camerino, Italy, May 21 to 24, 1985.
18. **Chong, M. T., Garrard, W. T., and Bonner, J.,** Purification and properties of a neutral protease from rat liver chromatin, *Biochemistry,* 13, 5128, 1974.
19. **Raj, N. B. K., Ro-Choi, T. S., and Busch, H.,** Nuclear ribonucleoprotein complexes containing U1 and U2 RNA, *Biochemistry,* 14, 4380, 1975.
20. **Rothbarth, K. and Werner, D.,** Amino-acid-transfer reactions in isolated nuclei of Ehrlich ascites tumor cells, *Eur. J. Biochem.,* 155, 149, 1986.
21. **Seeman, N. C., Rosenberg, J. M., and Rich, A.,** Sequence-specific recognition of double helical nucleic acids by proteins, *Proc. Natl. Acad. Sci. U.S.A.,* 73, 804, 1976.
22. **Cohen, P. P.,** Transport and processing of the precursor of mitochondrial carbamoyl phosphate synthetase I and ornithine transcarbamoylase, *Biochem. Soc. Trans.,* 12, 377, 1984.

Chapter 8

TISSUE SPECIFIC *IN VITRO* TRANSCRIPTION FROM MOUSE β-GLOBIN PROMOTER AND PREFERENTIAL BINDING OF ERYTHROID CELL NUCLEAR NONHISTONE PROTEINS TO UPSTREAM FLANKING SEQUENCES

J.-C. Lelong, G. Prevost, and M. Crepin

TABLE OF CONTENTS

I. MODULATION OF *IN VITRO* β GLOBIN TRANSCRIPTION BY CHROMOSOMAL NONHISTONE PROTEINS FROM MOUSE ERYTHROLEUKEMIA CELLS

A. Introduction

Expression of the β globin gene family in erythroid cells appears to be both tissue specifically determined[1-2] and inducible.[3-6] Mouse erythroleukemia (MEL) cells are Friend virus transformed erythroid cells, arrested at the proerythroblast stage of differentiation, that are easy to cultivate and may be induced with a variety of chemicals, including dimethyl sulfoxide (DMSO).[3] The accumulation of globin transcripts results from both transcriptional activation of the mouse globin genes and increased relative stability of their mRNAs.[7] Therefore, this system is a very suitable model to study the tissue-specific factors that are activated in order to promote the selectivity of transcription, leading to a particular type of terminal phenotypic expression.

The development of soluble cell free systems that mediate accurate transcription initiation on purified genes by RNA polymerase II *in vitro*[8-10] revealed a requirement for several additional proteins factors, probably common to all class II promoters,[11] but generally failed to allow the characterization of tissue-specific factors or combinations of factors, conferring promoter selectivity in the appropriate cell or tissue. One of the major difficulties in purifying these proteins has been the lack of a fast, quantitative, or semi quantitative assay, i.e., *in vitro* transcription products must be analyzed by gel electrophoresis in order to distinguish accurate from random initiation. We used agarose gel electrophoresis of radioactive nascent *in vitro* transcripts from an appropriate mixture of DNA fragments bearing different promoters[12] as a fast assay to characterize MEL specific nonhistone proteins and erythroid tissue-specific factors, which stimulate preferentially β globin transcription initiation.

B. Analysis of Transcription Initiation Complexes Formed from Mouse β Globin DNA and Phage λ DNA by Purified *E. coli* or Calf Thymus RNA Polymerase *In Vitro*

EcoR 1 restriction of the phage recombinant λ gtWESβMG2[13] allows the separation of the β major globin genomic insert (7 kb) from two λ DNA fragments: one of 14 kb, containing the early promoter, and one of 21 kb, lacking it (Figure 1A). These three DNA fragments have been transcribed under conditions favoring transcription initiation, as described by Chelm et al.[14] (see also the legends of Figure 1). The ternary complexes (DNA, proteins, and radioactive nascent RNA) were separated by agarose gel electrophoresis.[14] The amount of nascent transcript originated from each DNA fragment visualized by fluorography (Figure 1A) was revealed by autoradiography (Figure 1B).

When increasing concentrations of purified RNA polymerase (from either *Escherichia coli* or calf thymus) were used, the three DNA fragments were transcribed, and [α^{32}P]UTP incorporation in each band was found to be dependent on the length of the fragment (Figure 1B). Therefore, transcription by purified RNA polymerase II occurred at random on DNA templates, under these conditions *in vitro*, and did not allow any specific recognition of the β globin promoter, as compared to λ promoters (Figure 2a).

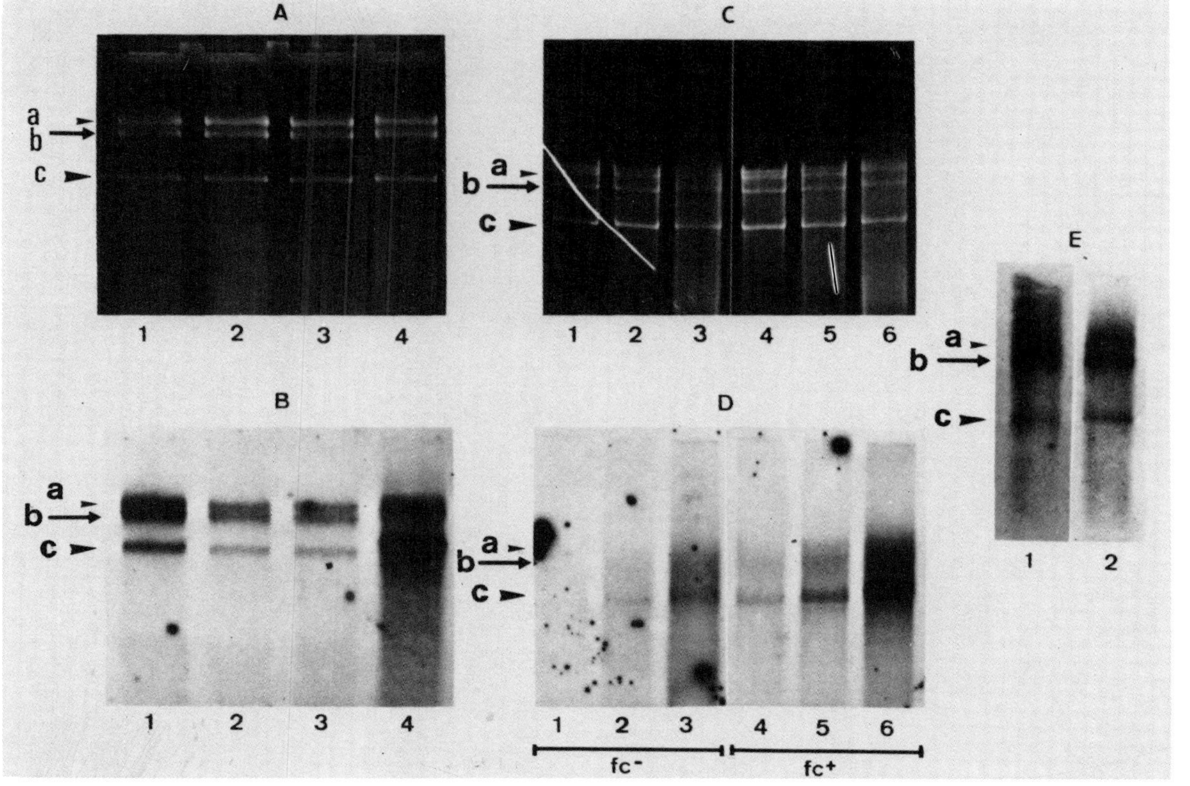

FIGURE 1.   Transcription initiation complexes formed on EcoR 1 cleaved λ globin DNA by purified calf thymus RNA polymerase II and the endogenous RNA polymerase of MEL cell and PCC3 crude extracts. [DNA/RNA - polymerase/RNA] ternary complexes were formed on purified DNA fragments obtained by EcoR 1 digestion of λ globin DNA from the λ gtWESβmG2 phage recombinant.[13] 1 μg EcoR 1 cleaved λ globin DNA was transcribed 10 min at 30°C under initiation conditions: 15 mM Tris-HCl pH 7.9; 7 mM MgCl$_2$; 2 mM EDTA; 1 mM dithiothreitol (DDT); 32 mM (NH$_4$)$_2$SO$_4$; 250 μM ATP, GTP; 25 μM CTP; 2.5 [α^{32}P]UTP (spec. act 400 Ci/mmol). RNA synthesis was stopped by addition of 10% glycerol, 20 mM EDTA. 100 μg/ml Heparin. Samples were loaded on 1% agarose gel (15 × 10 × 0.5 cm) in 40 mM Tris-acetate pH 7.9, 20 mM sodium acetate, 1 mM EDTA. Electrophoresis was performed at 4°C for 14 h at 40 V. After the dye front, containing the free ribonucleotides, had been removed, the gel was dried and autoradiographed. (A, B) Fluorography and autoradiography of the same agarose gel containing 12.5 ng, 25 ng, 100 ng calf thymus RNA polymerase II, purified as described,[14] in channels 3, 2, and 1, respectively. Channel 4 corresponds to 30 μg of crude extracts prepared as described[8] from uninduced MEL cells (FC−) clone 707-17-C. The positions of the two λ (a,b) and of the β globin (c) DNA fragments are indicated by arrows. (C, D) Fluorography and autoradiography of the same gel with increasing amount of uninduced MEL cell (FC−) extract (1 to 3) or Me$_2$SO treated cell (FC+) extract (4 to 6) prepared as described.[8] (From Triadou, P., Lelong, J. C., Gros, F., and Crepin, M., *BBRC*, 101, 45, 1981. With permission.)

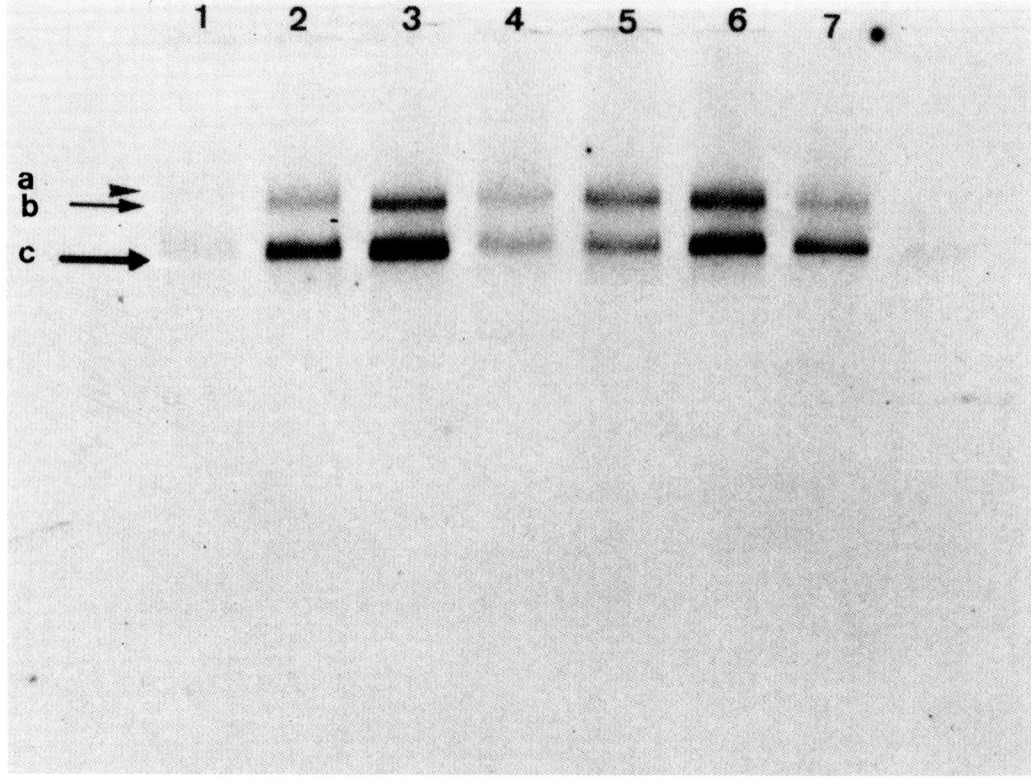

FIGURE 2.   Modulation of the initiation of transcription by the 50 mM phosphate NHP fraction from uninduced and DMSO-induced MEL cell nuclei. Three different concentrations of nuclear NHP eluted from an hydroxylapatite column at 50 mM phosphate[15] were dialyzed with 1 μg of EcoR 1 cleaved λ globin DNA against the transcription incubation mixture for 1 h at room temperature. DNA-protein complexes were then transcribed with 12.5 ng of *E. coli* RNA polymerase under similar conditions described in Figure 1. Lanes 1 to 7 of the autoradiography of the agarose gel electrophoresis correspond, respectively, to the transcription without NHP (1) with 1 μg, 2 μg, 3 μg of uninduced MEL cell NHP (2, 3, 4), with 1 μg, 2 μg, 3 μg of DMSO-induced MEL cell NHP (5, 6, 7). Arrows show the λ late (a), λ early (b), and β globin (c) DNA fragments. (From Triadou, P., Lelong, J.-C., Gros, F., and Crepin, M., *B.B.R.C.*, 101, 45, 1981. With permission.)

C. Specific Initiation of Transcription at the β Globin Promoter, in the Presence of Partially Purified MEL Nonhistone Proteins (NHP)

Such a transcription assay should be useful to investigate a possible regulatory function of nonhistone proteins (NHP). In order to test an NHP effect on specific β globin transcription, we have partially purifed NHP from uninduced and DMSO induced MEL cells on a hydroxylapatite column.[15] This purification leads to the isolation of a NHP chromatin subfraction free of histones and nucleic acids. Complexes, between various amounts of NHP and β globin fragments obtained by EcoR I digestion, were formed by dialysis against the transcription buffer and transcribed as described in the legends. Figure 2 shows the agarose gel electrophoresis of the transcription complexes with three concentrations of NHP from uninduced (lanes 2, 3, 4) and DMSO-induced (lanes 5, 6, 7) MEL cells. As compared to the naked DNA control (lane 1), NHP-DNA complexes are much more transcribed using both types of NHP. The overall transcription is increased to 100-fold depending on the NHP-DNA ratio. The two λ DNA fragments which migrate very closely are differently transcribed; whereas the fragment with late genes, which does not contain *in vitro* promoter, is not transcribed, the λ early gene DNA fragment with P_R and P_L promoters is significantly transcribed. Strikingly enough, the stimulatory effect of the NHP appears more or less specific for the β globin DNA fragment, depending on the NHP-DNA ratio. Both types of

NHP preparations, from either uninduced or induced MEL cells, are capable of specifically stimulating the β globin DNA transcription but at different NHP-DNA ratios, suggesting that several stimulating and inhibiting activities may coexist in the 50 m*M* phosphate NHP fractions, and that one or several proteins more abundant in uninduced MEL nuclei stimulate specifically the transcription of an eukaryotic gene.

D. Further Purification of MEL Nonhistone Proteins Suggests a Multiplicity of Protein Factors Involved in Specific β Globin Transcription

We have further fractionated the 50 m*M* phosphate protein fraction on a second hydroxylapatite column using a slightly different pH and a gradient elution (Figure 3A). The column fractions were tested for selective transcription at the optimal NHP:DNA ratio previously defined (Figure 2). Fractions Number 1 and 2 strongly stimulate nonspecific transcription, whereas fractions 4 to 6, 9, and 14 stimulate preferentially, if not exclusively, initiation of β globin transcription (Figure 3B). Other protein fractions, 10 to 13, inhibit nonspecifically λ DNA and β globin DNA transcription.

E. Tissue Specificity of β Globin Transcription in the Presence of Crude Extracts from MEL Cells or Mouse Erythroid Tissues

The next relevant question to be addressed is whether the observed β globin *in vitro* specific transcription is restricted to extracts of erythroid cells committed to express the β globin gene. Therefore, using the same *in vitro* system, we have compared the rate of β globin specific transcription initiation in crude extracts of erythroid and nonerythroid cells. The use of crude extracts,[8] instead of purified NHP, avoided the requirement of purified RNA polymerase II and ubiquitous transcription factors, since their endogenous RNA polymerase activities, when tested in initiation conditions on poly (dA-dT) as template, were not limiting (unpublished data). In contrast to the situation with purified RNA polymerase alone (Figure 1B), MEL cell extracts allow a preferential transcription of the β globin-containing fragment, when compared to the λ DNA fragments (Figures 1C,D). In addition, specific transcription of the β globin fragment occurs more efficiently with comparable amounts of proteins, in induced than in uninduced MEL cell extracts (Figure 1D). Crude extracts prepared from an early mouse embryonal carcinoma cell line (PCC3),[16] however, did not allow any specific recognition of the β globin promoter, as compared to λ promoters (Figure 1E), thus behaving, to some extent, like purified polymerase II (Figure 1B).

The same degree of selectivity of transcription initiation from the β globin promoter compared to λ promoters was also observed with erythroid tissue extracts prepared from anemic mice. Severe anemia was induced by phenylhydrazine injection.[17] In such mice, reticulocytes form up to 50 to 60% of the red circulating cells. In this case, only anemia induced mouse spleen extracts were capable of preferential β globin initiation of transcription (Figure 4). These results suggest the presence of factors in erythroid tissue which favor the recognition of the β globin promoter relative to other prokaryotic promoters (λ bacteriophage). These factors might be either absent or inactive, in extracts from cells which normally do not express globin genes (PCC3 cells from mouse teratocarcinoma, liver, thymus). One cannot exclude, however, that other factors in nonerythroid tissues might prevent specific transcription of the β globin gene. Further characterization might be facilitated by studying their binding to appropriate cloned DNA sequences.

II. *IN VITRO* CHARACTERIZATION OF TISSUE-SPECIFIC NUCLEAR PROTEINS PREFERENTIALLY BOUND TO THE MOUSE β GLOBIN GENE DURING MEL CELL TERMINAL DIFFERENTIATION

A. Introduction

Cis-acting DNA sequences that control gene transcription have been identified and dis-

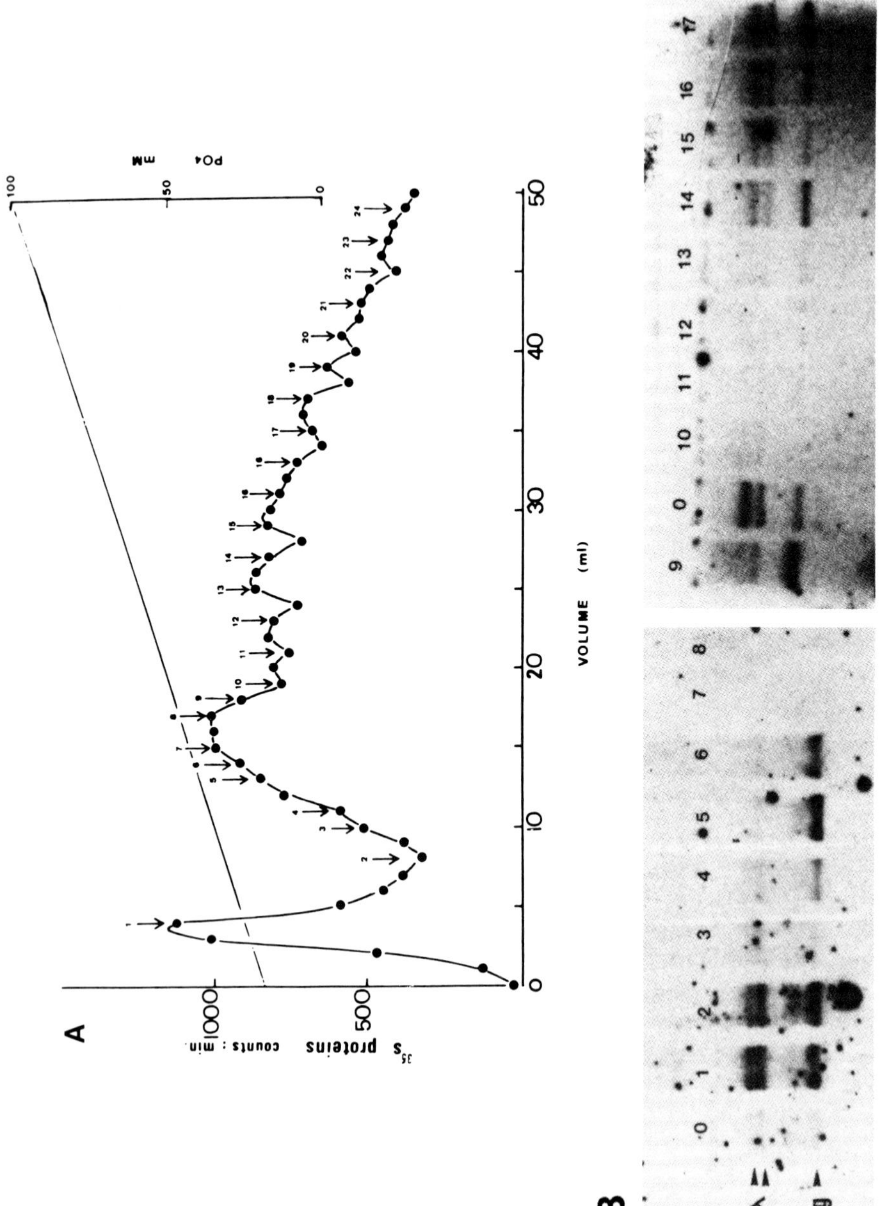

FIGURE 3. Modulation of β globin transcription initiation by partially purified nuclear NHP from MEL cell. (A) Hydroxylapatite elution profile of ^{35}S labeled NHP (50 m*M* phosphate fraction from uninduced MEL cell nuclei. specific activity 120 cpm/μg) with a gradient of potassium phosphate 1 to 100 m*M*. (B) 1 μg of protein from each fraction indicated by an arrow was added to the λ globin transcription initiation mixture under conditions described in Figure 2. Lanes 0 to 8 and lanes 9 to 17 correspond to two different test series of fractions. Lanes 0 are controls without NHP subfraction. λ and g indicate, respectively, the λ phage and β globin DNA fragments revealed by autoradiography of their nascent transcripts.[34] (From Triadou, P., Lelong, J. C., Gros, F., and Crepin, M.. *B.B.R.C..* 101. 45. 1981. With permission.)

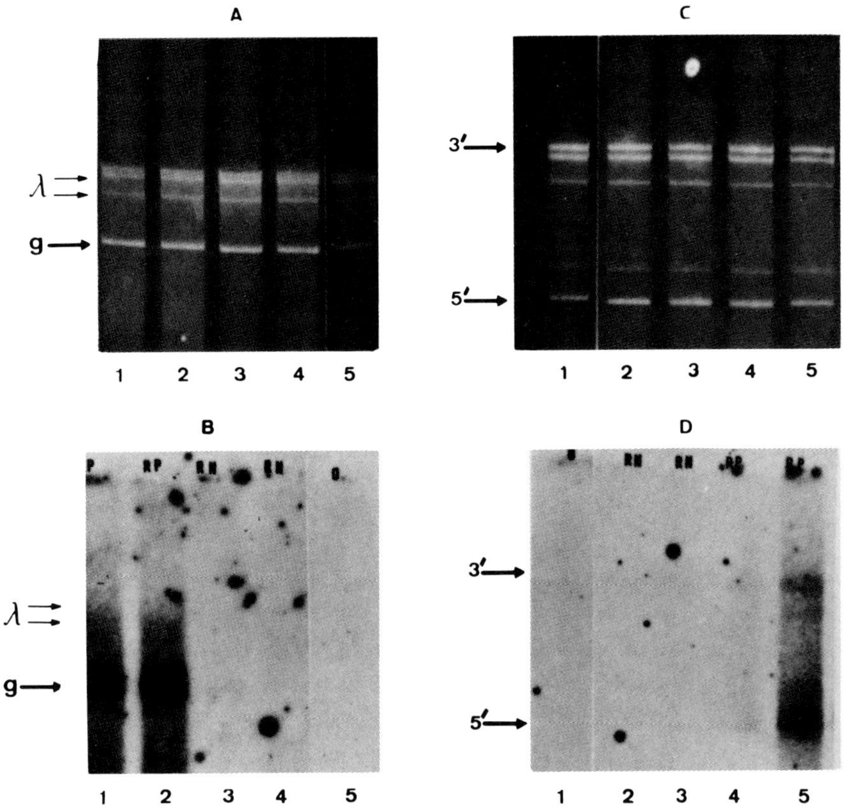

FIGURE 4.   Transcription initiation complexes formed on the mouse β globin DNA in the presence of erythroid tissue extracts. Spleen extracts were obtained from normal and phenylhydrazine induced anemic mice.[17] The tissue was cut into small-size pieces in phosphate buffered saline on ice, washed in the same buffer, and suspended in 10 mM Tris HCl pH 8, 2 mM MgCl$_2$, 3 mM CaCl$_2$, 1 mM EDTA, 1 mM dithrothreitol. Cells were lyzed by 10 strokes in a Dounce homogenizer (B pestle). The homogenate was then made to 0.3 M sucrose. Homogenization was subsequently pursued in the homogenizer. The lyzate was centrifuged at 10,000 g for 30 min. Aliquots of the supernatant were quickly frozen in liquid N2 and retained activity for at least 3 months. (A,B) Transcription-initiation complexes on λ globin DNA restricted with EcoR 1. (A) Fluorography of the 1% agarose gel. (B) Autoradiography of the same gel. 0.5 μg λ globin DNA restricted by EcoR I was transcribed under initiation conditions in the presence of spleen extracts from normal (3 and 4) and phenylhydrazine induced anemic mice (1 and 2); g: globin DNA fragment; λ: λ DNA fragments (14 kb and 21 kb). Amount of protein: (1) 25 μg (2) 50 μg (3) 50 μg (4) 25 μg. Lane 5 is a control without extract. (C,D) Transcription initiation complexes on λ globin DNA restricted with BamH1. (C) Fluorography and (D) autoradiography of the same 1% agarose gel. 5′ → indicates a 4.5-kb DNA fragment (containing the 5′ end of the globin gene and a λ sequence) and 3′ → a 21-kb fragment (containing the 3′ end of the globin gene and a λ sequence). (1) No extract; (2) and (3), respectively, 25 μg and 50 μg of normal spleen extract; (4) and (5), respectively, 10 μg and 25 μg of anemic mouse spleen extract. (From Triadou, P., Lelong, J.-C., Gros, F., and Crepin, M., *Eur. J. Biochem.*, 135, 163, 1983. With permission.)

sected by *in vitro* mutagenesis using mainly, so far, either *in vivo* assays of transient expression of introduced genes or *in vitro* transcription systems.[18-20] Since they are usually located at chromatin hypersensitive sites,[21-24] it is postulated that the function of these control DNA elements requires the coordinate binding of both general and promoter-specific transcription factors. The footprints of many such sequence-specific DNA binding proteins have been identified, but so far, only a few of them have been purified and thus are available for the study of their mechanism of action and their tissue-specific expression or activation. Most probably, the tissue and temporal regulation of mRNA synthesis in eukaryotic cells is controlled by cell-specific combinations of various promoter class specific factors.[25-29] Although a promoter-specific transcription factor need not be a sequence-specific DNA binding protein and vice versa, an important approach to the knowledge of the tissue-specific regulation of gene expression might be analyzing cell types and variations during differentiation, of the pattern of sequence specific nuclear DNA binding proteins using cloned fragments of a regulated gene as affinity probes.

Because protein blotting[30] is particularly suitable for a rapid and highly resolutive screening of cell crude extracts, we have used this technique to analyze MEL cell nuclear proteins which have been tested for (1) preferential binding to a cloned mouse β major globin gene; (2) erythroid cell specificity; (3) modulation of the DNA binding activity after induction of β globin expression.

We have characterized several DNA binding proteins which satisfy these criteria and describe also results suggesting that three of them may interact with specific sequences located more or less far upstream the transcription start.

B. Detection of High Affinity DNA Binding Proteins by the Protein Blotting Technique

Figure 5 shows the interaction of labeled β globin DNA with MEL cell nonhistone proteins as revealed by the protein transfer method described by Bowen et al.[30] Increasing concentrations of a crude MEL cell nonhistone protein preparation[10] were fractionated by SDS-gel electrophoresis, without prior denaturation (see legend of Figure 5), then transferred, after SDS removal in the presence of urea, onto nitrocellulose membranes, which were incubated with ^{32}P labeled β globin DNA, in the presence of variable excess of unlabeled competitor *E. coli* DNA. Clearly, both protein dilution and increasing competition by nonspecific DNA progressively abolish the radioactive signals given by the various DNA binding proteins of the extract and improves the specificity of the protein-DNA complexes. An appropriate choice of protein/specific DNA ratio and excess amount of unlabeled competitor DNA allows the selection of a restricted number of high affinity DNA binding proteins.[31] At the higher DNA/protein ratio used in this experiment, only one polypeptide ($\simeq$100 kDa) can be detected (Figure 5). Therefore, the specific DNA binding activity of nitrocellulose immobilized single polypeptides is preserved, and the stability of the protein-DNA complexes formed varies with the same parameters as those which influence the affinity of complexes measured by retention on nitrocellulose filters.[32]

C. Specific Binding of β Globin Genomic DNA to MEL Cell Nonhistone Proteins

The DNA sequence specificity of at least some of these erythroid nonhistone proteins is supported by the above competition experiment (Figure 5) and by the fact that preincubation of the nitrocellulose bound proteins with an excess of unlabeled native sonicated salmon sperm DNA (legend Figure 6) completely abolishes the binding of ^{32}P β globin DNA to the residual histones, but not to these DNA binding proteins (Figures 6 to 8) particularly to the 100 K polypeptide. In order to further investigate the degree of DNA sequence specificity of the 100-kDa protein, we explored its ability to bind three different, double-stranded DNAs: the 7 kb genomic β globin DNA fragment, the retroviral long-terminal repeat (5'LTR) of MMTV-DNA, and pBR322-DNA. Identical nitrocellulose replicates of increasing concentrations of SDS PAGE fractionated undifferentiated MEL cell proteins have been incubated separately with the three types of promoter containing cellular, viral, and bacterial labeled DNA fragments. Figure 6 shows a preferential binding of the 100 kDa to 95 kDa polypeptides to the β globin DNA (A) while no binding occurs to the LTR DNA (B) or to the pBR322 DNA (C). Thus, under stringent conditions, a group of high affinity DNA binding proteins (95 kDa to 100 kDa) recognizes sequences in the 7 kb β globin genomic DNA but has no affinity for a prokaryotic or even another type of eukaryotic promoter containing DNA: the MMTV-LTR-DNA. This is confirmed by the unspecific binding of

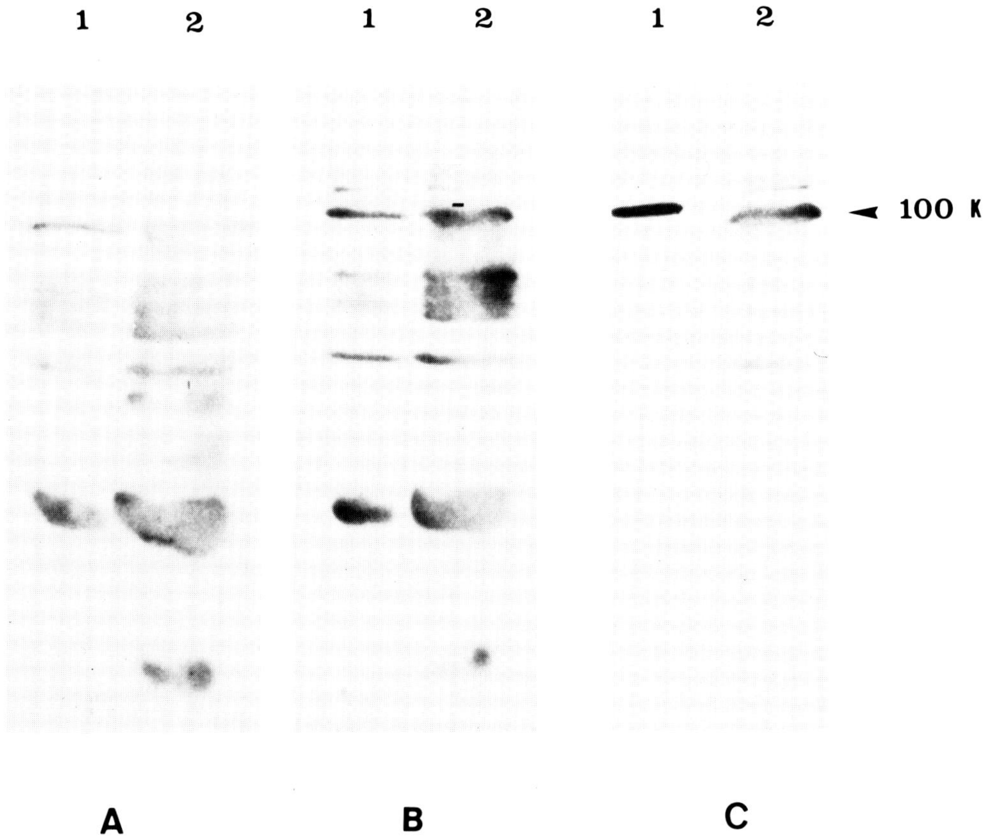

FIGURE 5.   Effect of protein concentration and nonspecific DNA competition on the formation of specific protein-DNA complexes on nitrocellulose. MEL cells (clone 707-17-C) were grown in MEM medium with 15% horse serum with or without 1.5% dimethylsulfoxide (DMSO). The crude extracts containing nonhistone proteins were prepared according to Manley et al.[10] Samples for SDS-polyacrylamide gel electrophoresis (acrylamide 12%, bisacrylamide 1%, SDS 0.1%) were prepared, without heating, in running buffer (Tris-glycine pH: 8.8, 25 mM; SDS 0.1%; β mercaptoethanol 2%). Protein renaturation and transfer to nitrocellulose (Schleicher and Schull BA 85) was done according to Bowen et al.;[30] Phenyl-methyl-sulfonyl fluorure (PMSF) 0.1 mM and DTT 0.2 mM were added to all buffer solutions. (1) 20 μg (2) 50 μg of protein from uninduced MEL cell extract were fractionated and transferred to nitrocellulose as identical replicates. Filters were incubated for 60 min at room temperature in 10 ml of binding buffer (Tris-HCl, pH 7.5, 10 mM; EDTA 1 mM; NaCl 0.1 M; PMSF 0.1 mM; DTT 0.2 mM; Bovine serum albumin (BSA) (fraction V) 0.02%; Ficoll 400: 0.02%; Polyvinyl-pyrollidone: 0.02%) containing 100 ng β globin DNA (7 Kb EcoR I fragment from the λ gtWESβmG2 recombinant), ^{32}P labeled by nick translation.[35] The filters were then washed several times, dried, and autoradiographed. The amount of competitor DNA added was (A) none (B) 1 μg (C) 5 μg of native sonicated unlabeled *E. coli* DNA. In subsequent experiments, *E. coli* DNA addition was replaced by preincubation of the filters for 10 min in binding buffer containing either 40 μg/ml of sonicated unlabeled salmon sperm DNA or 10 μg/ml Poly I-poly C (Böhringer).

the 50 kDa protein to all three DNA fragments indiscriminately, that can be taken as a useful internal control.

The DNA sequence specificity of the 100 kDa as well as a 35 kDa nonhistone MEL cell protein is also suggested by the comparison of their binding to labeled β globin DNA and repetitive poly d(AC)-poly d(GT) (Figure 7). The 95 kDa to 100 kDa proteins and the 35 kDa protein from undifferentiated MEL cells bind only to the β globin DNA, while the 50 kDa protein again binds also to the unspecific synthetic repetitive sequence (Figure 7b).

D. MEL Cell Specificity of β Globin DNA Binding Proteins

In order to analyze the cell specificity of the β globin DNA binding to these proteins, we have compared crude nonhistone protein preparations from various rodent cell types,

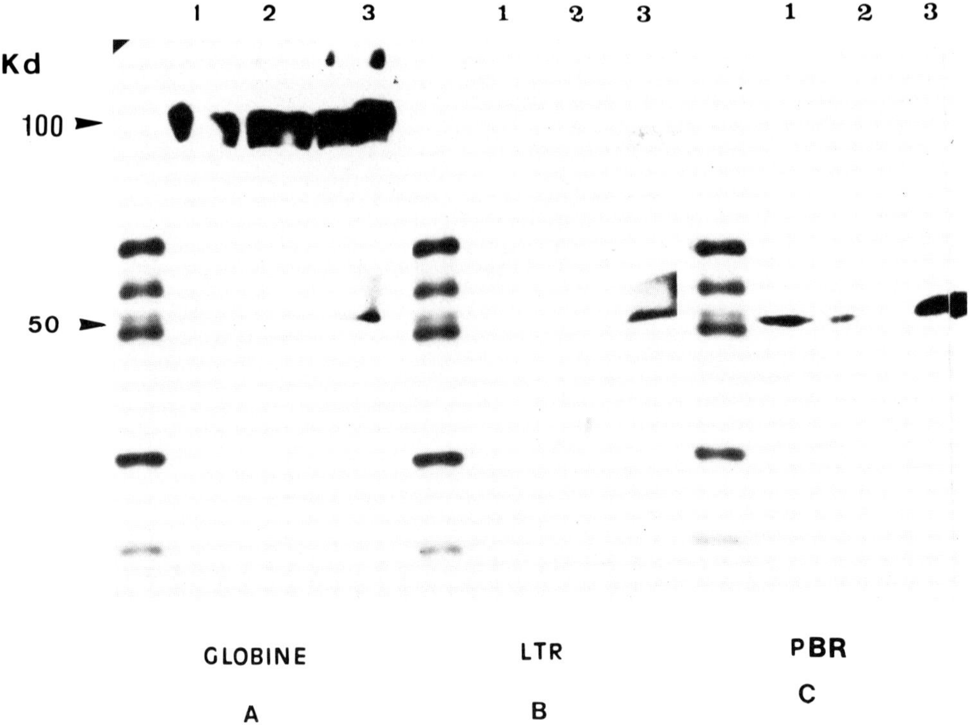

FIGURE 6.   DNA specificity of uninduced MEL cell NHP. Differential binding of β globin — MMTV-LTR and pBr322 DNA. Increasing concentrations (1) 15 μg (2) 30 μg (3) 50 μg of SDS-PAGE fractionated NHP from crude extracts of uninduced MEL cells were equally transferred on three nitrocellulose filters and allowed to bind, after preincubation with 50 μg/ml cold salmon sperm DNA to (A) 4.5 × 10⁶ cpm (10 ng/ml) EcoR I 7 kb β globin DNA fragment; (B) 4.10⁶ cpm (5 ng/ml) MMTV-LTR 5′ DNA (1,5 kb); (C) 5.5 × 10⁶ cpm (7 ng/ml) linear pBR322 DNA. Autoradiography of the washed and dried filters. The slots on the left of each filter contained ¹⁴C protein marker mixture as a control for the equal efficiency of protein transfer to the three nitrocellulose replicates A, B, C.

including immature or DMSO differentiated MEL cells, erythroblasts from embryonic liver, and epithelial mammary cells (Figure 8). Total proteins (A) and β globin DNA binding proteins (B) from MEL cells either uninduced (lanes 2, 3, 4) or DMSO induced (lanes 5, 6, 7) are shown after SDS gel electrophoresis (A) and incubation with ³²P-labeled 7 kb β globin DNA following nitrocellulose transfer (B). Total proteins stained with Coomassie blue are very similar in these three cell types (Figure 8A). By contrast, β globin DNA binding proteins are different in the three extracts (Figure 8B). At least five DNA binding proteins appear only in MEL cells (110 kDa, 100 kDa, 95 kDa, 75 kDa, 35 kDa). The major high molecular weight tissue-specific polypeptide of 100 kDa is also present in immature erythroblasts from rat embryonic liver (Figure 9, lanes 2 to 3), but absent from mammary cells (Figure 8B, lanes 8 to 10), and from other cultured cells as myoblasts (clone 10T984) or mink lung cells (Figure 9, lanes 8 to 10). One 55 kDa polypeptide binds DNA only in GR epithelial mammary cell extracts. Under the stringent conditions used in these experiments (high DNA/protein ratio, 0.1 *M* NaCl) only the 50 kDa polypeptide binds DNA in all cell types tested, according to its unspecific DNA binding (Figure 6B).

E. Effect of DMSO-Induced Terminal Differentiation on the Pattern of MEL Cell β Globin DNA-Binding Proteins

When MEL cells differentiate after 4 d of DMSO treatment, and the synthesis of β globin mRNA is maximum, the major DNA polypeptide of 100 kDa from immature cells no longer

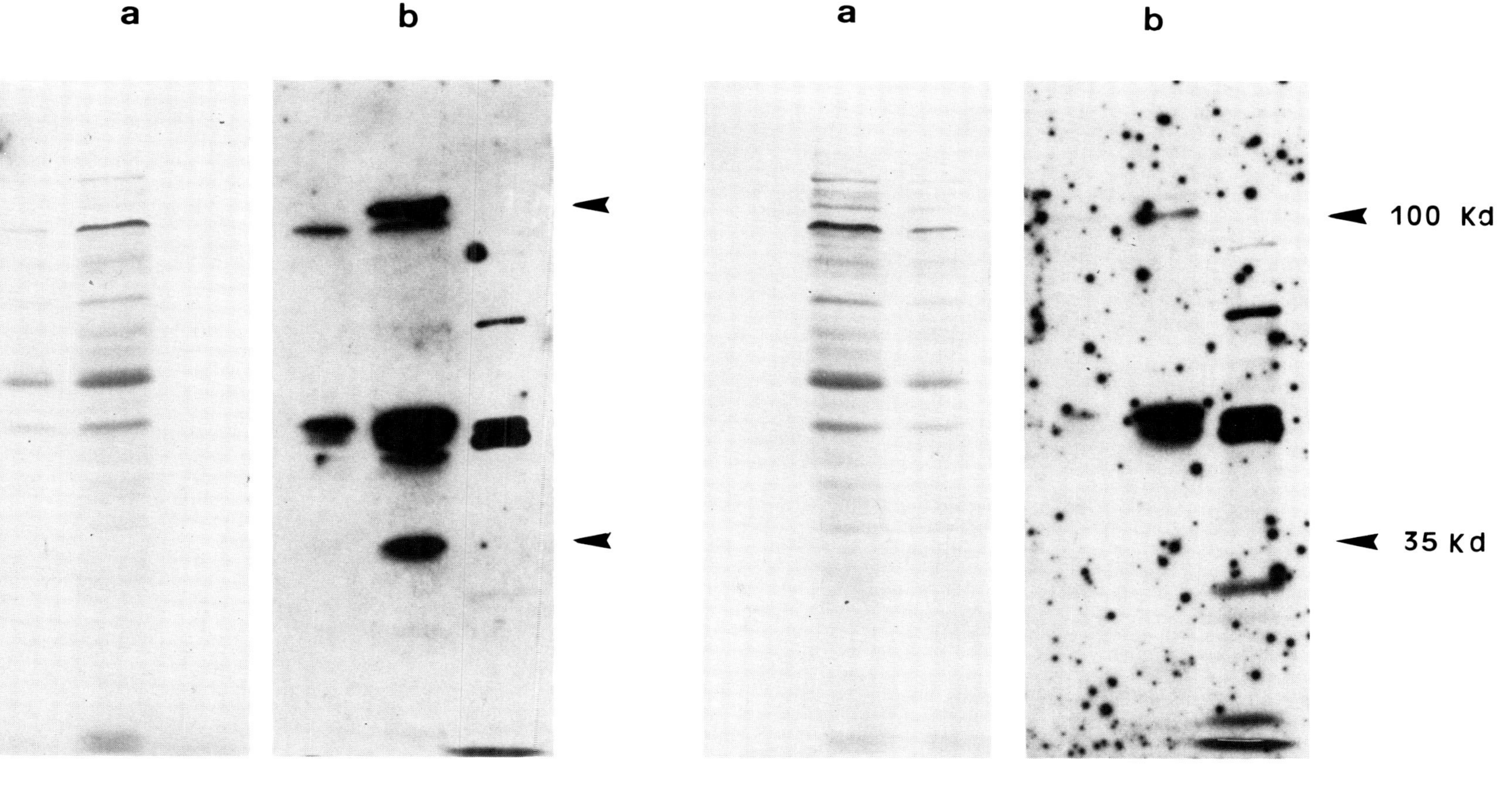

FIGURE 7. DNA sequence specificity of uninduced MEL cell NHP; differential binding of β globin DNA and repetitive poly d(AC)-poly d(GT). (a) Two identical nitrocellulose filters obtained by transfer of the same polyacrylamide gel of uninduced MEL cell NHP stained by China ink (Pelikan). (b) Autoradiography of the same nitrocellulose filters. (Left panel) filter incubated with poly I-poly C, then with ^{32}P labeled β globin DNA (EcoR 1 fragment); (Right panel) filter incubated with poly I-poly C, then with poly d(AC)-poly d(GT) (Böhringer) ^{32}P labeled by nicktranslation.[35]

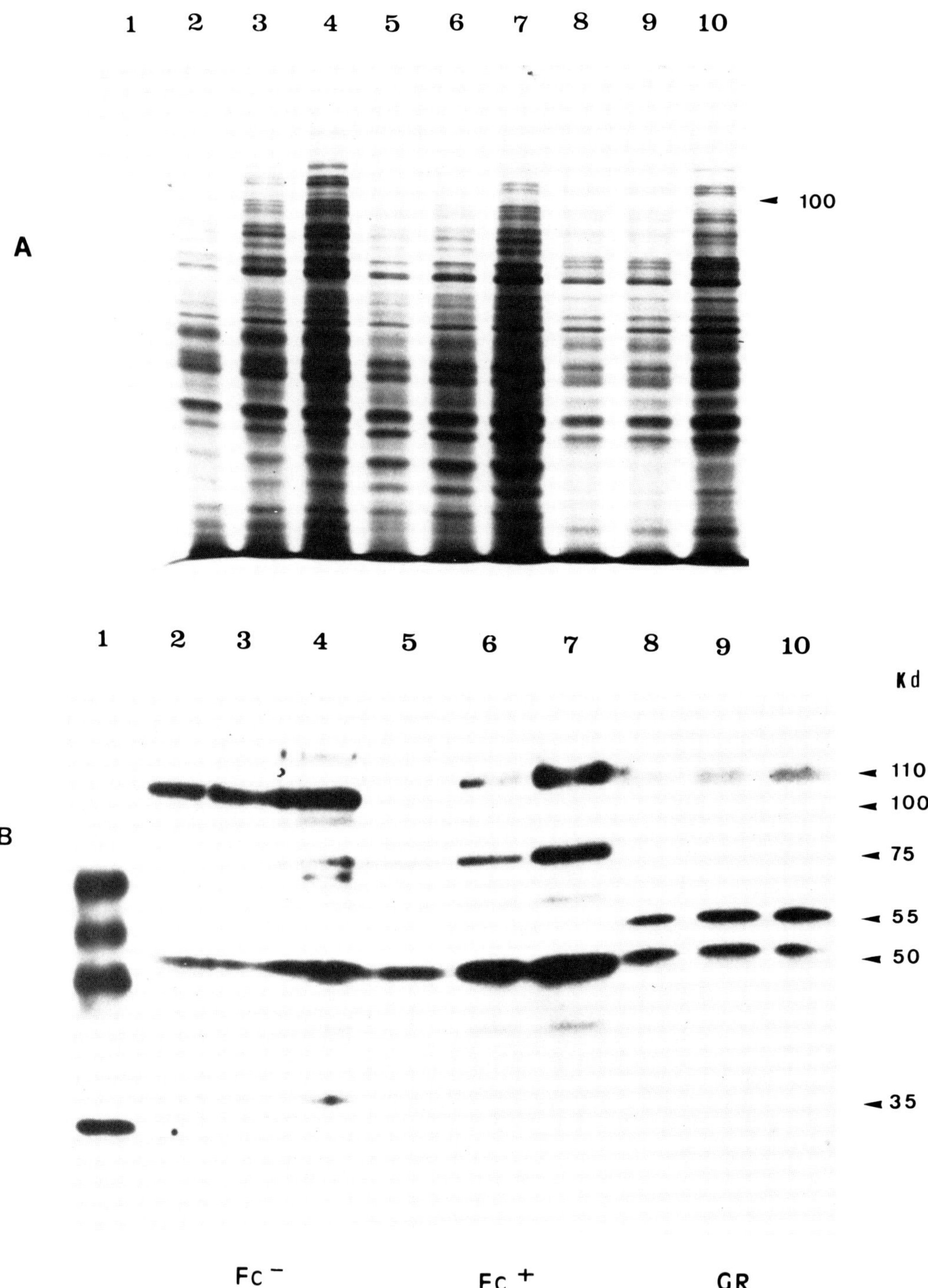

FIGURE 8.   Tissue specificity of β globin DNA binding proteins. (A) SDS-PAGE pattern of proteins from erythroid and nonerythroid cell extracts (Coomassie blue staining). (1) ¹⁴C MW marker proteins: BSA 69,000; IgG 55,000; ovalbumin 46,000, carbonic anhydrase 30,000. (2) 22 μg (3) 33 μg (4) 56 μg noninduced MEL cell NHP (5) 24 μg (6) 37 μg (7) 63 μg induced MEL cell NHP; (8) 28 μg (9) 39 μg (10) 56 μg GR mammary cell NHP. (B) Autoradiography of the same proteins transferred to a nitrocellulose filter, incubated in binding buffer, first with 40 μg/ml cold, native, sonicated salmon sperm DNA, then with 5 ng/ml ³²P labeled β globin DNA (EcoR 1 fragment) Spec. act: 7.5 × 10⁶ cpm/μg.

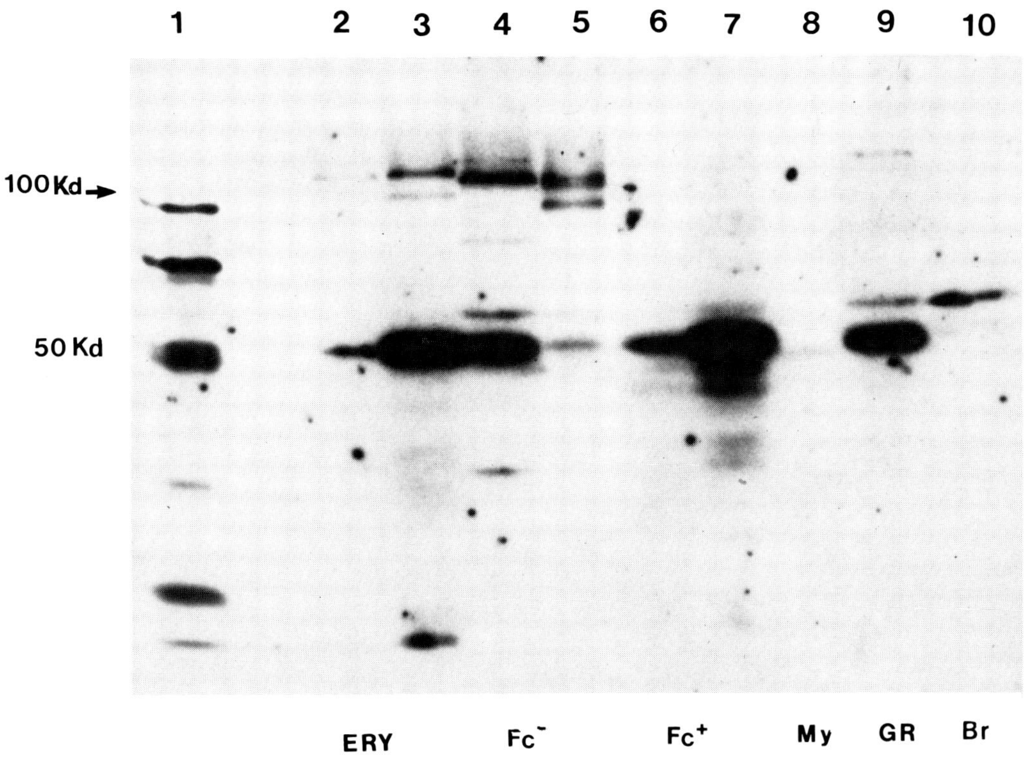

FIGURE 9.   Tissue specificity of β globin DNA binding proteins. Autoradiography of a nitrocellulose filter probed with [32]P Klenow end labeled [35] β globin EcoR 1-BamH1-1.8 kb. 5′ upstream DNA fragment — and containing the following proteins transferred from the same SDS-polyacrylamide gel: (1) ^{14}C MW marker proteins (2 and 3) increasing concentrations of NHP from embryonic liver erythroblasts (4) NHP from uninduced MEL cells (5) NHP from purified nuclei of uninduced MEL cells (6 and 7) increasing concentrations of NHP from DMSO-induced MEL cells (8) NHP from the myoblast cell line clone 10T984[16] (9) NHP from a mammary cell line (GR) (10) NHP from a mink lung cell line. NHP refers to extracts prepared as described.[10]

binds β globin DNA (Figure 8B, lanes 5, 6, 7). By contrast, two other protein-DNA complexes appear in differentiated MEL cells, which have been designated 110 kDa and 75 kDa according to their apparent molecular weight determined by reference to comigrating and cotransferred ^{14}C-labeled marker proteins (Figure 8, lane 1). The 110 kDa protein is clearly distinct from the major 100 kDa protein bound to β globin DNA in undifferentiated cells (lane 2 to 4) by its slightly different molecular weight and its immunoreactivity: the 100 kDa protein crossreacted on immuno-blots with an antibody raised against the 100 kDa DNA binding protein (nucleolin) described by Amalric et al.[33] while the 110 kDa protein does not.[39]

F. Further Characterization of MEL Cell β Globin DNA-Binding Proteins

1. Nuclear Localization

As expected from their DNA binding properties, we found that the proteins we have detected in undifferentiated MEL cells extracts are located exclusively in nuclei, with the exception of the ubiquitous 50 kDa protein that is also present in cytoplasmic extracts (Figure 10). Moreover, several proteins which bind to β globin DNA appear enriched in the nuclear extracts (95 kDa, 60 kDa, 35 kDa) (compare Figures 7 and 10).

2. Differential Affinity for Single- or Double-Stranded DNA

We have compared the binding affinity of the MEL cell DNA binding proteins for double-

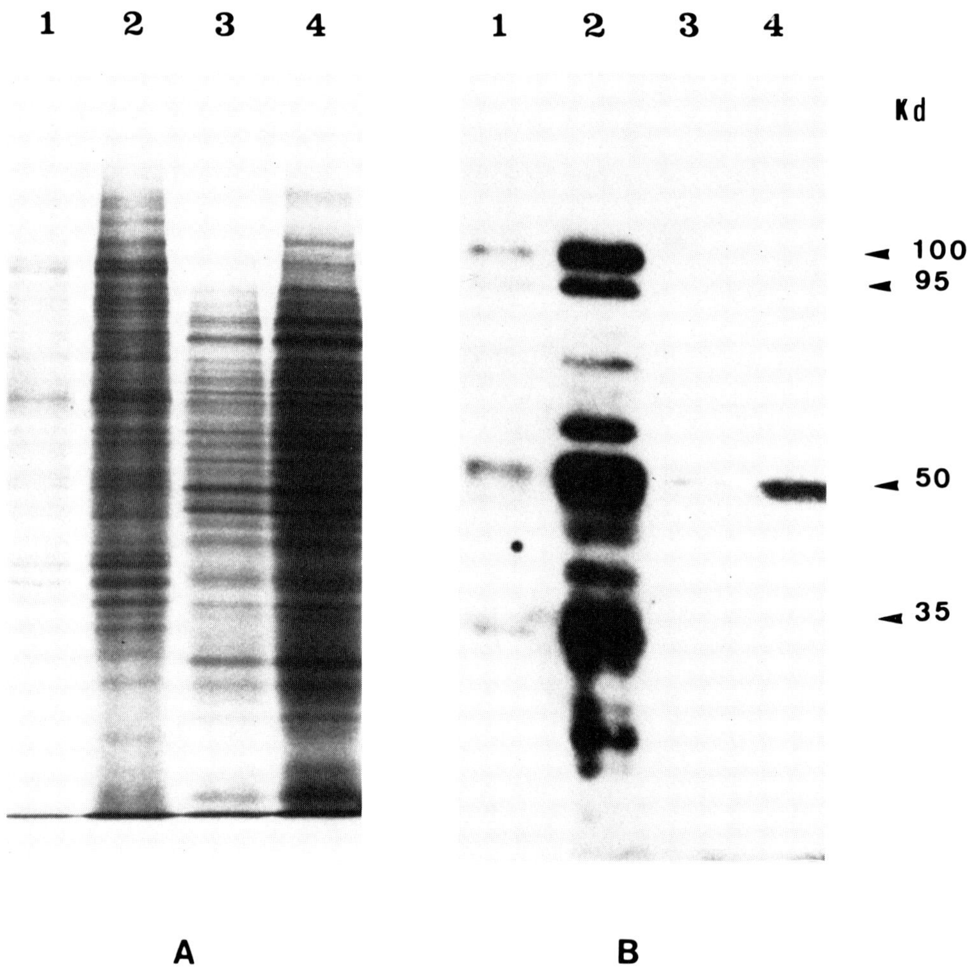

FIGURE 10.   Nuclear localization of MEL cell β globin DNA binding proteins. (A) SDS PAGE pattern of uninduced MEL cell extracts[10] stained with Coomassie blue; (1-2) 9 μg and 45 μg from purified nuclei (3-4) 25 μg and 125 μg from the cytoplasmic fraction. (B) Autoradiography of the same proteins transferred to a nitrocellulose filter, incubated first with 30 μg/ml cold native salmon sperm DNA, then with 2 ng/ml (5 × 10⁶ cpm) ³²P labeled β globin DNA (EcoR 1 7 kb fragment).

stranded native and single-stranded, heat-denatured β globin DNA. Interestingly, we found that the 100 kDa and 50 kDa proteins bind only to the native form while the 95 kDa, the 35kDa, and an additional 60 kDa protein are preferentially revealed by single-stranded β globin DNA (Figure 11B).

3. pH Dependent Transfer to Nitrocellulose Discriminates between Acidic and Basic DNA-Binding Proteins

In contrast with the results reported by Bowen et al.,[30] we found that protein transfer to nitrocellulose by diffusion after SDS removal was highly selective; proteins with no affinity for nitrocellulose are clearly not transferred while those which have a high affinity for it may be considerably enriched by transfer. This is the case of the 100 kDa and 95 kDa proteins since they become heavily stained on nitrocellulose, while hardly detected on Coomassie-blue-colored gels before transfer (compare Figures 7a and 10A). In addition, the ability for a protein to be transferred appears to be dependent from the pH of the transfer buffer: at pH 8.9, only the negatively charged acidic proteins are transferred and thus

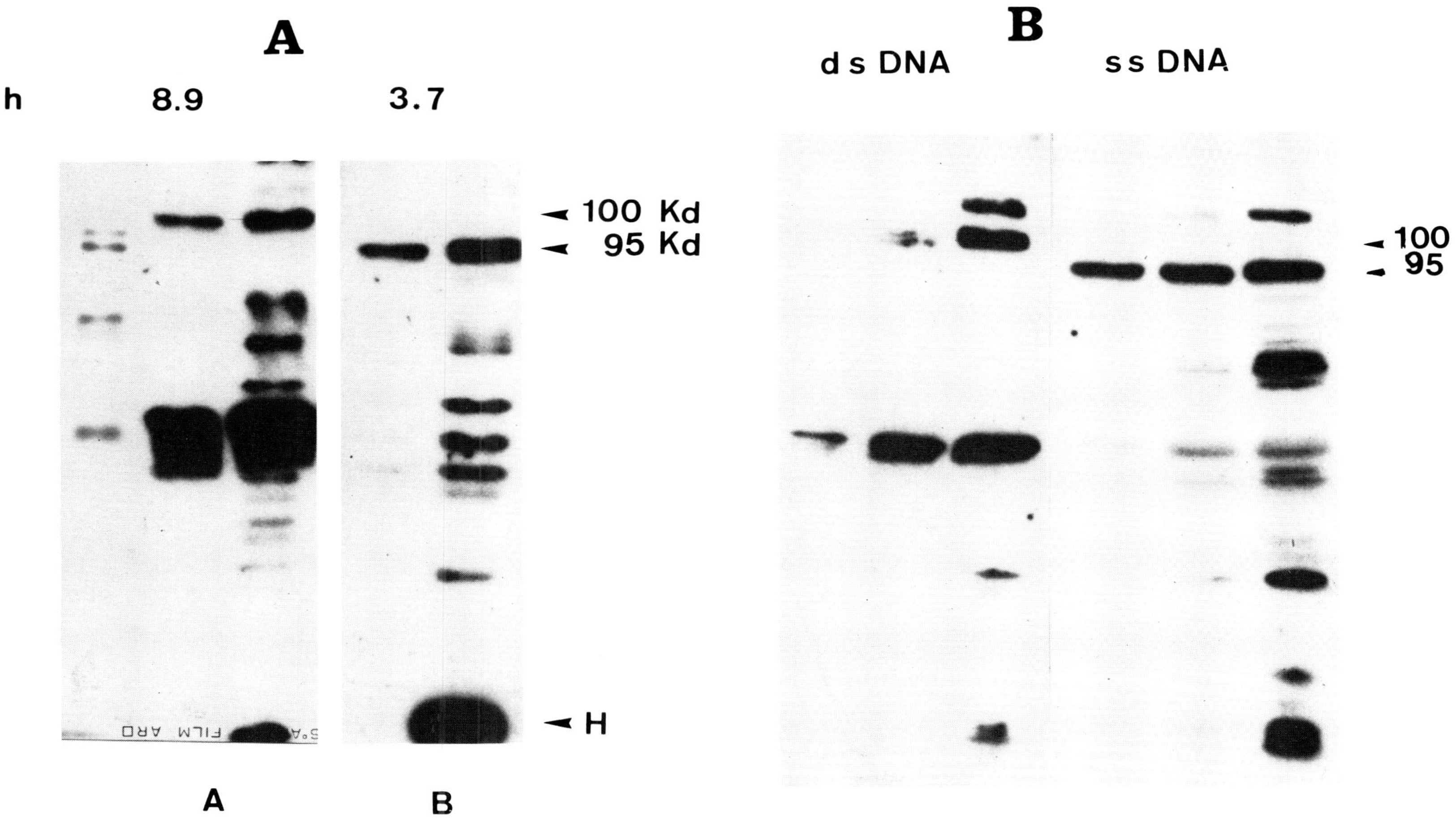

FIGURE 11. (A) Influence of the pH on nitrocellulose transfer of β globin DNA binding proteins. Identical increasing concentrations of nuclear proteins from uninduced MEL cells, fractionated by SDS-PAGE, were transferred to nitrocellulose filters: (A) using a transfer buffer at pH 8.9, (B) using the same buffer at pH 3.7; both filters were simultaneously probed by incubation with the same amount of ^{32}P-labeled β globin DNA. (Left channels) ^{14}C molecular weight proteins. (B) Differential affinity of β globin DNA-binding proteins for single- or double-stranded DNA. The same polyacrylamide gel loaded with increasing amounts of an uninduced MEL cell extract was equally transferred to two nitrocellulose filters: one (left) was subsequently probed with native ^{32}P-labeled β globin DNA and the other (right) was probed with the same amount of heat-denatured ^{32}P-labeled β globin DNA (EcoR 1 fragments).

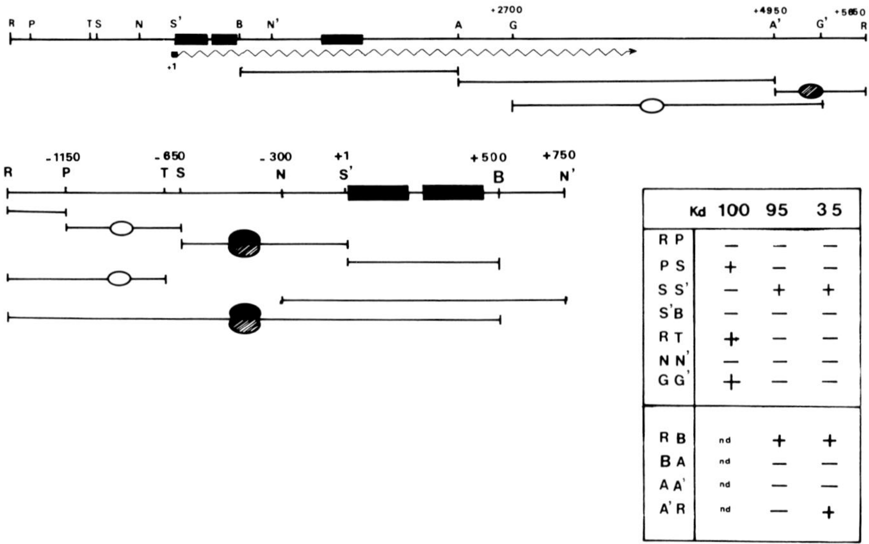

	Kd	100	95	35
R P		−	−	−
P S		+	−	−
S S'		−	+	+
S'B		−	−	−
R T		+	−	−
N N'		−	−	−
G G'		+	−	−
R B		nd	+	+
B A		nd	−	−
A A'		nd	−	−
A' R		nd	−	+

FIGURE 12.   Position of the various β globin DNA restriction fragments bound to the 100 kDa, 95 kDa, and 35 kDa NHP from uninduced MEL cells. Upper panel: the EcoR I β globin DNA fragment (7 kb) showing the restriction sites: R = EcoR I; P = Pst1; T = Sst1; S = Sau3A; N = Hind III; B = BamH1; A = HpAI; G = BglII, the exons (full rectangles), the noncoding regions (continuous thick line), and the distances in base pairs from the transcription initiation site (+1). The latter is indicated by a black square, followed by a zig-zag line covering the presumed length of the transcribed DNA according to Hofer et al.[7] The fragments used for protein-DNA binding experiments are delimited below. Those fragments occupied by an oval, are bound to the corresponding protein: (empty oval) represent the 100 kDa protein (full oval) represents the 95 kDa protein, and (dashed oval) represents the 35 kDa protein. For clarity, the 5'upstream region of the gene (fragment RN') was scaled up twice. The table summarizes the position of the binding sites of the three proteins using isolated DNA fragments (upper) or mixed DNA fragments (lower).

selectively enriched on nitrocellulose. Among these, are the 100 kDa protein and several other acidic DNA binding proteins (Figure 11A). By contrast, at pH 3.7, the latter proteins are not transferred while the positively charged basic proteins appear to be. The 95 kDa, the 35 kDa single-stranded DNA binding proteins, and the residual histones belong to this class (Figure 11A). This property might be used for selectively picking up rare polypeptides from a crude mixture of various proteins separated by SDS-polyacrylamide gel electrophoresis.

G. Selective Affinity of β Globin DNA-Binding Proteins for Upstream Flanking Sequences

By using restriction fragments from the 7-kb genomic β globin DNA, we have tentatively localized the specific binding sites of the best characterized nuclear DNA binding proteins from undifferentiated MEL cells. To perform these experiments, we have digested the EcoR 1 DNA fragment with Pst1, Sst1, Sau3A, Hind III, BamH1, HpaI, and/or BglII (Figure 12). The ^{32}P-labeled DNA subfragments were used as probes either alone or in combination. Using a single labeled subfragment, autoradiograph shows which protein binds the corresponding sequence. Using a mixture, we have eluted the unique labeled DNA fragment or fragments bound to a single polypeptide on nitrocellulose and analyzed the eluted DNA by agarose minigel electrophoresis: only the DNA fragments which have affinity for the individual protein are retained and will appear on agarose gel autoradiograph. Figures 13 and 14 show the pattern of the specific DNA-protein complexes obtained by probing with various isolated ^{32}P-labeled β globin DNA subfragments, four pairs of identical nitrocellulose rep-

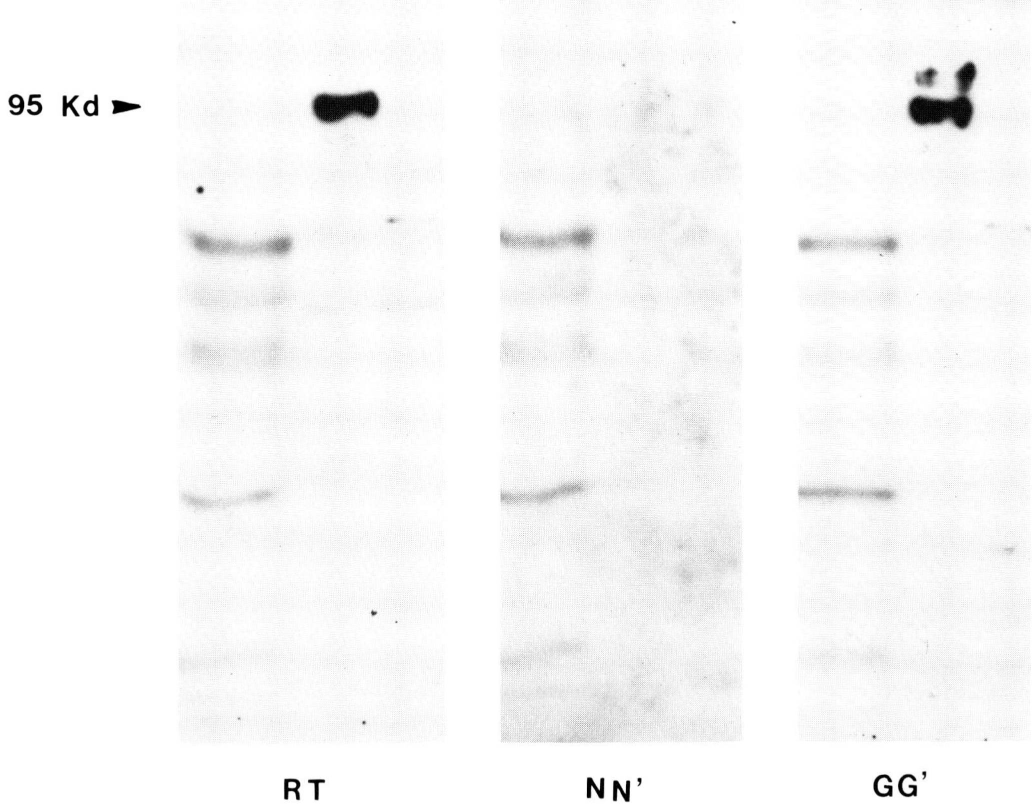

FIGURE 13.   Selective affinity of β globin DNA binding proteins for isolated genomic DNA subfragments. Autoradiography of identical nitrocellulose replicates obtained by transfer of uninduced MEL cell extracts fractionated by SDS-PAGE. The filters were probed with ^{32}P-labeled β globin restriction fragments EcoR 1-Sst1 (RT); Hind III-Hind III (NN') BGlII (GG') (Figure 12) as described in Figures 5 and 6. The slot on the left of each filter contains ^{14}C molecular weight proteins: BSA 69 kDa; IgG 55 kDa; ovalbumin 46 kDa; carbonic anhydrase 30 kDa; lactoglobuline A 18 kDa; lysozyme 14.5 kDa. The slot on the right contains MEL cell extract heated at 70°C for 10 min prior to SDS-PAGE.

licates from polyacrylamide gel electrophoresis of an undifferentiated MEL cell nonhistone protein preparation: the 100 kDa protein binds to the overlapping most upstream fragments EcoR I-Sst1 (RT) (Figure 13) and Pst1-Sau3A (PS) (Figure 15). The absence of interaction with fragments EcoR I-Pst1 (RP), Sau3A-Sau3A (SS'), Sau3A-BamH1 (S'B) (Figure 14) indicates that the 100 kDa protein may recognize a nucleotide sequence located between -1150 and -655 (Figure 12). Surprisingly enough, we observed an additional interaction of this protein downstream to $+2700$ (BglII-BglII, GG') (Figure 13).

By contrast, the 95 kDa and the 35 kDa proteins appear to bind specifically to the fragment Sau3A-Sau3A (SS') (Figure 14). Since they apparently do not bind to any other fragment tested, particularly fragments corresponding to the proximal promoter region (NN') (Figure 13) and (S'B) (Figure 14), we conclude that the latter two proteins recognize nucleotide sequences located between the two Sau3A restriction sites (SS') and upstream the Hind III site (N); therefore, in between -650 and -300 (SN) from the transcription initiation site (Figure 12).

The labeled DNA retained on the nitrocellulose filter by the 95 kDa and the 35 kDa proteins from a (BamH1 + HpaI) digest mixture of the entire 7 kb β globin DNA fragment was eluted after excision of the corresponding radioactive spots and analyzed by agarose gel electrophoresis. The results of such experiments are shown on Figure 15. Among the

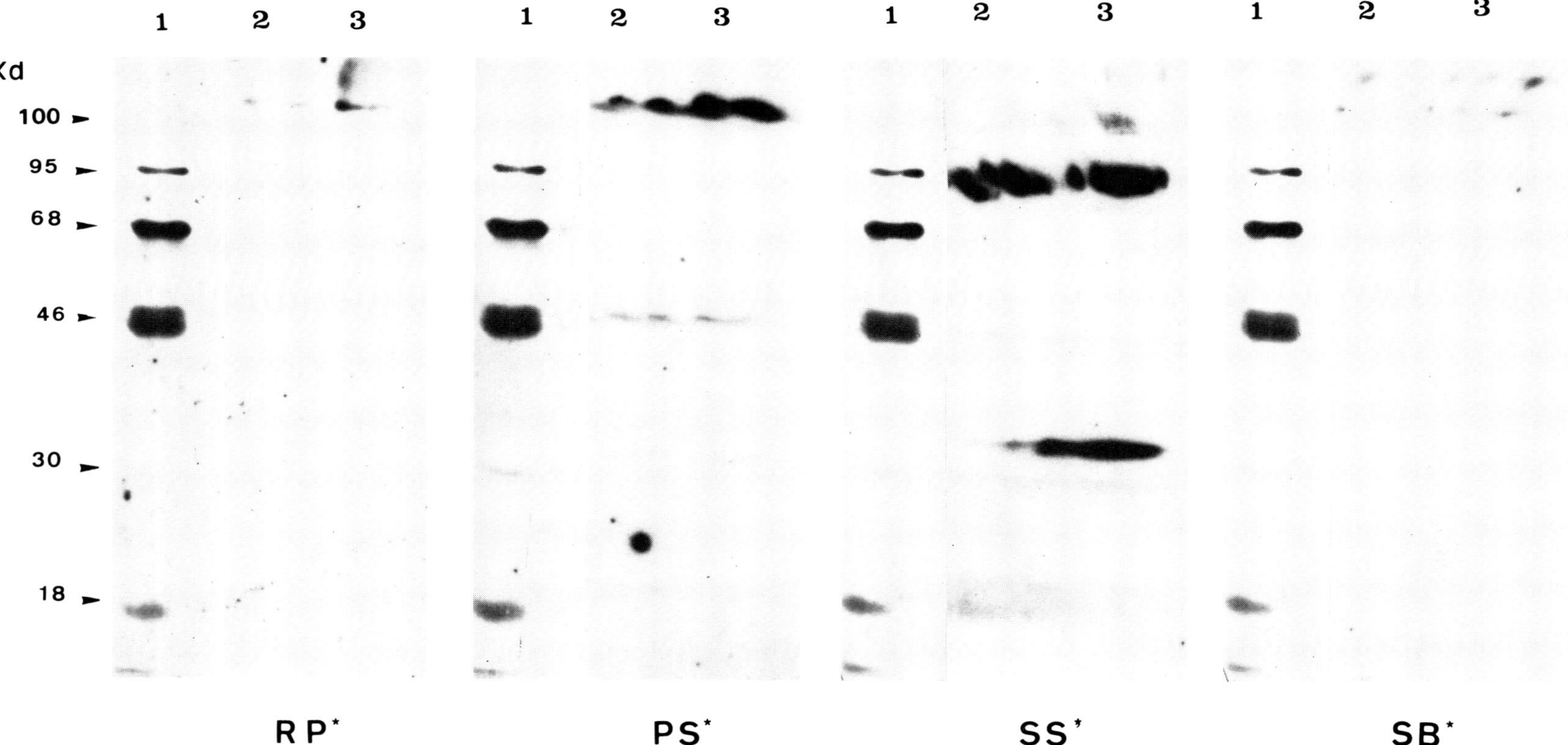

FIGURE 14. Selective affinity of β globin DNA binding proteins for isolated genomic DNA subfragments. Autoradiography of identical nitrocellulose transfer replicates of uninduced MEL cell extracts fractionated by SDS-PAGE. The filters were probed with ^{32}P Klenow end-labeled[35] β globin restriction fragments EcoR 1-Pst1 (RP), Pst1-Sau3A (PS) Sau3A-Sau3A (SS′) Sau3A-BamH2 (SB) (Figure 12). The experimental conditions were as described in Figures 5, 6, and 13. (1) ^{14}C molecular weight proteins: phosphorylase B 92, 5 kDa; BSA 69 kDa, ovalbumin 46 kDa, carbonic anhydrase 30 kDa, lactoglobuline A 18 kDa, cytochrome C 12 kDa, (2) and (3) increasing concentrations of uninduced MEL cell extracts.

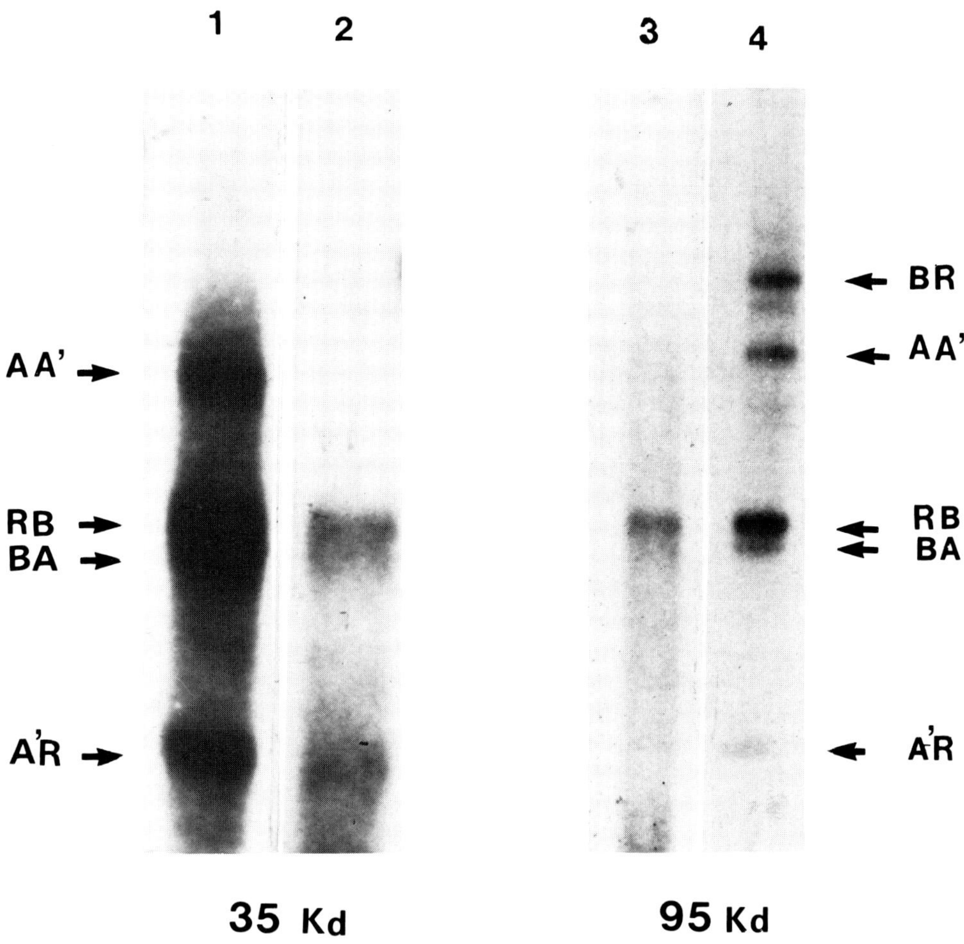

FIGURE 15.   Selective affinity of the 95 kDa and 35 kDa DNA binding proteins for β globin upstream flanking sequences: discrimination between mixed genomic β globin DNA subfragments. Uninduced MEL cell NHP transferred to nitrocellulose after SDS-PAGE were probed by incubation with a mixture of [32]P Klenow end-labeled [35] β globin DNA restriction fragments indicated by arrows and positioned in Figure 12. Autoradiography of 1.2% agarose gels loaded with (1) (4) mixture of ^{32}P labeled β globin DNA fragments, and (2) (3) with the ^{32}P DNA fragments eluted from the 35 kDa (left panel) and 95 kDa (right panel) radioactive spots of the nitrocellulose filter (not shown).

four DNA subfragments covering the entire length of the β globin genomic fragment, and one additional subfragment resulting from partial digestion at the 3′ HpaI sites (Figure 12), only the 1.9 kb EcoR I-BamH1 subfragment (R-B), including the 5′ upstream flanking sequences and the two first exons, appears to be retained in the eluate from the excised 95 kDa radioactive nitrocellulose spot. Moreover, it is selectively bound to this protein, since DNA fragments including the second intron, the third exon, and the 3′ flanking sequences are not bound (Figure 15, lanes 3 and 4). A similar result is obtained with the DNA fragments bound to the 35 kDa protein. The major fragment retained is again the 5′EcoR 1-BamH1 1.9-Kb fragment (Figure 15, lanes 1 and 2). There is no retention of the 3′HpaI-HpaI large fragment (AA′) nor of the BamH1-HpaI (BA) fragment. Interestingly, the small 3′ extreme distal 0.7 kb HpaI-EcoR 1 fragment (A′R) which was previously shown to contain an *in vitro* initiation site of transcription,[12] is also retained by the 35 kDa protein. Using isolated DNA fragments, another protein, the 100 kDa, has been shown to interact also with the 3′ region of the gene (Figure 13).

III. CONCLUDING REMARKS

The advantages of the protein-blotting procedure for the detection of DNA-binding proteins has been already extensively discussed in previous reports.[30,31,36] We find this technique particularly useful to screen various differentiating cell types, of normal or tumoral origin, for tissue-specific DNA-binding proteins harboring a preferential affinity *in vitro* for a given type of purified genomic DNA fragment. No prior purification steps are necessary and the size of the protein is directly observable.

However, one has to be aware of the existence of nonspecific DNA-protein interactions that may be reduced by carefully optimizing the DNA-protein ratio as well as the amount and nature of the competitor DNA used. Figure 12 summarizes the results we have obtained, which suggest the existence of several high affinity DNA-binding proteins interacting specifically with upstream sequences of the β-globin gene. Noteworthy is the fact that the protein-binding sites we have observed reside in a region previously shown to possess a transcriptional "dehancer" activity *in vivo*.[37] The apparent erythroid cell specificity and the variations of the binding activity of these proteins during terminal differentiation, imply a possible role in the modulation of β-globin gene expression, by positive or negative regulatory mechanisms. However, further experiments are needed to correlate the *in vitro* DNA binding of these polypeptides with the modulation of the initiation of *in vitro* transcription we described in the Section I. We have also no evidence as yet that the sites bound by these proteins *in vitro* are occupied *in vivo* or reflect a physiologically relevant association. In other respects their nuclear localization does not rule out a nucleolar[33] or nuclear matrix[38] origin.

Finally, although the β globin DNA binding proteins we have detected show an obvious degree of DNA sequence specificity, they may recognize DNA sequences present in multiple copies in the genome, which could be involved in regulation of a class of genes that are coordinately expressed. In this respect, they might be similar to the DNA-binding proteins detected by the same procedure in a human tumor cell line with the transferrin receptor gene promoter.[36]

ACKNOWLEDGMENTS

This work was supported by grants from the Centre National de la Recherche Scientifique, the Ministère da l'Industrie et de la Recherche, the Institut National de la Santé de la Recherche Médicale, and the Fondation pour la Recherche Médicale Française.

REFERENCES

1. **Marks, P. A. and Rifkind, R. A.,** Protein synthesis: its control in erythropoiesis, *Science,* 175, 955, 1972.
2. **Groudine, M. and Weintraub, H.,** Activation of globin genes during chicken development, *Cell,* 24, 393, 1981.
3. **Ross, J., Gielen, J., Packman, S., Ikawa, Y., and Leder, P.,** Globin gene expression in cultured erythroleukemic cells, *J. Mol. Biol.,* 87, 697, 1974.
4. **Chao, M. V., Mellon, P., Charnay, P., Maniatis, T., and Axel, R.,** The regulated expression of β-globin genes introduced into mouse erythroleukemia cells, *Cell,* 32, 483, 1983.
5. **Charnay, P., Treisman, R., Mellon, P., Chao, M., Axel, R., and Maniatis, T.,** Differences in human α and β globin gene expression in mouse erythroleukemia cells; the role of intragenic sequences, *Cell,* 38, 251, 1984.
6. **Wright, S., de Boer, E., Grosveld, F. G., and Flavell, R. A.,** Regulated expression of the human β-globin gene family in murine erythroleukemia cells, *Nature,* 395, 333, 1983.

7. **Hofer, E., Hofer-Warbinek, R., and Darnell, J. E.**, Globin RNA transcription: a possible termination site and demonstration of transcriptional control correlated with altered chromatin structure, *Cell,* 29, 887, 1982.
8. **Weil, P. A., Luse, D. S., Segall, J., and Roeder, R. G.**, Selective and accurate initiation of transcription at the Ad2 major late promoter in a soluble system dependent on purified RNA polymerase II and DNA, *Cell,* 18, 469, 1979.
9. **Tsuda, M. and Suzuki, Y.**, Faithful transcription initiation of fibroin in a homologous cell free system reveals an enhancing effect of 5′flanking sequence far upstream, *Cell,* 27, 175, 1981.
10. **Manley, J. L., Fire, A., Cano, A., Sharp, P. A., and Gefter, M. L.**, DNA dependent transcription of adenovirus genes in a soluble whole cell extract, *Proc. Natl. Acad. Sci. U.S.A.,* 77, 3855, 1980.
11. **Sawadogo, M. and Roeder, R. G.**, Factors involved in specific transcription by human RNA polymerase II: analysis by a rapid and quantitative *in vitro* assay, *Proc. Natl. Acad. Sci. U.S.A.,* 82, 4394, 1985.
12. **Crépin, M., Triadou, P., Lelong, J.-C., and Gros, F.**, Identification of transcription initiation sites for bacterial RNA polymerase and eukaryotic RNA polymerase II on the 5′end of the mouse β-globin gene, *Eur. J. Biochem.,* 118, 371, 1981.
13. **Tilghman, S. M., Tiemeier, D. C., Polsky, F., Edgell, M. H., Seidmann, J. G., Leder, A., Enquist, L. W., Norman, B., and Leder, P.**, Cloning specific segments of the mammalian genome: bacteriophage λ containing mouse β-globin and surrounding genes sequences, *Proc. Natl. Acad. Sci. U.S.A.,* 74, 4406, 1977.
14. **Chelm, B. K. and Geiduschek, E. P.**, Gel electrophoresis separation of transcription complexes: an assay for RNA polymerase selectivity and a method for promoter mapping, *Nucleic Acids Res.,* 7, 1851, 1980.
15. **McGillivray, A. J., Cameron, A., Krauze, R. J., Rickwa, D., and Paul, J.**, The non-histone proteins of chromatin, their isolation and composition in a number of tissues, *Biochim. Biophys. Acta,* 277, 384, 1972.
16. **Jakob, H., Buckingham, M. E., Cohen, A., Dupont, L., Fiszman, M., and Jacob, F.**, A skeletal muscle cell line isolated from a mouse teratocarcinoma undergoes apparently normal terminal differentiation *in vitro, Exp. Cell Res.,* 114, 403, 1978.
17. **Tambourin, P. E., Wandling, F., Gallien-Lartigue, O., and Huaulme, D.**, Phenylhydrazine induced anemia in mice, *Biomedicine,* 19, 112, 1973.
18. **Breathnach, R. and Chambon, P.**, Organization and expression of eukaryotic protein-coding nuclear split genes, *Annu. Rev. Biochem.,* 50, 349, 1981.
19. **Dierks, P., Van Ooyen, A., Cochran, M. D., Dobkin, L., Reiser, J., and Weissmann, C.**, Three regions upstream from the cap site are required for efficient and accurate transcription of the rabbit β-globin gene in mouse 3T6 cells, *Cell,* 32, 695, 1983.
20. **Myers, R. M., Tilly, K., and Maniatis, T.**, Fine structure genetic analysis of a β-globin promoter, *Science,* 232, 613, 1986.
21. **Cart Wright, I. L., Abmayr, S. M., Fleischmann, G., Lowenhaupt, K., Elgin, S. CR., Keene, M. A., and Howard, G. C.**, Chromatin structure and gene activity: the role of non histone chromosomal proteins, *CRC Crit. Rev. Biochem.,* 13, 1, 1983.
22. **Elgin, S. C. R.**, Anatomy of hypersensitive sites, *Nature,* 309, 213, 1984.
23. **Cohen, R. B. and Sheffery, M.**, Nucleosome disruption precedes transcription and is largely limited to the transcribed domain of globin genes in murine erythroleukemia cells, *J. Mol. Biol.,* 182, 109, 1985.
24. **Wu, C.**, Activating protein factor binds *in vitro* to upstream control sequences in heat shock gene chromatin, *Nature,* 311, 81, 1984.
25. **Augereau, P. and Chambon, P.**, The mouse immunoglobulin heavy-chain enhancer: effect on transcription *in vitro* and binding of proteins present in Hela and lymphoid B cell extracts, *EMBO J.,* 5, 1791, 1986.
26. **Schlokat, U., Bohmann, D., Schöler, H., and Gruss, P.**, Nuclear factors binding specific sequences within the immunoglobulin enhancer interact differentially with other enhancer elements, *EMBO J.,* 5, 3251, 1986.
27. **Sen, R. and Baltimore, D.**, Multiple nuclear factors interact with the immunoglobulin enhancer sequences, *Cell,* 46, 705, 1986.
28. **McKnight, S. and Tjian, R.**, Transcriptional selectively of viral genes in mammalian cells, *Cell,* 46, 795, 1986.
29. **Cohen, R. B., Sheffery, M., and Chul-Geun-Kimn**, Partial purification of a nuclear protein that binds to the CCAAT box of the mouse α1 globin gene, *Mol. Cell. Biol.,* 6, 821, 1986.
30. **Bowen, B., Steinberg, J., Laemmli, U. K., and Weintraub, H.**, The detection of DNA binding proteins by protein-blotting, *Nucleic Acids Res.,* 8, 1, 1980.
31. **Jack, R. S., Gehring, W. J., and Brack, C.**, Protein component from Drosophila larval nuclei showing sequence specificity for a short region near a major heat shock protein gene, *Cell,* 24, 321, 1981.
32. **Riggs, A. D., Suzuki, H., and Bourgeois, S.**, Lac repressor-operator interaction. I. Equilibrium studies, *J. Mol. Biol.,* 48, 67, 1970.

33. **Buggler, B., Cairzergues-Ferrer, M., Bouche, G., Bourbon, H., and Amalric, F.,** Detection and localization of a class of proteins immunologically related to a 100 kd nucleolar protein, *Eur. J. Biochem.,* 128, 475, 1982.
34. **Germond, J. E., Hirt, B., Oudet, P., Gross-Bellard, M., and Chambon, P.,** Folding of the DNA double helix in chromatin-like structures from Simian Virus 40, *Proc. Natl. Acad. Sci. U.S.A.,* 72, 1843, 1975.
35. **Maniatis, T., Fritsch, E. F., and Sambrook, J.,** *Molecular Cloning, A Laboratory Manual,* Cold Spring Harbor Laboratory, Cold Spring Harbor, NY, 1982.
36. **Miskimins, W. K., Roberts, M. P., McClelland, A., and Ruddle, F. H.,** Use of a protein blotting procedure and a specific DNA probe to identify nuclear proteins that recognize the promoter region of the transferrin receptor gene, *Proc. Natl. Acad. Sci. U.S.A.,* 82, 6744, 1985.
37. **Gilmour, R. S., Spandidos, D. A., Vass, K., Gow, J. W., and Paul, J.,** A negative regulatory sequence near the mouse β-major globin gene associated with a region of potential Z-DNA, *EMBO J.,* 3, 1263, 1984.
38. **Hentzen, P. C., Rho, J. H., and Bekhor, I.,** Nuclear matrix DNA from chicken erythrocytes contains β-globin gene sequences, *Proc. Natl. Acad. Sci. U.S.A.,* 81, 304, 1984.
39. **Lelong, J.-C., Prevost, G., Crepin, M.,** unpublished results.

Chapter 9

QUANTITATION OF TWO-DIMENSIONAL POLYACRYLAMIDE GEL FLUOROGRAMS OF mRNA TRANSLATIONAL PRODUCTS BY SOFT LASER DENSITOMETRY

Isaac Bekhor

TABLE OF CONTENTS

I. ABSTRACT

Messenger RNA was isolated from cataractous lens of normal rats and rats maintained on a 50% galactose diet for 32 d and from rats fed a diet of Purina® Chow for 12 d following the 50% galactose period. The total mRNA from the three experimental groups was translated *in vitro* in the presence of ^{35}S-methionine. The labeled products were separated by a two-dimensional polyacrylamide gel electrophoresis, processed for fluorography, and exposed to X-ray film to localize the labeled products. The film was scanned by computer-driven two-dimensional soft laser scanner for quantitation. This type of quantitation demonstrates that a highly significant detailed analysis of the fluorograms is now possible.

II. INTRODUCTION

The visual inspection of mRNA translational products on a fluorogram yields qualitative data that could be more significant if quantitated. The present contribution utilizes the 2-D laser scanner to detect and quantitate changes in mRNA population occurring as a result of changes in the environment of a specific tissue. We used rat lenses from which we isolated messenger RNA for translation *in vitro,* and separated the [35]-methionine labeled products by two-dimensional polyacrylamide gel electrophoresis. The lenses were obtained from either normal rats or rats maintained on a diet of Purina® Chow/galactose at a ratio of 1:1, then reversed to a normal diet. Galactose induces formation of cataracts in the latter group of animals.[1-5] Removal of galactose from the diet reverses cataracts.[6] Reversal of cataracts is expected to lead to mRNA synthesis of various classes and levels as found in normal lens.[7] Specifically, the present data demonstrates that quantitation of mRNA products can elaborate our analysis on changes occurring at the genetic level. This methodology can easily be applied to studies on nonhistone chromosomal protein population variability in specific tissues.

III. MATERIALS AND METHODS

Induction of cataracts was carried out as previously described.[7] Messenger RNA isolation was done by precipitation from 3 *M* Na-acetate (pH 6.0) as described in our previous report.[7] Translation of mRNA was done with a translation kit from BRL (Bethesda, MD) using ^{35}S-methionine. The products were separated by two-dimensional polyacrylamide gel electrophoresis essentially as described by O'Farrell[8] and according to the methodology described by the manufacturer of the electrophoresis unit (BioRad, Richmond, CA). Following electrophoresis, samples of known cpm (250 to 10,000) were spotted on the gel to serve as internal standards for conversion of integrated volume values (see below) to cpm when desired, and also to obtain a standard curve to insure that the values or the intensities of the various protein products did not exceed the sensitivity limits of the film. The gel was later stained with Coomassie blue to localize the markers, then washed in deionized water, immersed into Fluoro-Hance (Research Products International, Mt. Prospect, IL) at about 200 ml per 16 × 20-cm gel, and processed as described[7] for fluorography. Exposure to the XAR-5 X-ray film was for 4 d at −70°C.

A. Scanning the 2-D Fluorogram

Computer-assisted analysis of two-D polyacrylamide gels has been discussed in detail by Garrels[9] and by Lester et al.[10] The soft laser densitometry technique was described by Zeineh et al.[11] for quantitation of tube isoelectric focusing gels and by Zeineh[12] for computer-driven scanning densitometry of 2-D polyacrylamide gels. Briefly, the fluorogram was placed on the sample holder of Zeineh soft laser Scanning Densitometer, model SL-2DUV, and menu driven by an IBMpc. The 2-D PCI program was loaded into computer memory. From the main menu, the "scan & save" mode was selected. In the present study, the scan length of 9.3 cm was entered. Scan width of 5.8 cm was calculated by computer. The frame dimensions were proportional to the monitor screen of 320 × 200 pixels. Scan speed was set at 1, equivalent to 1 cm/s speed in the X-axis direction. The laser beam of 30-μm long and 10-μm wide emerged from below and passed through the sample to the photodetector located on the lid of the scanner. With the laser beam incident on a region clear of dark spots, the Zero knob on the front panel of the scanner was offset so that the optical density reading on the screen was 5. The sample was electrically moved so the laser beam was incident on the darkest spot on the fluorograph and the gain knob was adjusted so the screen reading was 210. The zero and gain settings assured the reading of optical density range of 0 to 255. The sample was electrically moved to the starting point, upper left corner of the fluorograph, and scanning was started. At the end of the scan, the 2-D data was saved on the diskette in Drive B. Similarly, the control fluorograph was scanned under the same conditions. For amplification of the data in selected areas of the fluorograph, smaller regions — one of 5 cm in length and another of 1.2 cm for either the test or control data — were scanned and saved for comparison purposes.

B. Three-D and Two-D Graphics

From the main menu, the "view and process" submenu was loaded, and contour mapping was applied to the various computer-saved scanned data. Selected areas on the test fluorographs were compared to those of the control both by integration of volume of each product and superimposition. Visual inspection of the superimposed contour maps displayed qualitative differences, presence or absence of a spot at a glance, and also revealed semiquantitative changes as indicated by color and spread of contours. Pictorial displays, or 3-D graphics, aid the quantitative evaluation (height of the 3-D peak proportional to the intensity or optical density of the spot) as well as the analysis of shape and overlapping of spots.

C. Quantitation of the Labeled Protein Products by Integration

From the contour-graphic mode, integration with automatic or manual mode was performed, and results were recorded as shown below.

IV. RESULTS

The data in Figure 1 is a fluorograph of a two-D polyacrylamide gel electrophoretic profile of translational products from about 5 μg of total RNA from lenses of rats maintained on Purina® Chow alone. Figure 2A shows the products obtained from lenses of rats fed a diet containing 50% galactose for 32 d. At that time the lenses appear to be totally cataractous.[1-5] This was compared with the results obtained from rats fed a Purina® Chow diet for 12 d following the 32 d galactose feeding period (reversal of cataracts, Figure 2B). The boxed areas show the products obtained from the lens minor mRNAs (noncrystallins); those products below the boxed area represent the crystallins (products A through S, except for I). Qualitatively, there appears to be an increase in the number of the minor mRNA products with little change in the crystallins following reversal of the cataracts. Product I, as shown in the boxed area, was chosen as a reference point for the two-D scan analysis. In the data to follow we show detailed analysis on specific products to signify the extent of quantitation by computer analysis of the two-D scan data.

The data shown in Plate 1A* (9.3-cm scan) represent computer generated analysis of the laser scan as seen in the upper half of the 32-d-cataractous-lens-fluorogram of Figure 2A (boxed area, noncrystallin mRNA translational products), including both low and high

---

* Plates 1 to 4 appear following page 120.

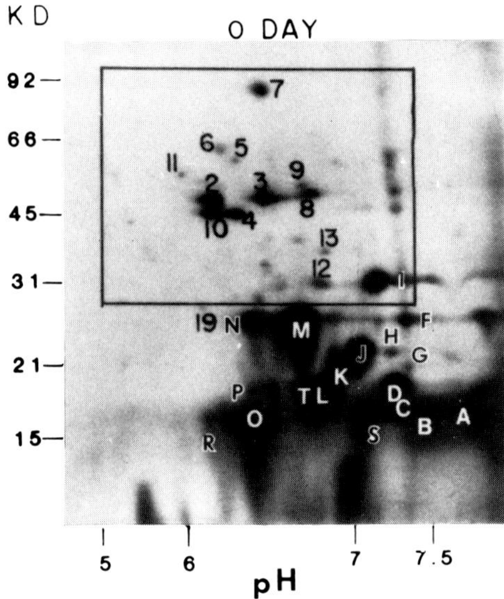

FIGURE 1.   Fluorograph of a two-D polyacrylamide gel electrophoretic profile of translation products from about 5 μg of RNA from lens of 50-g female Sprague Dawley rats fed a diet of Purina® Chow alone. All of the assays were done as described.[7] Proteins are numbered for reference only. The crystallins were located as described.[7] Alpha crystallin, product O; beta crystallins, products C,D,J,K,L,M; gamma crystallins, products A,B. About 400,000 cpm were layered on the IEF gel. Exposure to the film was for 4 d at −70°C.

intensity radiospots. By manipulation of the levels of the baseline (with the "edit contour" mode of the PCI program), the major products can be highlighted as shown in Plate 1B. Similar analysis was done on the lower half of the gel (the crystallins) showing the calculated data obtained with low (Plate 1C) and high (Plate 1D) baseline settings. The demarcation of the crystallin products is clearly seen, and the volumes of these spots can be easily determined by integration (Table 1). The three-dimensional presentation of the scans is shown in Plate 1E (equivalent to Plate 1B), and in Plate 1F (equivalent to Plate 1D). The three-D data can be amplified to verify the presence or absence of minor mRNA translational protein products (data not shown).

The scan obtained from the 12-d reversed cataracts is shown in Plate 2A (noncrystallins, boxed area of Figure 2B) and in Plate 2C (crystallins, lower half of Figure 2B) at a low baseline setting, while Plates 2B and 2D show the same data at a high baseline setting, highlighting the higher intensity (major) products only. A portion of the scans was amplified as shown in Plate 3A (a 5-cm scan of noncrystallins, 32-d cataractous lens mRNA products, boxed area in Figure 3B), and in Plate 3B (a 5-cm scan of noncrystallin mRNA translational products from a 12-d reversed cataracts, boxed area in Plate 2B). The data from these figures were superimposed for visual examination of the changes that may have occurred as a result of reversal of the diseased state (Plate 3C). Further amplification can also be done on a specific set of proteins, e.g., as was done with those proteins shown in the boxed area in Plates 1B and 2B. Amplification is done by reducing the area that is scanned. Plates 4A (from Plate 1B) and 4B (from Plate 2B) show those results at a scan of 1.2 cm. The three-D presentation of the data of Plates 4A and 4B is shown in Plates 4C and 4D. The data suggest that in the reversed cataracts state, a new protein appears to be synthesized which may be suggestive of changes occurring in response to the new state of recovery. This protein was not detected in the 32-d cataractous state, nor was it clearly seen in the 9.3-cm scan of the original fluorogram for the reversed 12-d cataractous state (Figure 2B). Therefore, the methods described in this communication signify the strength of these analytical pro-

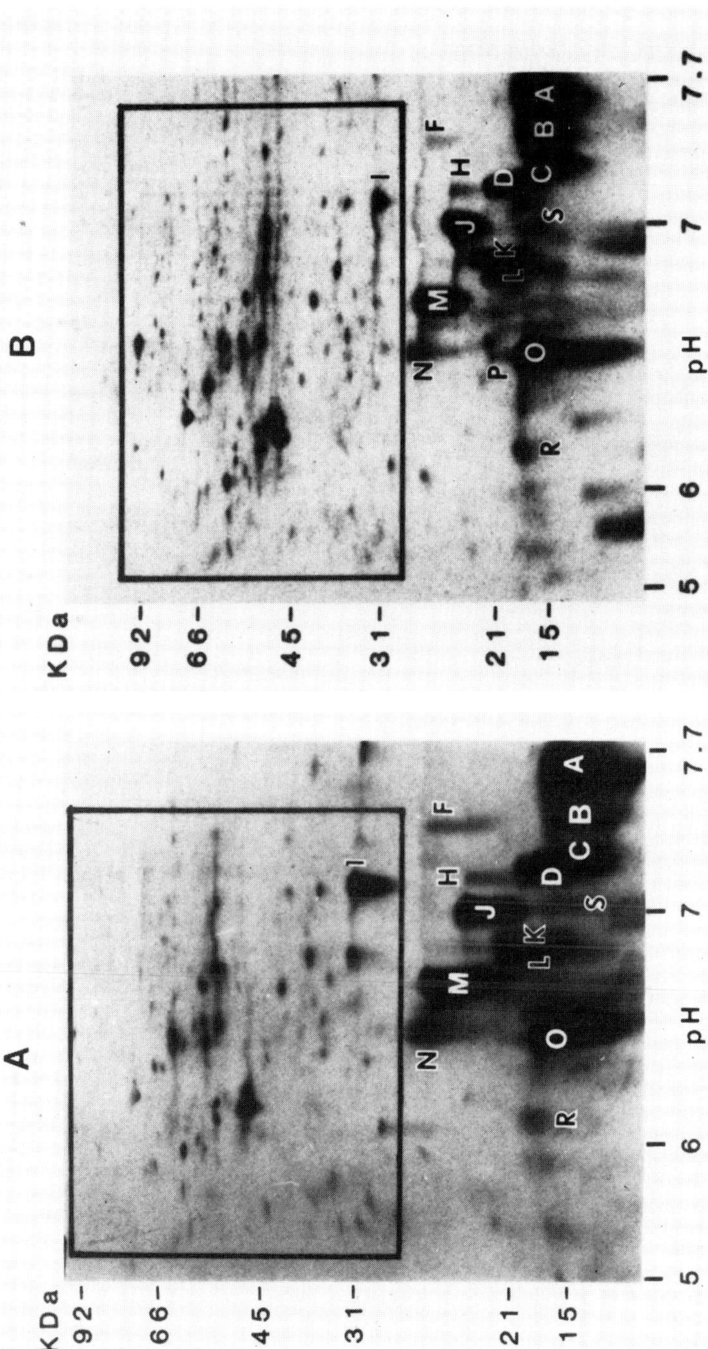

FIGURE 2. Fluorograph of a two-D polyacrylamide gel electrophoretic profile of translation products from about 5 μg of RNA from lens of 50-gm female rats fed 1:1 galactose:Purina® Chow for 32 d (A), followed by feeding with Purina® Chow alone for 12 d (B). Details of the methodology may be found in previous publications.[7,14] The boxed area designates the noncrystallin mRNA products (minor mRNAs); the products marked 3 to 6 designate those proteins scanned in Plates 4A and B; those marked from A to S represent the major crystallin mRNA products. The protein designated I is used as a reference point for the two-D gel scans as shown below. About 400,000 cpm were layered on the isoelectrofocusing (IEF) gel. Exposure to the X-ray film was for 4 d at −70°C.

Table 1
RELATIVE ABUNDANCE OF SPECIFIC
NONCRYSTALLIN PROTEIN PRODUCTS
FROM mRNA OF 32-d CATARACTOUS
LENS AND LENS UNDERGOING
REVERSAL OF CATARACTS

| Protein product | Percent[a] | |
(#, Figures 2A, 2B)	In Figure 2A	In Figure 2B
I	37.2	9.2
3	19.5	18.1
4	0.99	0.57
5	0.83	0.71
6	0.00	0.57

[a]   Percent (or abundance) is calculated by measuring the relative volume of a specific protein product vs. the total volume of all the products as done by the "I" mode of the PCI program.

Table 2
PERCENT OF ALPHA-, BETA-, AND GAMMA-
CRYSTALLIN mRNA TRANSLATION PRODUCTS

State of lens	Percent of total translation products			
	Alpha	Beta	Gamma	Alpha + beta + gamma
Normal range[a]	8—11	40—45	24—27	70—83
(average)	(9)	(42)	(25)	(76)
11-d gal[b]	17	59	ND	76
32-d gal	10	42	25	77
32-d gal + 12-d p.c.[a]	8	41	28	77
32-d gal + 28-d p.c.	10	48	25	83
32-d gal + 61-d p.c.	21	48	9	78
32-d gal + 84-d p.c.	21	50	8	80

Note: As determined from Figures 1 and 2 by computer analysis following 2-D scanning of the respective radioautograms in a soft laser densitometer.

[a]   Data obtained from previous results[7,14] on decapsulated lens from 50-g female Sprague Dawley rats showing a range between three different measurements.
[b]   Gal, galactose; d, day; p.c., Purina® Chow; ND, not detectable; e.g., 32-d gal + 12-d p.c., means rats were fed a diet of 50% galactose for 32 d then switched to a diet of Purina® Chow for an additional period of 12 d.

cedures for examining data obtained from 2-D gel electrophoretic profiles of radiolabeled mRNA translational products.

Table 1 presents typical calculation of data on a sample of selected noncrystallin protein products; e.g., product I which is relatively in high abundance in the cataractous lens (Figure 2A) was reduced significantly in lens undergoing reversal of cataracts (Figure 2B). Table 2 shows the percentage of various crystallin-protein mRNA products found in normal and cataractous lens and those lenses undergoing reversal of cataracts (calculated from Plates 1 and 2). The percent is calculated from the volume of each product determined by integration following scanning by computer-driven laser densitometry. Volume is related to the intensity and area of each product which, in turn, is proportional to the cpm as is concluded from

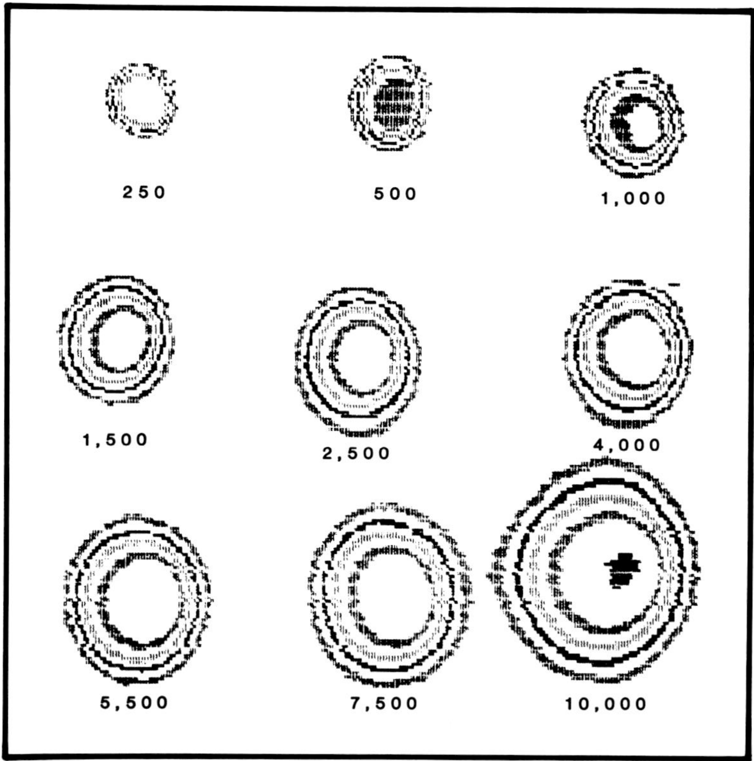

FIGURE 3.   Computer generated contours of a two-D scan of known amounts of cpm spotted on a gel and processed as described.[7] Data demonstrate the relationship of area (area of film exposed to the radioactive source) and intensity (change in color from the periphery to the center of the contour) vs. cpm applied to the gel. [^{35}S]-methionine labeled protein was applied at a constant volume of 1 µl per spot; cpm values are shown below each contour. Exposure to the film was for 3 d.

the scan of the standard cpm data shown in Figure 3, and plotted as volume vs. cpm in Figure 4. The initial analysis demonstrated that there was a linear relationship between cpm applied to the gel and volume of each standard spot (Figure 4). Further, our measurements showed that, on a fluorogram, the observed linearity between volume and cpm may also be due to the linear increase in the area of the exposed film (not volume) as a function of cpm (in a constant volume of 1 µl) applied to that spot (Figure 5). The procedure, therefore, demonstrates that limits of sensitivity of exposure of the film (shown as peak heights in Figure 6) do not appear to limit the accuracy with which one can quantitate cpm levels spotted on the gel. The perfect linear relationship between area and volume is shown in Figure 7. In other experiments we spotted as much as 100,000 cpm of labeled protein and confirmed that this value, when scanned and integrated, still remained within the linear region of this curve (Figure 7), although approaching higher values. Thus, there appears to be no limitation to number of counts that one can detect by the procedures described in this communication. It is clear then, that the area and the volume of the labeled product become a function of the cpm detected by the X-ray film, as measured by the present technique of computer-driven densitometry. Volume values are obtained by integration using the ''I'' mode of the PCI program. From these volumes (being a function of the area of each product) one can calculate the relative percent or the abundance of a specific protein in that population.

Table 2, therefore, displays relative abundance (or percent) of the various crystallin products seen in Figures 1 and 2. These data are expected to reflect gene activity or abundance

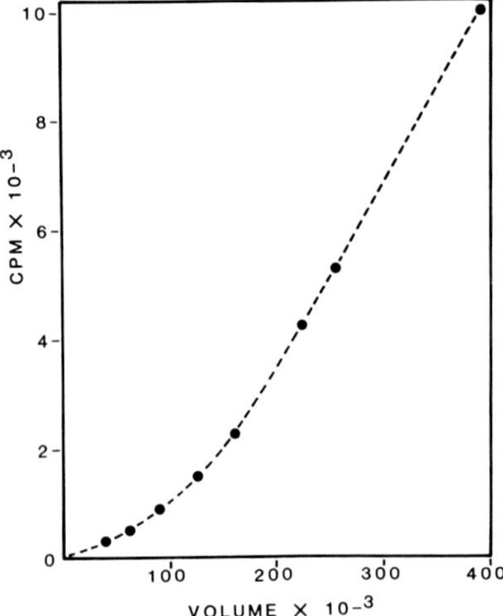

FIGURE 4.   Relationship between integrated volume of a radioactive spot on a radioautograph vs. cpm applied to that spot (Figure 3). Volume refers to arbitrary units proportional to the area and intensity of the spot being integrated, based on a scale of isobars ranging from 0 to 255. Zero refers to no intensity, and 255 refers to maximum intensity detected by a laser beam 30 μm-long and 10-μm wide. A semi-linear relationship between volume and cpm is observed. A 10,000 cpm per spot falls within this region. All of the spots integrated in our experimental gels fall within this semi-linear relationship. Each experimental gel contained its own internal standard protein spots, thus allowing for no errors in determining cpm found per protein product in that gel.

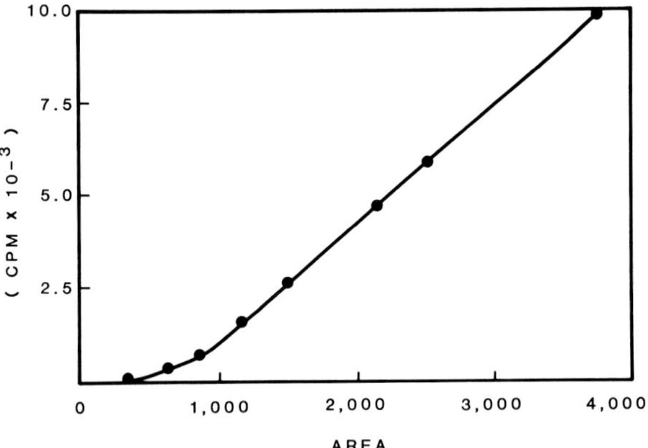

FIGURE 5.   Relationship of area (arbitrary units) vs. cpm applied to each spot (Figure 3). Area appears to show a similar relationship to cpm as found with volume to cpm (Figure 4). Units are based on an assigned value of a maximum scanner reading of 255 as described under Materials and Methods.

of a specific mRNA in the whole population of the isolated mRNA. For example, one of the gamma-crystallins, designated by B in Figure 2B, is increased significantly. The latter finding is in agreement with previously published data on the effects of cataractogenesis on gamma-crystallin synthesis.[13]

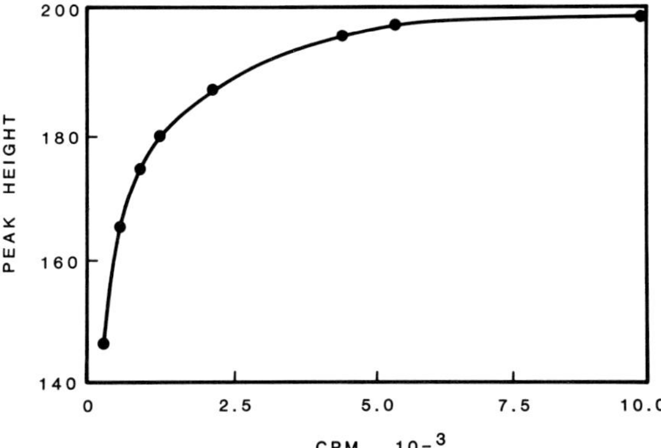

FIGURE 6.   The relationship of maximum intensity (peak height) of each radioactive spot vs. cpm. This number is dependent on the sensitivity of the X-ray film being used. The data establish that maximum exposure is reached at about 2500 cpm per spot. Values are arbitrary units based on optical density units assigned at maximum reading of 255 as described under Materials and Methods.

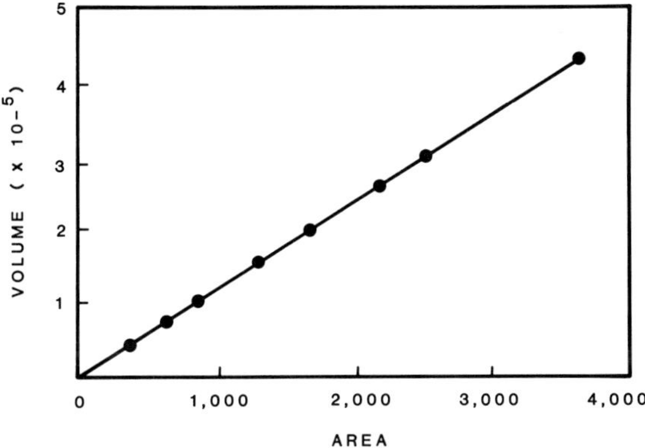

FIGURE 7.   The relationship between area of a radioactive spot vs. its volume as measured by the methods described in this communication. The data demonstrate that the observed linear relationship between these two parameters makes possible quantitation of cpm incorporated per protein product independent of the sensitivity of the film being used.

V. DISCUSSION

The methods of laser scanning and computer analysis of the data clearly simplifies analysis of 2-D gel electrophoretic profiles which were not possible previously. In addition to the results shown in Tables 1 and 2, it is also possible to calculate cpm incorporated into each protein product by entering the factor of cpm vs. area or intensity units as discussed above and calculated with the aid of a computer. The standard cpm vs. intensity can be determined on the same gel by spotting known amounts of cpm on the gel prior to exposure to the X-ray film (data not shown).

The results presented in this communication confirm that one of the gamma crystallins is resynthesized during the process of reversal of the cataracts in agreement with other data.[7,13] In addition, it is also possible to document and quantitate synthesis of new mRNAs (reflected by the appearance of new protein products on the gel) or decrease in mRNA synthesis as a result of the disease.

Similar analysis can be carried out on the abundance of specific nonhistone protein mRNA in a specific tissue or abundance of a specific protein in the cell nucleus. Scanning of Coomassie blue or silver-stained gels can also be easily done by the two-D laser scanner, and data calculated as shown in this communication.

ACKNOWLEDGMENT

This work was supported in part by the National Institutes of Health (EY05406 to IB), and in part by Biomed Instruments, Inc., Fullerton, CA.

REFERENCES

1. **Kinoshita, J. H., Merola, L. O., and Dikmak, E.,** *Exp. Eye Res.,* 1, 405, 1962.
2. **Kador, P. F., Zigler, J. S., and Kinoshita, J. H.,** *Invest. Ophthalmol. Vis. Sci.,* 18, 696, 1979.
3. **Unakar, N. J., Genyea, C., Reddan, J. R., and Reddy, V. N.,** *Exp. Eye Res.,* 26, 123, 1978.
4. **Unakar, N. J., Tsui, J. Y., and Harding, C. V.,** *Ophthalmic Res.,* 13, 20, 1981.
5. **Kuwabara, T., Kinoshita, J. H., and Cogan, D. G.,** *Invest. Ophthalmol.,* 8, 133, 1969.
6. **Unakar, N. J., Weinseider, A., and Reddan, J. R.,** *Ophthalmic Res.,* 9, 296, 1977.
7. **Hsu, M. Y., Unakar, N. J., and Bekhor, I.,** *Lens Res.,* 3, 305, 1986.
8. **O'Farrell, P. H.,** *J. Biol. Chem.,* 250, 4007, 1975.
9. **Garrels, J. I.,** *J. Biol. Chem.,* 254, 7961, 1979.
10. **Lester, E. P., Lemkin, P., Lipkin, L., and Cooper, H. L.,** *Clin. Chem.,* 26, 1392, 1980.
11. **Zeineh, R. A., Nijm, W. P., and Al-Azzawi, F. H.,** *Am. Lab.,* 7, 51, 1975.
12. **Zeineh, R. A.,** *Am. Biotech. Lab.,* 4, 25, 1986.
13. **Bours, J.,** *Interdiscip. Top. Gerontol.,* 12, 196, 1978.
14. **Hsu, M. Y., Jaskoll, T. F., Unakar, N. J., and Bekhor, I.,** *Exp. Eye Res.,* 44, in press.

Chapter 10

NONHISTONE PHOSPHOPROTEINS WITHIN THE REALM OF NUCLEAR STRUCTURE AND FUNCTION: AN OVERVIEW

M. J. Halikowski and C. C. Liew

TABLE OF CONTENTS

I. INTRODUCTION

The regulatory genetic mechanisms in eukaryotic organisms in contrast to prokaryotic organisms is far more difficult to elucidate. Unlike their prokaryotic counterparts, the eukaryotic genome has been demonstrated to be extremely complex. This dynamic nucleoprotein structure is composed of DNA, histones, and nonhistone proteins.[1-4] In recent years, the major role of histones in packaging DNA into discrete nucleosomes has been well established.[5-7] However, the precise role of the nonhistone chromatin proteins (NHCP) in general has remained a more challenging problem due in large part to the extreme heterogeneity of this fraction, the relative insolubility of many of its components, and the lack of *in vitro* functional assays which would reflect *in vivo* conditions.[8-10] Despite these limitations, progress has been made in demonstrating that NHCP do indeed take part in several diverse functions such as chromatin packaging, transcriptional regulation, and various enzymatic activities.[11-14] The NHCP fraction, like its histone counterpart, is subject to numerous post-transcriptional modifications which may be involved in the regulation of its effects on chromatin structure and gene activity.[15-18] Of these numerous modifications, the most frequently encountered by far is phosphorylation. The protein-bound phosphate content of the NHCP fraction accounts for four to five phosphorylated amino acids for every 100 amino acids present giving this protein fraction the highest phosphate content of any protein component in the cell nucleus.[11,19]

These observations have resulted in the isolation and identification of individually phosphorylated NHCP in order that their specific role within the realm of nuclear organization and/or function can be elucidated. Investigations into NHCPs has shown that indeed a large variety of these proteins are phosphorylated and thus the potential diversity of function within this fraction is probably more complex than previously thought.[19] In earlier work, these phosphorylated proteins were considered to be associated with transcriptional regulation.[8] Subsequently, numerous studies have revealed that phosphorylated NHCP could also be associated with hnRNP (heterogeneous nuclear ribonucleoprotein) particles[20] and the nuclear envelope lamina.[21,22] Therefore, phosphorylation of NHCP has been inferred to involve several regulatory processes in gene expression, RNA transcription, transport of gene products, RNP assembly, cell cycle control, and nuclear substructure regulation. This review will attempt to provide an overview of the properties of several of these groups of phosphorylated NHCP (since the review of Olson[17]). We will also try to tie in the possible functional significance of each phosphoprotein within the realm of the nucleus.

II. GENERAL PROPERTIES OF NONHISTONE PHOSPHOPROTEINS

A. Tissue and Species Specificity

During the late 1960s, a large body of evidence was accumulated to demonstrate the presence of a highly heterogeneous population of phosphoproteins in the nucleus of various cell types under a wide range of physiological conditions. Due to their characteristics in heterogeneity, species specificity, and accessibility to covalent modifications such as phosphorylation, some putative role of NHCPs in the regulation of gene expression was inferred,[23] based upon the following observations: (1) changes in nuclear protein phosphorylation has been demonstrated during differentiation in perinatal rat hepatocytes[24] and in crown gall tumor;[25] (2) changes in specific DNA binding of NHC phosphoproteins have been identified in bovine lymphocytes[26] and in *Physarum polycephalum*;[27] (3) changes in phosphorylation state of nuclear proteins during the cell cycle in HeLa[28] and *Xenopus laevis*;[29] (4) phosphorylation of NHCP can be altered by physiological conditions as seen in aging rat liver[15] and regenerating rat liver;[2] (5) there are differences in the content of phosphoproteins as revealed in Morris hepatoma, Walker tumor, and Novikoff hepatoma;[30,31] and (6) phosphorylation can be induced by hormone administration as demonstrated in rat liver (T3 and glucocorticoid)[32,33] and kidney[34] and by chemical agents such as heparin (pig testis)[35] and phorbol diester (HL-60 leukemic cells).[36]

With the advent of the two-dimensional gel electrophoresis, the NHCPs have been shown to comprise several hundred distinct species in marked contrast to the evolutionary conserved histones.[37,38] The use of developing tissue systems (e.g., sea urchin embryos, rat hepatocytes, etc.) has been especially instrumental in demonstrating this tissue specificity.[9] These proteins may indeed serve as either repressor or activators of specific genes, and phosphorylation may act as a "fine" control of this system. However, the tissue specificity and specific phosphorylation of NHCP are only suggestive of their role in gene regulation and does not provide conclusive support of this theory.

B. NHCP and the Cell Cycle

The role of nuclear proteins in various stages of the cell cycle has largely been centered around the determination of phosphorylation with changes in the rates of transcription. In particular, phosphorylation events occurring during mitosis (when RNA synthesis is greatly reduced), late G1 and early S phases (when histone mRNA synthesis is maximal) have to date been the most extensively investigated. Indeed, most studies have concluded that the rate of protein phosphorylation is highest when the rate of RNA synthesis is maximal.[39] Studies are now being focused on the participation of individual phosphoprotein species since it has also been demonstrated that some of these proteins lose phosphate much more rapidly than others, according to the particular phase of the cell cycle.

A strong correlation between phosphorylation of histone H1 and H3 and the onset of chromosome condensation and mitotic initiation has been demonstrated.[40,41] However, it is generally believed that H1 superphosphorylation is by itself not sufficient to bring about chromosomal condensation since phosphorylation can occur without any condensation.[42] Attention has now shifted to the role of NHCP phosphorylation in this scheme of mitotic related events. For example, phosphorylation of HMG proteins has been suggested to be responsible for decreased gene transcription during mitosis.[43] In addition, increased intermediate filament protein (i.e., vimentin) phosphorylation[44] as well as phosphorylation/dephosphorylation of lamina proteins[45] has been implicated in the dissolution/reformation of the nuclear envelope. Phosphorylation of ribosomal protein S6, along with increased phosphorylation of other NHCP (i.e., maturation promoting factor) have been shown to be associated with meiotic maturation of *Xenopus luevis* oocytes.[46] More recent work has demonstrated that phosphorylation/dephosphorylation is intimately involved during oocyte maturation brought about by HeLa cell mitotic factors.[29] Those phosphorylated NHCP that are extractable with 0.2 M NaCl are in fact casually related with the entry of cells into mitosis while their selected dephosphorylation occurs during exit from mitosis. Specific monoclonal antibodies have been shown to react with these phosphoproteins only during mitosis while during G1 and S they are dephosphorylated and lose antigenicity.

During meiosis of amphibian oocytes, another major change in nuclear protein phosphorylation has been demonstrated.[47] Earlier studies revealed a two- to threefold increase in ^{32}P phosphate incorporation into total nuclear protein with most of it localized in one nuclear protein, nucleoplasmin. This is the most abundant protein of the oocyte nucleoplasm (5 to 8 mg/ml in the nucleus) and this level of phosphorylation represents a net increase in nuclear phosphate content approximately 6000 times greater than the total phosphate content of the oocytes chromosomal DNA. Such a level of phosphorylation associated with the abundance of nucleoplasmin in vertebrate somatic nuclei suggests that phosphorylation of this protein may play some key role during the mitotic cycle.

C. DNA Binding of NHC Phosphoproteins

Since NHCPs have been postulated to play some role in this selective transcriptional control, it is likely that they do so by means of their association with the DNA itself. This has been particularly well documented in prokaryotes (e.g., lac repressor inhibits lac operon transcription).[48] Earlier studies demonstrated that the greatest proportion of DNA-binding proteins were phosphorylated and, in general, this heterogeneous population of phosphoproteins accounted for approximately 1% of the total NHCP in chromatin.[49] The characterization of individual DNA binding NHC phosphoproteins was first analyzed using rat liver nuclear proteins which were soluble at low ionic strength. Several of these (designated B33, C18, CN′) were in fact found to be phosphorylated.[50] Bluthmann[26] has characterized a DNA binding phosphoprotein (M_r: 30 kDa) from bovine lymphocytes which exhibits preferential binding to lymphocyte DNA as opposed to *E. coli* DNA. However, this protein has since been shown to recognize DNA with a low degree of selectivity, binding to as much as 30% of lymphocyte DNA (one binding site: 1700 base pairs). This specificity together with the high content of acidic amino acids led to further speculation that this protein may in fact be a member of the HMG family. However N-terminal analysis has demonstrated an arginine residue ruling this out as a possibility.[17]

Though DNA-binding proteins have long been suspected of being involved in the regulation of gene transcription, they may also play a role in DNA replication and recombination. DNA-binding phosphoproteins which may play such roles have recently been characterized from mouse ascites cells, rat spermatocytes, and animal virus infected cells.[51,52] Phosphorylation of a DNA-binding protein from mouse ascites cells reduces its binding to single-stranded DNA but not duplex DNA and abolishes the ability of this protein to stimulate DNA polymerase activity (suggestive of a "helix-destabilizing" role). Similarly a DNA-binding protein from adenovirus and the T antigen of SV 40 (both necessary in the initiation of DNA synthesis) have been found to be phosphoproteins.[53] When protein kinase is added to cultured rat liver cells, DNA synthesis is stimulated (and subsequently abolished if kinase inhibitor is present) suggesting that specific phosphorylation of DNA binding proteins may play a key role in the initiation of DNA synthesis.[54,55] Rat spermatocyte meiotic cells have also been shown to contain a phosphorylatable helix-destabilizing DNA-binding protein which is present mainly between the S-phase of the meiotic cell cycle and the termination of chromosome pairing.[27] Its high level of phosphorylation during this period suggests a possible function in genetic recombination.

An alternative approach to the characterization of specific DNA:NHC phosphoprotein complexes formed *in vivo* has been developed by Bekhor and co-workers.[56,57] Following high salt extraction of chromatin and ultracentrifugation, a nucleoprotein complex containing 1% of the total nuclear DNA complexed to a specific NHCP fraction has been isolated. In almost all tissues investigated, the DNA in this complex is enriched 20-fold in transcribed sequences. In addition, some of these proteins have been shown to be highly phosphorylated.[2] Since these proteins bind with such high affinity to specific DNA sequences, rearrangement during isolation is unlikely, and this suggests the possibility of some kind of regulatory role.

D. Effects of Hormones (and Other Chemicals)

It is now well documented that many hormones and drugs alter the expression of specific genes in their target tissues. A correlation between quantitative changes in NHCP and/or their phosphorylation and increases in transcription induced by the target agents[58] has been reported. This level of phosphorylation affects a broad range of NHCPs and strongly suggests a complex control mechanism involving activation of several genes and/or processes within the nucleus. However, controversy exists concerning the phosphorylation of NHCPs and their involvement at the pretranscriptional level under the influence of hormone activation. Earlier studies by Klyzejko-Stefanowicz et al.[59] showed that the binding of 5 α-dihydroxytestosterone complex of rat prostate chromatin was enhanced by phosphorylation of chromatin, while conversely Nielsen et al.[60] have reported that the glucocorticoid receptor protein from mouse fibroblasts and rat liver is inactivated by incubation with alkaline phosphatase.

With the development of newer receptor purification methods, it has been shown conclusively that steroid receptors themselves are phosphoproteins and, when unoccupied, reside primarily in the nucleus. The purified avian progesterone receptor (subunit A and B) can also be readily phosphorylated by a cAMP dependent protein kinase *in vitro.*[55] Additional *in vivo* studies utilizing avian progesterone (oviduct tissue slices) or glucocorticoid receptors (mouse fibroblast L cells) following incubation with ^{32}P clearly demonstrated receptor subunit phosphorylation following autoradiography.[61] The question remains, however, as to what role this phosphorylation may play. Evidence now points to receptor phosphorylation as a necessary precondition for steroid binding. Migliaccio et al.[62] have shown that pretreatment of purified estrogen receptors from calf uterus with a nuclear phosphatase abolished estrogen receptor binding ability which could be restored later by treatment with protein kinase. It now appears that protein kinase(s) themselves may be an inherent component of the receptor complex.[63] In the future, efforts will be centered upon the elucidation of what role phosphorylation/dephosphorylation plays in receptor-hormone mediated binding at nuclear sites (where the complex may act to either promote or inhibit specific gene transcription).

III. DISTRIBUTION AND CHARACTERISTICS OF NONHISTONE PHOSPHOPROTEINS

In the late 1960s, most investigators defined chromatin components into either active or inactive fractions and attempts were then made to identify those nonhistone proteins in relation to specific gene expression.[1] The further establishment of the nucleosome structure and the more detailed analysis of nuclear components have also revised our recent approaches in examining those nonhistone phosphoproteins involved in selective aspects of gene expression. The following discussion is thus based on the location and characteristics of nonhistone phosphoproteins in the subnuclear structure with emphasis on their possible functional significance (listed in Table 1).

A. Nucleosome Associated
1. HMG Proteins
The high mobility group (HMG) proteins are among the most extensively characterized group of NHCPs.[64,65] They have been readily identified on the basis of their high solubility in mineral acids (i.e., 2% TCA, PCA), their lack of tissue or species specificity, and their high concentration in the nucleus. Four main proteins (HMG-1, 2, 14, and 17) are present in isolated nucleosomes. Attention has grown in recent years over the selective localization of HMG proteins in regions of "active" genes (DNase I-sensitive regions of chromatin) suggesting their participation in some facet of selective gene control.[66] However, these findings have also raised doubts due to the fact that HMG protein release has not always been carefully quantified along with the tremendous difficulty in comparing HMGs from different organisms.[8]

HMG proteins have been shown to undergo a large variety of post-synthetic modifications including acetylation, methylation, phosphorylation, and ADP ribosylation.[67] In particular, phosphorylation has come under very close scrutiny due in large part to its suspected role in altering some HMG protein interaction with DNA and/or other chromatin proteins.[43] Earlier studies using Ehrlich ascites cells clearly showed that ^{32}P phosphate incorporation occurred in HMG 14 and 17, but not HMG 1 and 2.[68] In addition, the overall kinetics of labeling and turnover rates were close to that of histone H1. Studies utilizing CHO cells have also shown preferential phosphorylation of HMG-17 during interphase (80%).[82] However, during metaphase, this level of HMG-17 phosphorylation fell to half of the interphase value suggesting a possible cell-cycle dependent phosphorylation. It now appears that specific protein kinases are involved in this scheme of phosphorylation. Nuclear cAMP independent

Table 1
SPECIFIC PHOSPHOPROTEINS IDENTIFIED IN NUCLEAR ORGANIZATION AND/OR FUNCTION

Species	Comments	Ref.
Nucleosome associated		
High mobility group proteins (HMG 14 and 17)	HMG 14 and 17 bind to two specific sites on the core nucleosome (i.e., 160 bp DNA). May play key role in chromosome condensation and/or dispersal prior to mitosis	65—68
P1 (HMG-like)	M_r: 48—53,000, highly acidic pH: 5.0. Highly phosphorylated in a variety of proliferating and nonproliferating mammalian cells. May bind to satellite DNA sequences	69
B$_2$ phosphoprotein	M_r: 68,000, pI: 6.5—8.2. Associated with mononucleosomes of actively transcribed chromatin as well as a tightly bound fraction associated with the nuclear matrix	70—84
a (p41), *b* (p31)	*a* associated with monomer particles from hepatoma cells while *b* is present in larger dimer and trimer particles (possibly linker regions). Specific role of each unknown	71
D-55	D-55 (55 kDa) associated with nucleosomes from calf thymus. Phosphorylation abolishes any increase in transcription brought about by initial D-55 binding to DNA	72
A-24	Conjugate protein formed by linkage of carboxy terminus of ubiquitin to H2A. Found to be phosphorylated in Novikoff hepatoma cells on serine sequence of H2A but role unknown	73
Nucleolus associated		
C23, B23	C23 (M_r: 100 kDa), B23 (M_r: 40 kDa) both are highly phosphorylated and may be involved in the increased synthesis and output of ribosomes	74
Nuclear envelope/lamina associated		
Lamina A, B, and C phosphorylation	Lamina protein associated with the disassembly of interphase nuclear structure nucleus prior to the onset of mitosis	21, 45
RNP particle associated		
40S particle	M_r: 30—40 kDa, phosphorylation may play role in association with and/or processing of hnRNA	75
Low M_r chromatin associated proteins		
p11 (I-NHP) p29	p11 (Ehrlich ascites tumor chromatin) p29 (calf thymus chromatin) bind to DNA inhibiting transcription at the point of RNA chain initiation	17
p13	p13 (mouse spleen chromatin) phosphorylation stimulated by binding to DNA but not core particle. Specific role unknown	17
Enzymes		
DNA topoisomerase I and II	Nicking-closing enzymes involved in DNA relaxation and reassociation. In most cases phosphorylation results in stimulation of activity.	76
DNA polymerase α	Phosphorylation may be involved in the modulation of activity during different stages of cell cycle	77
RNA polymerase II	Associated with core particle from transcribed genes. Phosphorylation believed to enhance activity *in vivo*	78
Poly(A) polymerase	Phosphorylation enhances addition of poly(A) to newly processed mRNA	79
Nuclear oncogene products		
c-myc, c-myb, c-fos, c-ski, p53	All appear to be intimately involved in cellular transformation and are phosphorylated but their specific functional roles are unknown	80, 81

protein kinase NII appears to preferentially phosphorylate HMG-17 (and to a lesser extent HMG-14), while a cytoplasmic kinase phosphorylates HMG 1 and 2 to a higher degree than HMG 14 and 17.[43]

The precise functional role of HMG phosphorylation is currently not known. However, recent studies mapping the phosphorylation sites of HMG-14 have shed some light on this.[83] When HMG-14 was phosphorylated by c-GMP dependent kinase at serine-6, it eluted from a ssDNA-agarose column at a lower ionic strength (i.e., 0.27 M NaCl) as compared to the unphosphorylated native protein (i.e., 0.3 M NaCl). Conversely, when HMG-14 was incubated with nuclear protein kinase II which catalyzes phosphorylation of serine-89 (i.e., carboxyl-terminal region), the binding to DNA was not effected. Thus, it appears that various regions of HMG-14 have different functions, with the basic amino terminal region possessing DNA-binding properties, with phosphorylation in this region possibly playing some critical role in chromosome condensation and/or dispersal.

2. Other Phosphoproteins

The presence of phosphorylated NHCP associated with the particulate subunit of chromatin, the nucleosome, has been known for some time. However, the specific cause/effect functional relationship which their interaction with this unit may have is yet to be fully elucidated. Liew and Chan[84] were among the first to demonstrate the presence of specific nonhistone phosphoproteins associated with mononucleosomes from nuclease-digested rat liver chromatin. One of these proteins designated B_2 (M_r: 68 kDa; pI: 6.5 to 8.2) has been purified to homogeneity.[70] Its amino acid composition has been shown to possess a nearly one-to-one ratio of acidic-to-basic amino acids with nearly half of its amino acids containing charged side chains. As such, this may confer both protein- and DNA-binding capability. This type of pattern appears to be a common feature of most nuclear phosphoproteins isolated thus far.

Defer et al.[71] have also demonstrated the presence of phosphorylated nonhistone proteins with mononucleosomes of rat liver nuclei. Further analysis using a hepatoma cell line revealed the presence of two main polypeptides a (M_r: 41 kDa) and b (M_r: 31 kDa) which differed in their relative distribution on nucleosome particles. Phosphoprotein a was found to be present on monomer particles released by micrococcal nuclease, while phosphoprotein b was present in larger monomers and dimers suggesting an association of these proteins with different regions of chromatin and thus possibly possessing different functional and/or organizational roles within the nucleus.

Safer and Coleman[72] have also isolated and partially characterized a phosphorylated nucleosomal binding protein from calf thymus designated D-55 (M_r: 55 kDa). The unphosphorylated form of this protein has been shown to bind to DNA, histones, and nucleosomes. Upon phosphorylation, D-55 binding to histones and nucleosomes has been shown to be enhanced, while its DNA-binding properties were largely unaffected. D-55 by itself upon nucleosomal binding enhances transcription of nucleosomal DNA some 100-fold but this effect can be completely abolished by its selective phosphorylation. Even though a specific role for this protein is not known at this time, it appears likely that it plays some role in facilitating transcription.

B. Nucleolar Phosphoproteins

The nucleolus has long been known to be the site of synthesis of ribosomal rRNA as well as the location for ribosome precursor assembly and processing.[85,86] The nucleolus contains several distinct classes of proteins. Among these are histones, nonhistones (from chromatin), and ribosomal proteins (from preribosomal particles). In addition, several distinctive nucleolar enzymes are present including RNA polymerase I, methylases (which act upon preribosomal RNA), and specific preribosomal RNA processing nucleases.[87]

Using two-dimensional electrophoresis, almost 40 nucleolar phosphoprotein fractions have been analyzed under *in vivo* labeling conditions.[88,89] Two polypeptides designated B23 (M_r: 40 kDa) and C23 (M_r: 100 kDa) are particularly highly phosphorylated. These two proteins were found to be components of the nucleolar preribosomal particles, suggesting that their phosphorylation may be intimately involved in the maturation process of these particles.

Nucleolar chromatin itself has been analyzed for the presence of phosphoproteins with a protein designated C18 being the major phosphorylated species.[88] In addition, C18 appears to be localized primarily in the nucleolus and to possess physical and chemical characteristics identical to protein C23. This in itself suggests a dual form for C18/C23 with one form tightly bound to chromatin while the other nonchromatin component complexes itself in part with RNA (perhaps aiding in transport and/or processing).

Tsutsui and Oda[90] have purified a high molecular weight (p110, M_r: 110 kDa) acid-soluble phosphoprotein from mouse ascites sarcoma cell nucleoli. It has been recently found to be present in several cell types including Ehrlich ascites, mouse leukemias, and mouse/rat liver. Interestingly, it has been shown that p110 levels are highest in more rapidly growing tissues, and it appears on the basis of pI and amino acid composition to be similar to C23. This phosphoprotein can form a complex of 280 kDa with a nonphosphoprotein component (M_r: 32 kDa),[91] and this protein complex has been found to preferentially bind to nucleolar DNA nucleosome complexes and to possess protein kinase activity capable of *in vitro* autophosphorylation. Phosphorylation via this process occurs primarily on serine but to a smaller degree on tyrosine residues. This holoenzyme can be immunoprecipitated using certain oncogene protein antibodies suggesting a possible role in transformation.

The study of nucleolar protein phosphorylation has taken on more attention recently due to the fact that their phosphorylation has been shown to vary with the physiological state of the cells from which they were derived. In the case of Novikoff hepatoma cells compared to normal rat liver, the uptake of ^{32}P into total acid soluble nucleolar proteins *in vivo* and *in vitro* was at least twofold greater, with C23 showing the greatest difference.[92] The use of the regenerating liver model has also demonstrated a correlation of phosphorylation changes in the nucleolar protein fraction *in vitro*. Uptake of ^{32}P into total acid soluble nucleolar proteins was found to increase 22 times after 24 h liver regeneration (with C23 again accounting for most of this difference).[85] The dynamic nature of this regeneration system is reflected by the fact that several other proteins increased in ^{32}P uptake, while others decreased. These increases in ^{32}P uptake during proliferation have also been shown to correlate with increases in protein kinase activity found in regenerating liver and Novikoff hepatoma nucleoli.

Despite the fact that the functional significance of these phosphorylation changes remains unclear, there does seem to be a general correlation between the level of phosphorylation of nucleolar proteins, enhanced levels of RNA Polymerase I, and the rate of output of ribosomes.

C. Nuclear Matrix Phosphoproteins

Interest has been growing in recent years in a particular nucleoskeletal complex that is obtained by extracting nuclei with high salts and nucleases (known alternatively as the nuclear cage or nuclear protein matrix). Its potential structural role in attaching chromatin at various points in the nuclear architecture has now been associated with a large number of functional roles ranging from DNA replication, post-translational RNA processing, and intranuclear translocation.[93-95] It is because of these diverse functions that the potential role of phosphorylation on this particular class of NHCP has been actively investigated. Allen et al.,[96] using the regenerating rat liver system as a model, have shown that ^{32}P incorporation into nuclear matrix proteins was two- to threefold higher than that of control livers. In addition, 12 h following partial hepatectomy, nuclear matrix protein phosphorylation reached a maximum level. Since the onset of DNA synthesis occurs 16 to 18 h after partial hepatectomy, it is quite conceivable that this matrix phosphorylation may play some key role in gene regulation and/or onset of DNA synthesis.

Sevaljevic et al.[97] have in addition shown that nuclear matrix protein phosphorylation may play a central role in cellular differentiation. Using sea urchin embryos, they were able to demonstrate that more phosphorylation occurred in the earlier blastula than pluteus stage, suggesting that matrix protein phosphorylation may act as a trigger "signal" to initiate further cell division.

Additional cell cycle studies in HeLa cells have shown an increase in phosphorylation of nuclear matrix proteins in S phase and premitosis, with a lower level of phosphorylation in G1 phase. Song and Adolph[28] have further defined the components of this system and identified a protein of molecular weight 119 kDa as the major phosphorylated species in the matrix of dividing cells. Henry and Lodge[98] were also able to identify phosphoserine as the major phosphorylated amino acid in this fraction together with significant levels of phosphothreonine and phosphotyrosine, and suggested an inherent tyrosine protein kinase activity in the matrix (which has recently been substantiated[99]). These findings all point to the relative complexity of nuclear matrix protein phosphorylation/dephosphorylation as cells prepare for and go through mitosis.

D. Nuclear Envelope Phosphoproteins

The nuclear envelope has been shown to consist of two concentric membranes separated by a 40- to 60-nm perinuclear space.[22,100] Both of these membranes and the perinuclear space are transversed by so-called nuclear pore complexes which itself is embedded in a proteinaceous detergent and salt insoluble lamina which acts to "anchor" the inner nuclear membrane with the peripheral chromatin. This structural lamina consists of three predominant polypeptides designated lamina A, B, and C, having molecular weights of 70, 67, and 60 kDa, respectively.[21,101] As the nuclear envelope disappears during mitosis, there is a concurrent reversible disassembly of the lamina complex. It has now been documented that during interphase the lamina remains in a polymerized form but becomes monomeric during mitosis with lamina A and C dissociating from the membranes and becoming soluble. Attention has been centered upon the precise role of phosphorylation in this sequence of events. Gerace and Blobel[45] have shown that the incorporation of ^{32}P into lamina A and C was three times more during interphase, while lamina B was six times more. In addition, during early G1 lamina, phosphorylation is intermediate between mitosis and S-G2 (exponentially growing cells) suggestive of the role of phosphorylation in regulation of envelope disassembly.

More recent work has demonstrated that the phosphorylation of various nuclear envelope polypeptides may be under the control of specific protein kinases present in the membrane itself. Kasper and Lam[102] demonstrated the phosphorylation of lamina B when purified envelope was incubated with γ-^{32}P ATP. Despite the proposed role of lamina B phosphorylation in nuclear disassembly, they suggest that lamina B and its associated kinase activity may play some role in nucleocytoplasmic transport. Nuclear envelope associated protein kinase activity has also been demonstrated by several other groups[103,104] which again points out the relative complexity and dynamic nature of this substructure.

E. Nuclear RNP Particle Proteins

Ribonucleoproteins associated with heterogeneous nuclear RNA (hnRNA) to form ribonucleoprotein complexes (hnRNP), have long been believed to be involved in the packaging and processing of hnRNA.[105,106] Based on morphological and sedimentation studies of material release from nuclei by mild sonication or RNase digestion, current models of hnRNP structure are believed to be a "beads on a string" structure with the 40S particles forming these beads. Evidence now suggests that the 40S particle is associated with sequence specific hnRNA.[107] Biochemical analysis has demonstrated that the major components of 40S particles are proteins with molecular weights in the 30 to 40 kDa range classified as A, B, and C.[105,108] In addition, other proteins can be isolated *in vitro* with hnRNA but are not present on the 40S particle. These have included a protein which is preferentially associated with polyadenylate, poly(A).[109,110]

Previous work demonstrated the presence of phosphorylated proteins in rat brain hnRNP particles.[111] Labeling with ^{32}P has revealed that the radioactivity was associated with predominantly two polypeptides in the 30 to 40 kDa molecular weight range. The preferential phosphorylation of four 40S hnRNA proteins from HeLa cells[75] has also been shown. In addition, endogenous protein kinase and phosphoprotein phosphatase activities were found associated with these particles.[112] The endogenous kinase activity (cAMP independent) was found to preferentially phosphorylate two proteins of M_r: 28 kDa and 37 kDa, respectively. Schweiger and Schmidt[113] have also noted that when rat liver 30S nuclear RNA particles are incubated with γ-^{32}P-ATP a M_r: 39 kDa polypeptide can be identified with RNA binding ability.

Additional work has now been carried out on rat liver 40S RNP phosphorylation. Two proteins of the designated C class (C1, M_r: 42 kDa and C2, M_r: 44 kDa) have been shown to be phosphorylated both *in vivo* and *in vitro*.[106] Among the six major proteins in the 40S particle, the two C group proteins are the most tightly bound to RNA, suggesting that phosphorylation may play some role in their association with and possibly processing of hnRNA. By the use of an exogenous kinase, rat liver 40S RNP complexes were studied before and after treatment with increasing concentrations of NaCl.[114] The phosphorylation pattern prior to the increase in NaCl concentrations showed a major group of labeled proteins in the 30- to 40-kDa range. As the salt concentration was raised (i.e., up to 1.2 *M*), polypeptides 37, 33, 26, 25, and 17 kDa remained in the complex, suggesting that their tight association with the hnRNA chain forms the latent backbone of the RNP particle. Phosphorylated polypeptides 30 and 27 kDa were, however, released by the increasing NaCl concentration suggestive of their presence on the outside of the particle (exhibiting a weak binding property).

In regions of transcriptionally active hnRNA gene loci (i.e., salivary gland cells) the phosphorylation of hnRNPs 42, 33, 30, and 25 kDa could be directly correlated with this transcriptional activity.[101] Consequently, when hnRNP genes were inactivated at the level of chain initiation by nucleoside analogues, the transcriptional block coincided in time with the inhibition of phosphorylation of these same hnRNPs. These results suggest that phosphorylation of hnRNPs may play some crucial role in the initiation of hnRNA synthesis and perhaps turnover.

Small nuclear ribonucleoprotein particles (U-snRNPs) and their associated small nuclear RNAs (U-snRNAs) have recently received increasing attention due to their potential role in processing and/or transport of hnRNA.[20] Six U-snRNA species have been localized in the nucleoplasm, are uridylic acid-rich, and are designated U1 to U6 snRNAs. During interphase, U-snRNPs are associated with the nuclear matrix, while the onset of mitosis results in a breakdown of these nuclear protein clusters and their redistribution between condensing chromosomes. Upon completion of mitosis with chromosome decondensation cluster formation occurs. Phosphorylation of snRNPs appears to parallel the phosphorylation of the lamina proteins with higher levels of ^{32}P incorporation occurring during mitosis while dephosphorylation triggers the return to the interphase state.[20] However, the actual role of this modification in the disassembly/reassembly of snRNPs is currently unknown.

F. Specific Nuclear Enzymes

Along with the numerous structural and regulatory NHC phosphoproteins, a variety of enzymes whose activity can be regulated by phosphorylation/dephosphorylation have also been characterized. A few of the more pertinent enzymes which catalyze nucleic acid synthesis and post-synthetic modifications of nucleic acids and proteins will be outlined.

1. RNA Polymerases

Three distinct RNA polymerases are present in nuclei. RNA polymerase I is involved in the synthesis of large ribosomal RNA precursors; RNA polymerase II synthesizes hnRNA and some small nuclear RNAs; and RNA polymerase III synthesizes transfer RNA precursor and 5S ribosomal RNA.[78] The actual process involved in synthesizing any RNA molecule is highly complex involving several stages, including correct initiation, elongation, and termination. One of the limitations involved in studying the effects of phosphorylation on any of these enzymes is the fact that most of the assay systems used are nonselective for aberrant types of RNA synthesis that may have predominated.[18]

The possibility of regulation of eukaryotic RNA polymerases via phosphorylation/dephosphorylation has been suggested by the findings that all three RNA polymerases from various sources are phosphorylated.[78,115] For example, in yeast, all three RNA polymerases are phosphorylated *in vivo*. However, in rat liver, when isolated nuclei were incubated with γ-^{32}P-ATP,[116] a localization of ^{32}P on serine and threonine was found on nucleolar RNA polymerase. Most phosphorylation studies to date have been extended to systems where isolated RNA polymerase is phosphorylated by exogenous protein kinase(s) and the relative effect on enzyme activity measured. Kranias et al.[117] have shown phosphorylation of a M_r: 25 kDa subunit of calf thymus RNA polymerase II using a homologous nuclear cAMP-dependent protein kinase. This resulting phosphorylation resulted in a stimulation of activity by as much as threefold over the control, while dephosphorylation of the polymerase by *E. coli* alkaline phosphatase lowered activity. Dahmus[118] has also observed a five- to sevenfold stimulation of Novikoff hepatoma ascites cell RNA polymerase II due to phosphorylation by a protein factor with kinase activity. Recent work has also demonstrated the activation of RNA polymerase II in wheat embryo following phosphorylation of a M_r: 220 kDa subunit.[120] This phosphorylation could occur *in vitro* when the polymerase was incubated with a homologous kinase preparation or *in vivo* at the onset of germination. In fact, it has been hypothesized that some RNA polymerases may contain endogenous protein kinases among their subunits which may act to regulate the entire holoenzyme. In fact, Rose et al.[120] have copurified casein kinase II with yeast RNA polymerase I. However, this finding has been challenged by others.[121]

Despite the suggestive role between *in vitro* RNA polymerase phosphorylation and enzyme activation, the precise functional role *in vivo* for this modification remains unclear as is whether the *in vitro* results may parallel the actual *in vivo* situation.

2. DNA Polymerase

Most mammalian cells have been shown to contain multiple forms of DNA polymerases. Three distinct polymerases (α, β, γ) have been extensively characterized with respect to size, template specificity, and immunological probing. From these investigations, it has been demonstrated that DNA polymerase α is the major fraction of cellular DNA polymerase activity. Its main role is to incorporate deoxyribonucleoside monophosphate into both leading and lagging strands of the developing replication fork. Studies dealing with the effects of post-translation modifications (i.e., phosphorylation) on α-polymerase activity have not been extensively carried out due to its low abundance and also its tendency to aggregate in large oligoenzyme complexes. Earlier studies[122] using avian myeloblastosis virus DNA polymerase α did show that when a phosphoprotein of the complex was in the phosphorylated form, the activity of the enzyme itself could increase by as much as tenfold (over the effect brought about by the unphosphorylated form). However, as to whether this protein was a subunit of the enzyme itself was not fully investigated. More recently, phosphorylation studies on the polymerase enzyme itself have been carried out.[77] When rat embryonic fibroblast DNA polymerase α was isolated and incubated with alkaline phosphatase, DNA polymerase activity was almost completely inhibited. However, upon incubation with ATP, DNA polymerizing activity could be restored suggesting that phosphorylation/dephosphorylation could serve to modulate the activity of this polymerase, perhaps during different stages of the cell cycle. The α polymerase phosphorylation has recently been demonstrated in human KB cells[123] suggesting this modification may be a universal control mechanism in eukaryotic replication systems.

3. Poly(A) Polymerase

Along with the regulatory and enzymatic proteins associated with DNA transcription and replication, several other nuclear enzymes and proteins have been shown to be phosphorylated by protein kinase(s). One of these is chromatin associated poly(A) polymerase which is known to be involved in the post-transcriptional addition of 100 to 200 poly(A) residues to newly synthesized RNA chains.[19] Rose et al.[79] have demonstrated that protein kinase-mediated phosphorylation of purified poly(A) polymerase is associated with activation of the enzyme with the degree of activation dependent upon the amount of protein kinase added to the assay. Phosphorylation of poly(A) polymerase increased the affinity of the enzyme for its RNA substrate thereby stabilizing the messenger RNA and enhancing the rate of poly(A) synthesis. It is particularly interesting to note that the poly(A) polymerase of rapidly growing liver tumors has now been shown to be more highly phosphorylated *in vivo* than the enzyme of normal liver, suggesting some role for it in the neoplastic state.[124]

4. Histone Deacetylase

Histone deacetylase first isolated from calf thymus by Vidali et al.[125] catalyzes the hydrolysis of the ε-N-acetylysine linkage on histones 2A and 3. The isolated protein (M_r: 160 kDa) has an acidic isoelectric point (pI: 4.5) and contains 1.3% phosphorous by weight suggestive of multiple phosphorylation sites. Currently, the precise relationship between its degree of phosphorylation and enzymatic activity is not known.

5. Topoisomerases

DNA topoisomerases catalyze the *in vitro* conversion of one DNA topological isomer to another. These include relaxation of superhelical DNA, reassociation of complementary single-stranded DNA circles, knotting/unknotting, and catenation/decatenation of circular DNA.[76,126] These reactions have all been explained on the basis of a transient breakage of phosphodiester bonds with topoisomerase I catalyzing transient single-strand breaks while type II catalyzes transient double-strand breaks. These enzymes appear to be ubiquitous in eukaryotic organisms and in most cases are found together in the nucleus. However, the role of topoisomerases in *in vivo* systems has been difficult to clarify. Topoisomerase I appears to play some indirect role in DNA replication, transcription, and transposition, while type II topoisomerases appear to be more involved in DNA replication (as to which specific stage of replication, e.g., DNA chain elongation, termination, etc. is not known).[127]

Despite the relative confusion regarding specific functions, it is believed that regulation of topoisomerase enzyme activity may result in alterations of gene expression. Phosphorylation of topoisomerases by protein kinases may in fact represent one such regulatory mechanism. In support of this theory, it has been shown that in Novikoff ascites cells, phosphorylation results in an increase in the activity of topoisomerase I[128] while, conversely, in calf thymus, tyrosine phosphorylation by pp60[src] brings about topoisomerase I inhibition.[129] In the case of topoisomerase II, it has been shown that purified preparations of this enzyme from Drosophila contains an endogenous protein kinase and the enzyme itself occurs as a phosphoprotein.[130] Topoisomerase II is also subject to serine phosphorylation by casein kinase II leading to a threefold increase in enzyme activity.[131]

More interestingly, it has recently been shown that Drosophila DNA topoisomerase II can be phosphorylated by subnanogram amounts of protein kinase.[132] This phosphorylation was present solely on serine residues and resulted in marked activation of topoisomerase II (based on unknotting and relaxation enzyme activities). In addition, if topoisomerase II activity was blocked by specific inhibitors, it was found that phorbol ester-induced differentiation (initiated by protein kinase C stimulation) of leukemic HL-60 cells was also blocked. This raises the possibility that phosphorylation of topoisomerase II may in fact be involved in the mediation of nuclear effects by one or more second messenger systems.

6. Ornithine Decarboxylase

Ornithine decarboxylase has been shown to catalyze the initial rate-limiting step in the synthesis of the polyamines spermidine and spermine in eukaryotes.[133] Its activity is clearly regulated by means of reversible phosphorylation and dephosphorylation. While phosphorylated, the enzyme is inactive, but the kinase required to accomplish this is inhibited by putrescine, an intermediate in the pathway forming polyamines. The polyamines themselves stimulate the kinase, thus providing feedback inhibition for the synthesis of polyamines.

Ornithine decarboxylase itself (when in the phosphorylated form) has been shown to stimulate transcription of the ribosomal RNA genes in *Physarum polycephalum*.[134] These genes are found on DNA molecules (rDNA-"mini" chromosomes) located in the nucleolus (approximately 60 kb). Each of these DNA molecules contains two copies of the genes for rRNA arranged palindromically near each end and interrupted with a large spacer region. It has been shown that the phosphorylated form of the enzyme can bind to this central spacer region and stimulate rRNA gene transcription but does not appear to do so by stimulating RNA polymerase I itself.[135]

7. Nuclear Protein Kinases

A large variety of different nuclear protein kinases are known, and currently they are being isolated and characterized with greater scrutiny. For histones alone, each species has been shown to have its own specific kinase (or kinases in the case of H_1 and H_4) while for nonhistones two major forms of nonhistone protein kinase (which are cyclic nucleotide independent) have been isolated and purified (i.e., NI and NII).[19,136] However, the exact number of different protein kinases present within the entire nucleus itself (including all associated nuclear subfractions) has been difficult to pinpoint, but at least 18 distinct types of enzymes have been categorized. It appears that this large number of essentially the same enzyme allows for the phosphorylation of different substrates to be "fine" controlled. Nearly all nuclear protein kinase preparations retain some endogenous substrate that can be phosphorylated upon the addition of ATP and Mg^{2+}. The possibility that some of this substrate could be the kinase itself has gained greater credence largely due to the fact that purified cytoplasmic protein kinases are capable of phosphorylating themselves as a means of regulating their activity.[137] Recently, a cyclic nucleotide independent serine-threonine protein kinase (molecular weight >100 kDa) has been purified from HeLa cell nuclei which can be separated into several subunit species in the range of 25 to 42 kDa.[138] Two of these major subunits are autophosphorylated. However, questions still remain concerning the exact effect phosphorylation has on the enzyme activity, its precise location in the nucleus (matrix or otherwise), and its functional role. Previous investigations have shown that protein kinase NII could be autophosphorylated.[121,139] Kruh et al.,[140] using a new procedure for the isolation of rat liver nuclear protein kinases NI and NII, demonstrated that a proteolytic degradation fragment of NII (M_r: 27 kDa) could be phosphorylated, and when phosphorylated, stimulates the activity of the NII holoenzyme complex by nearly 40%. Nevertheless, this autophosphorylation may be only one of several other mechanisms involved in the regulation of this enzyme which includes the interaction of other proteins with the holoenzyme and possibly the partial proteolysis of the native enzyme itself.

G. Nuclear Oncogene Products

It has been established that many oncogene proteins are found in association with the plasma membrane (i.e., ras, abl, erb-B, etc.) and appear to affect the interaction of growth factors and growth factor receptors.[141,142] However, four viral oncogenes (v-myc, v-myb, v-fos, and v-ski) and their cellular counterpart encoded proteins have been localized in the nucleus.[143] A fifth protein p53, which was first recognized due to its association with the Simian virus 40 (SV 40) large T antigen, is also of nuclear origin and appears to function as an oncogene. Despite the fact that the kinetics and mechanism of induction of each of these proteins tend to vary, both the mRNAs and the proteins encoded for by c-myc, c-myb, c-fos, and p53 are all characterized by having short half-lives (i.e., 20 to 25 min).[81,144] This in turn has been used to suggest that these nuclear proteins function as specific effectors to link events at the cell surface to nuclear events (i.e., changes in gene transcription, initiation of DNA synthesis, etc.). As to how each of these nuclear oncogene products may render a cell tumorigenic is not clear, but the current theory is the possibility of their rendering cells more sensitive to the effects of growth factors.

Several of these nuclear oncogene proteins are phosphorylated and as a result can be classified as nonhistone phosphoproteins.

1. c-myc

The protein products of the c-myc gene (along with the related N-myc and L-myc genes) in vertebrate cells appear to be several highly related nuclear phosphoproteins (M_r: 64, 65, 67, 68 kDa).[142,144,145] It has been shown that the synthesis, turnover, and modification of c-myc proteins is in fact constant throughout the cell cycle, although transient increases can occur upon serum stimulation of resting cells. Only a small percentage of intranuclear c-myc protein is associated with chromatin, although *in vitro* DNA binding has been demonstrated. The majority of c-myc protein has been shown to be tightly associated to the nuclear matrix-lamina in interphase cells (although this localization has recently been challenged[146]) but show a cytoplasmic distribution in mitotic cells (possibly due to the dissolution of the nuclear envelope/lamina during mitosis).[147] The precise role of c-myc in normal cells is unknown, but it is believed to play some role in RNA processing due to its association with snRNP complexes.[20,148,149] In addition, its expression appears to be involved in maintaining the cell in a proliferative state by allowing at least a partial bypass of growth factor requirements for S phase entry.[81] However, the importance of its phosphorylation/dephosphorylation state is not known.

2. c-myb

The retroviral transforming gene v-myb has been shown to encode for a 45 kDa nuclear transforming protein (p45 v-myb). This p45 v-myb has since been shown to be a truncated and mutated version of a normal cellular M_r: 75 kDa phosphoprotein (p75 c-myb).[150] The c-myb protein, unlike c-myc or c-fos, is not continually expressed throughout the cell cycle, but is more closely associated with the onset of the S phase.[81] It has been demonstrated that approximately 80 to 90% of the total p45 v-myb and p75 c-myb present in the cell nucleus can be released by low salt, exhibits a DNA binding capability, and is released attached to nucleosomes after digestion with DNase I. A minor portion (approximately 10 to 20%), however, remains tightly attached to the nuclear matrix-lamina fraction. The possible functions for each subpopulation along with the role of phosphorylation/dephosphorylation remains obscure, but due to their rapid turnover (< 1 h) a regulatory role in such processes as replication and/or transcription appears likely. In addition, the rapid release of the bulk of v- and c-myb protein following nuclease digestion suggests a preferential association with transcriptionally active genes.

3. c-fos

The oncogene fos is the transforming gene of the FBJ murine osteosarcoma virus. The gene has been found to encode for a 55 kDa DNA binding phosphoprotein whose levels are markedly increased in several cell types including murine leukemia and human promyelocytic cells.[151] Interestingly, its expression can be increased very early after stimulation by growth factors in fibroblasts suggesting its induction is a primary event during cell growth. However, in certain cases, its expression can be correlated with cell transformation.[152] Recent studies (using serum stimulated mouse fibroblasts) demonstrated that the c-fos protein could form a noncovalent complex with another nuclear protein (M_r: 39 kDa).[153] The c-fos/p39 complex can be almost completely released from nuclei following either micrococcal nuclease or DNase I digestion suggesting an association with active genes. In addition, the complex demonstrates DNA binding capability (e.g., binds to either single- or double-stranded calf thymus DNA) but is bound to chromatin by only mild electrostatic forces (e.g., 0.4 M NaCl could elute >90% of the complex). The importance of its phosphorylation and the precise role it plays currently remains unknown. It has been theorized that increased levels of c-fos protein could lead cells to acquire sensitivity to growth factors to which they were previously unresponsive, thus increasing the probability of uncontrolled proliferation.[81]

4. p53

The nuclear oncogene phosphoprotein p53 (M_r: 53 kDa) has been shown to be tightly complexed to Simian virus 40 T (SV40 T) antigen in SV 40 transformed mammalian cells. It has also been found complexed with adenovirus 58K protein (E1b) in adenovirus-transformed mouse cells.[154,155] p53 is in very low abundance and has also been shown to be very unstable in established but nontransformed cell lines (i.e., 3T3 cells) with a half-life of approximately 20 min.[156] However, when p53 is complexed to T antigen, its half-life increases to 24 h along with a concurrent 1000-fold increase in level as compared to the parent 3T3 cells from which they were originally established.[157] Increases in p53 level appear to be a general characteristic of transformed and tumor cell lines.[158] The normal cellular role of p53 phosphorylation along with the importance of its complexing to viral oncogene products in virus-transformed cells remains unclear. Nevertheless, recent studies have strongly suggested that p53 may play a crucial role in the control of cell growth. For example, it has been demonstrated that microinjection of monoclonal antibodies to p53 into the nucleus of quiescent fibroblasts can abolish the RNA synthesis response of these cells to serum stimulation.[159] When the cloned p53 gene itself is attached to a strong viral promoter and then transfected into certain established cell lines, tumorgenesis resulted.[69]

IV. CONCLUSION

This brief review has attempted to provide a coherent background into the many facets of nonhistone phosphoprotein function. When first discovered and characterized in the 1960s,[160,161] these proteins were believed to play a key role in the regulation of gene expression. Despite the fact that this hypothesis has remained essentially a theory during this interim, indeed, additional functional roles have emerged. Thus phosphoproteins and phosphorylation itself appears to be a mechanism for controlling the activities of nuclear enzymes (e.g., RNA polymerases, topoisomerases, etc.), chromatin and nuclear structure, the functional aspects of hormone receptors, RNA packaging, etc. The ultimate challenge in the future will be to further unravel details of these processes as well as the structure of each specific phosphoprotein — a task that will be made more feasible with the identification and purification of all the components involved. This information will in turn provide further insights into a more complete understanding of how phosphorylation/dephosphorylation of nonhistone chromatin proteins is involved in gene regulation.

ACKNOWLEDGMENTS

The generous support by MRC, the Ontario Heart and Stroke Foundation, and the National Institutes of Health are greatly appreciated. The major portion of research on the B_2 nuclear phosphoprotein was primarily supported by MRC. M. J. Halikowski is supported both by the University of Toronto Open Fellowship and the Ontario Graduate Scholarship Program. This review constitutes part of the fulfillment of his Ph.D. thesis.

REFERENCES

1. **Liew, C. C.,** Non-histone proteins, in *Concepts of the Structure and Function of DNA, Chromatin and Chromosome,* Dion, A. S., Ed., Year Book Medical Publishers, Chicago, 1979, 153.
2. **Liew, C. C., Halikowski, M. J., and Zhao, M. S.,** Two specific groups of NHC proteins involved in gene expression, in *Progress in Nonhistone Protein Research,* Volume I, Bekhor, I., Ed., CRC Press, Boca Raton, 1984, 29.
3. **Burgoyne, L. A.,** DNA, chromatin and chromosomal levels of nuclear organization, *Cytobios,* 43, 141, 1985.
4. **Cartwright, I. L., Eissenberg, J. C., Thomas, G. H., and Elgin, S. C. R.,** Selected topics in chromatin structure, *Annu. Rev. Genet.,* 19, 485, 1986.
5. **von Holt, C.,** Histones in perspective, *BioEssays,* 3, 120, 1985.
6. **Klug, A., Richmond, T. J., and Finch, J. T.,** The structure of the nucleosome core particle, *Biochem. Soc. Trans.,* 14, 221, 1986.
7. **Wu, R. S., Panusz, H. T., Hatch, C. L., and Bonner, W. M.,** Histones and their modifications, in *CRC Critical Review in Biochemistry,* Vol. 20, Fasman, G. D., Ed., CRC Press, Boca Raton, FL, 1986.
8. **Stein, G., Stein, J., and Kleinsmith, L. J.,** Nonhistone proteins and gene regulation, in *Eukaryotic Genes,* Maclean, N., Gregory, S. P., and Flavell, R. A., Eds., Butterworths, London, 1983, 31.
9. **Stein, G. S., Plumb, M., Stein, J. L., Phillips, I. R., and Shephard, E. A.,** Nonhistone proteins in genetic regulation, in *Nonhistone Chromosomal Proteins,* Vol. I, Hnilica, L. S., Ed., CRC Press, Boca Raton, FL, 1983, 127.
10. **Yaniv, M. and Cereghini, S.,** Structure of transcriptionally active chromatin, in *CRC Critical Review in Biochemistry,* Vol. 21, Fasman, G. D., Ed., CRC Press, Boca Raton, FL, 1986, 1.
11. **Kleinsmith, L. J.,** Phosphorylation of nonhistone proteins, in *The Cell Nucleus,* Vol. VI, Busch, H., Ed., Academic Press, New York, 1978, 222.
12. **Liew, C. C., Jackowski, G., Ma, T., and Sole, M. J.,** Nonenzymatic separation of myocardial cell nuclei from whole heart tissue, *Am. Physiol. Soc.,* C3-C10, 1983.
13. **Kumar, S. and Leffack, M.,** Assembly of active chromatin, *Biochemistry,* 25, 2055, 1986.
14. **Bustin, M.,** Immunochemical analysis of the structure and function of chromosomal proteins, *Cytometry,* 8, 251, 1987.
15. **Liew, C. C. and Gornall, A. G.,** Covalent modification of nuclear proteins during aging, *Fed. Proc.,* 34, 186, 1975.
16. **Surie, D. and Liew, C. C.,** Characterization of proteins associated with nuclear ribonucleoprotein particles by two dimensional gel electrophoresis, *Can. J. Biochem.,* 57, 32, 1979.
17. **Olson, M. O. J.,** Nonhistone nuclear phosphoproteins, in *Chromosomal Nonhistones Proteins,* Vol. III, Hnilica, L. S., Ed., CRC Press, Boca Raton, FL, 1984, 129.
18. **Waterborg, J. H. and Matthews, H. R.,** Reversible modifications of nuclear proteins and their significance, in *The Enzymology of Post-Translational Modification of Proteins,* Vol. 2, Freedman, R. B. and Hawkins, H. C., Eds., Academic Press, New York, 1985, 126.
19. **Kleinsmith, L. J. and Mitchell, S. J.,** Nuclear protein kinases, in *Chromosomal Nonhistone Proteins,* Vol. III, Hnilica, L. S., Ed., CRC Press, Boca Raton, FL, 1984, 132.
20. **Spector, D. L. and Smith, H. C.,** Redistribution of U-snRNPs during mitosis, *Exp. Cell Res.,* 163, 87, 1986.
21. **Gerace, L., Comeau, C., and Benson, M.,** Organization and modulation of nuclear lamina structure, *J. Cell Sci. Suppl.,* 1, 137, 1984.
22. **Maul, G. G. and Schatten, G.,** The nuclear envelope and its associated structures, in *UCLA Symposia on Molecular Biology and Cellular Biology,* Vol. 26, Smuckler, E. A. and Clawson, G. A., Eds., Alan R. Liss, New York, 1986, 13.

23. **Phillips, I. R., Stein, G. S., Stein, J. L., and Shepard, E. A.,** Role of nonhistone chromosomal proteins in selective gene expression, in *Eukaryotic Gene Regulation,* Vol. II, Kolodny, G. M., Ed., CRC Press, Boca Raton, FL, 1980, 113.
24. **Kruh, J., Guguen-Guillouzo, C., and Tichonicky, L.,** Hepatocyte chromosomal non-histone proteins in developing rat, *Eur. J. Biochem.,* 95, 235, 1979.
25. **Schafer, W. and Kahl, G.,** Phosphorylation of chromosomal proteins changes during the development of Crown Gall tumors, *Plant Cell Physiol.,* 25, 1187, 1984.
26. **Bluthmann, H.,** Specific binding of a nonhistone chromosomal protein from lymphocyte to DNA, *Eur. J. Biochem.,* 70, 233, 1976.
27. **Hotta, Y. and Stern, H.,** The effect of dephosphorylation on the properties of a helix-destabilizing protein from meiotic cells and its partial reversal in Physarum Polycephalum, *Eur. J. Biochem.,* 90, 29, 1978.
28. **Song, M.-K. H. and Adolph, K. W.,** Phosphorylation of nonhistone proteins during the HeLa cell cycle, *J. Biol. Chem.,* 258, 3309, 1983.
29. **Rao, P. N., Adlakha, R. C., Sahasrabuddhe, C. G., Wright, D. A., and Bigo, H.,** Role of nonhistone protein phosphorylation in the regulation of mitosis in mammalian cells, in *Experimental Biology and Medicine,* Skehan, P. and Friedman, S. J., Eds., Humana Press, Clifton, N.J., 1985, 59.
30. **Thomson, J. A., Chiu, J.-F., and Hnilica, L. S.,** Nuclear phosphoprotein kinase activities in normal and neoplastic tissues, *Biochim. Biophys. Acta,* 407, 114, 1975.
31. **Ganpath, N., Prestayko, A. W., and Busch, H.,** Comparison of nuclear nonhistone phosphoproteins of rat liver and Novikoff hepatoma, *Cancer Res.,* 37, 1290, 1977.
32. **Metlas, R., Kanazir, D. T., Trajkovic, D. P., Ribarac-Stepic, N., and Popic, S. D.,** Cortisol dependent acute metabolic responses in rat liver cells, *J. Steroid Biochem.,* 9, 467, 1978.
33. **Coleoni, A. H. and DeGroot, L. J.,** Liver nuclear protein phosphorylation in vivo and the effect of triiodothyronine, *Endocrinology,* 106, 1103, 1980.
34. **Liew, C. C., Suria, D., and Gornall, A. G.,** Effects of aldosterone on acetylation and phosphorylation of chromosomal proteins, *Endocrinology,* 93, 1025, 1973.
35. **Katoh, N., Kimura, K., and Sakurada, K.,** Effects of heparin and polyamine on the phosphorylation of high mobility group proteins by cyclic nucleotide-independent phosvitin kinase from pig testis, *J. Biochem.,* 97, 859, 1985.
36. **Macfarlane, D. E.,** Phorbol diester-induced phosphorylation of nuclear matrix proteins in HL-60 promyeloctes. Possible role in differentiation studied by cationic detergent gel electrophoresis systems, *J. Biol. Chem.,* 261, 6947, 1986.
37. **Suria, D. and Liew, C. C.,** Isolation and analysis of non-histone chromosomal proteins from rat liver nuclei by three different methods, *Can. J. Biochem.,* 52, 1143, 1974.
38. **Peterson, J. L. and McConkey, E. H.,** Non-histone chromosomal proteins from HeLa cells. A survey by high resolution, two-dimensional electrophoresis, *J. Biol. Chem.,* 251, 548, 1976.
39. **Rao, P. N., Adlakha, R. C., and Davis, F. M.,** Role of phosphorylation of nonhistone proteins in the regulation of mitosis, in *Control of Animal Cell Proliferation,* Vol. I, Boynton, A. L. and Leffert, H. L., Eds., Academic Press, Orlando, 1986, 485.
40. **Hohmann, P., Tobey, L. R., and Gurley, L. R.,** Phosphorylation on distinct regions of f1 histone. Relationship to the cell cycle, *J. Biol. Chem.,* 251, 3685, 1976.
41. **Gurley, L. R., D'Anna, J. A., Halleck, M. S., Barham, S. S., Walters, R. A., Jett, J. J., and Tobey, R. A.,** Relationships between histone phosphorylation and cell proliferation, in *Protein Phosphorylation,* Book B, Rosen, O. M. and Krebs, E. G., Eds., Cold Spring Harbor Laboratory, New York, 1981, 1093.
42. **Jungmann, R. A., Laks, M. S., Harrison, J. J., Suter, P., and Jones, C. E.,** Modulation of nuclear protein kinases at times of gene activity, in *Protein Phosphorylation,* Book B, Rosen, O. M. and Krebs, E. G., Eds., Cold Spring Harbor Laboratory, New York, 1981, 1109.
43. **Bhorjee, J. S.,** Phosphorylation of the high mobility group nonhistone proteins, in *Progress in Nonhistone Protein Research,* Vol. II, Bekhor, I., Mirell, C. J., and Liew, C. C., Eds., CRC Press, Boca Raton, FL, 1985, 57.
44. **Evans, R. M. and Fink, L. M.,** An alteration in the phosphorylation of vimentin-type intermediate filaments is associated with mitosis in cultured mammalian cells, *Cell,* 29, 43, 1982.
45. **Gerace, L. and Blobel, G.,** The nuclear envelope lamina is reversibly depolymerized during mitosis, *Cell,* 19, 277, 1980.
46. **Nielsen, P. J., Thomas, G., and Maller, J. L.,** Increased phosphorylation of ribosomal protein S6 during meiotic maturation of Xenopus oocytes, *Proc. Natl. Acad. Sci. U.S.A.,* 79, 2937, 1982.
47. **Laskey, R. A.,** Phosphorylation of nuclear proteins, *Phil. Trans. R. Soc. Lond.,* 302, 143, 1983.
48. **Adler, K., Beyreuther, K., Fanning, E., Geisler, N., Gronenborn, B., Mueller-Hill, B., Phahl, M., and Schmitz, A.,** How lac repressor binds to DNA, *Nature (London),* 237, 322, 1972.
49. **Kleinsmith, L. J.,** Specific binding of phosphorylated non-histones chromatin proteins to deoxyribonucleic acid, *J. Biol. Chem.,* 248, 5648, 1973.

50. **Prestayko, A. W., Crane, P. M., and Busch, H.,** Phosphorylation and DNA binding of nuclear rat liver proteins soluble at low ionic strength, *Biochemistry,* 15, 414, 1976.

51. **Knippers, P., Otto, B., and Baynes, M.,** A single-strand specific DNA-binding protein from mouse cells that stimulates DNA polymerase. Its modification by phosphorylation, *Eur. J. Biochem.,* 73, 17, 1977.

52. **Henning, R. and Monteharh, M.,** Simian virus 40 T-antigen phosphorylation is variable, *FEBS Lett.,* 114, 107, 1980.

53. **Collins, J. K., Tegtmeyer, P., and Rundell, K.,** Modification of simian virus 40 protein A, *J. Virol.,* 21, 647, 1977.

54. **Whitfield, J. F. and Boynton, A. L.,** A possible involvement of type H cAMP-dependent protein kinase in the initiation of DNA synthesis by rat liver cells, *Exp. Cell Res.,* 126, 477, 1980.

55. **Weigel, N. L., Tash, J. S., Means, A. R., Schrader, W. T., and O'Malley, B. W.,** Phosphorylation of hen progesterone receptor by cAMP dependent protein kinase, *Biochem. Biophys. Res. Commun.,* 102, 513, 1981.

56. **Bekhor, I. and Mirell, C. J.,** Simple isolation of DNA hydrophobically complexed with presumed gene regulatory proteins (M3), *Biochemistry,* 18, 609, 1979.

57. **Norman, G. L. and Bekhor, I.,** Enrichment of selected active human gene sequences in the placental DNA fraction associated with tightly bound nonhistone chromosomal proteins, *Biochemistry,* 20, 3568, 1981.

58. **Toft, D. O., Puri, R. K., and Dougherty, J. J.,** Phosphorylation of steroid receptors, *TIBS,* 6, 83, 1985.

59. **Klyzejko-Stefanowicz, L., Chiu, J. F., Tsai, P.-H., and Hnilica, L. S.,** Acceptor proteins in rat androgenic tissue chromatin, *Proc. Natl. Acad. Sci. U.S.A.,* 73, 1954, 1976.

60. **Nielsen, C. J., Sando, J. J., and Pratt, W. B.,** Evidence that dephosphorylation inactivate glucocorticoid receptors, *Proc. Natl. Acad. Sci. U.S.A.,* 74, 1398, 1977.

61. **Dougherty, J. J., Puri, R. K., and Toft, D. O.,** Phosphorylation in vivo of chicken oviduct progesterone receptor, *J. Biol. Chem.,* 257, 14226, 1982.

62. **Migliaccio, A., Rotondi, A., and Auricchio, F.,** Calmodulin-stimulated phosphorylation of 17-estradiol receptor on tyrosine, *Proc. Natl. Acad. Sci. U.S.A.,* 81, 5921, 1984.

63. **Garcia, T., Tuohimaa, P., Mester, J., Buchou, T., Renoir, J.-M., and Baulieu, E.-E.,** Protein kinase activity of purified components of the chicken oviduct progesterone receptor, *Biochem. Biophys. Res. Commun.,* 113, 960, 1983.

64. **Ishida, H. and Ahmed, K.,** Studies on chromatin-associated protein phosphokinase of submandibular gland from isoproterenol-treated rats, *Exp. Cell Res.,* 84, 127, 1974.

65. **Johns, E. W.,** HMG, history, definitions and problems, in *The HMG Chromosomal Proteins,* Johns, E. W., Ed., Academic Press, New York, 1982, 1.

66. **Weisbrod, S., Groudine, M., and Weintraub, H.,** Interaction of HMG 14 and 17 with actively transcribed genes, *Cell,* 19, 289, 1980.

67. **Allfrey, V. G.,** Postsynthetic modifications of HMG proteins, in *The HMG Chromosomal Proteins,* Johns, E. W., Ed., Academic Press, New York, 1982, 123.

68. **Inoue, A., Tei, Y., Hasuma, T., Yukioka, M., and Morsawa, S.,** Phosphorylation of HMG 17 by protein kinase NII from rat liver cell nuclei, *FEBS Lett.,* 117, 68, 1980.

69. **Ostvold, A. C., Holtlund, J., and Laland, S. G.,** A novel, highly phosphorylated protein, of the high-mobility group type, present in a variety of proliferating and non-proliferating mammalian cells, *Eur. J. Biochem.,* 153, 469, 1985.

70. **Chan, P. K. and Liew, C. C.,** Purification of a phosphoprotein from chromatin of rat liver, *Biochem. J.,* 183, 143, 1979.

71. **Defer, N., Kitzis, A., Levy, F., Tichonicky, L., Sabatier, M.-M., and Kruh, J.,** Presence of non-histone proteins in nucleosomes, *Eur. J. Biochem.,* 88, 583, 1978.

72. **Safer, J. D. and Coleman, J. E.,** Reversible phosphorylation of a nucleosome binding protein that stimulates transcription of nucleosome and deoxyribonucleic acid, *Biochemistry,* 19, 5874, 1980.

73. **Goldknopf, I. L., Rosenbaum, F., Sterner, R., Vidali, G., Allfrey, V. G., and Busch, H.,** Phosphorylation and acetylation of chromatin conjugate protein A24, *Biochem. Biophys. Res. Commun.,* 90, 269, 1979.

74. **Olson, M. O. J., Orrick, I. R., Jones, C. E., and Busch, H.,** Phosphorylation of acid-soluble nucleolar proteins of Novikoff hepatoma ascites cells in vivo, *J. Biol. Chem.,* 249, 2823, 1974.

75. **Blanchard, J. M., Brunel, C., and Jeanteur, P.,** Characterization of an endogenous protein kinase activity in ribonucleoprotein structure containing heterogeneous nuclear RNA in HeLa cell nuclei, *Eur. J. Biochem.,* 79, 117, 1977.

76. **Wang, J. C.,** DNA Topoisomerases, *Annu. Rev. Biochem.,* 55, 665, 1985.

77. **Donaldson, R. W. and Gerner, E. W.,** Phosphorylation of a high molecular weight DNA polymerase, *Proc. Natl. Acad. Sci. U.S.A.,* 84, 759, 1987.

78. **Sentenac, A.,** Eukaryotic RNA polymerases, in *CRC Critical Reviews in Biochemistry,* Vol. 18, Fasman, G. D., Ed., CRC Press, Boca Raton, FL, 1985, 31.

79. **Rose, K. M., Roe, F. J., and Jacob, S. T.,** Two functional states of poly(adenylic acid) polymerase in isolated nuclei, *Biochim. Biophys. Acta,* 478, 180, 1977.

80. **Baserga, R., Mercer, W. E., and Avignol, C.,** Role of p53 protein in cell proliferation as studied by microinjection of monoclonal antibodies, *Mol. Cell. Biol.,* 4, 276, 1984.

81. **Eisenman, R. N. and Thompson, C. B.,** Oncogenes with potential nuclear function: myc, myb and fos, *Cancer Sur.,* 5, 309, 1986.

82. **Arfmann, H.-A., Haase, E., and Schroter, J.,** High mobility group proteins from CHO cells. Their characterization and modifications during cell cycle, *Biochem. Biophys. Res. Commun.,* 101, 137, 1981.

83. **Maenpaa, P. H., Palvimo, J., and Linnala-Kankkunen, A.,** Phosphorylation alters the affinity of high mobility group protein HMG 14 from single-stranded DNA, *Biochem. Biophys. Res. Commun.,* 133, 343, 1985.

84. **Liew, C. C. and Chan, P. K.,** Identification of non-histone chromatin proteins in chromatin subunits, *Proc. Natl. Acad. Sci. U.S.A.,* 73, 3458, 1976.

85. **Busch, H.,** Nucleolar proteins: purification, isolation and functional analyses, in *Chromosomal Nonhistone Proteins,* Vol. IV, Hnilica, L. S., Ed., CRC Press, Boca Raton, FL, 1984, 233.

86. **Fakan, S. and Hernandez-Verdun, D.,** The nucleolus and nucleolar organizer regions, *Biol. Cell,* 56, 189, 1986.

87. **Busch, H., Busch, R. K., Chan, P.-K., Kelsey, D., and Takahashi, K.,** Nucleolar antigens of human tumors, in *Methods in Cancer Research,* Vol. XIX, Busch, H., Ed., Academic Press, New York, 1982, 110.

88. **Olson, M. O. J., Ezrailson, E. G., Guetzow, K., and Busch, H.,** Localization and phosphorylation of nuclear, nucleolar and extranucleolar nonhistone proteins of Novikoff hepatoma ascites cells, *J. Mol. Biol.,* 97, 611, 1974.

89. **Jackowski, G., Suria, D., and Liew, C. C.,** Fractionation of nucleolar proteins by two-dimensional gel electrophoresis, *Can. J. Biochem.,* 54, 9, 1976.

90. **Tsutsui, K. and Oda, T.,** Isolation and characterization of a high molecular weight acid soluble nuclear protein from mouse ascites-sarcoma cells, *Eur. J. Biochem.,* 108, 497, 1980.

91. **Izawa, M., Sato, S., and Kawashima, K.,** Properties and functions of a nucleolus-specific phosphoprotein of mouse ascites sarcoma cells, *Cell Struc. Funct.,* 9, 291, 1984.

92. **Kang, Y.-J., Olson, M. O. J., Jones, C., and Busch, H.,** Nucleolar phosphoproteins of normal rat liver and Novikoff hepatoma ascites cells, *Cancer Res.,* 35, 1470, 1975.

93. **Berezney, R.,** Organization and functions of the nuclear matrix, in *Chromosomal Nonhistone Proteins,* Volume IV, Hnilica, L. S., Ed., CRC Press, Boca Raton, FL, 1984, 119.

94. **Coffey, D. S., Nelson, W. G., Pienta, K. J., and Barrack, E. R.,** The role of the nuclear matrix in the organization and function of DNA, *Annu. Rev. Biophys. Chem.,* 15, 457, 1986.

95. **Kaufmann, S. H., Fields, A. P., and Shaper, J. H.,** The nuclear matrix-current concepts and unanswered questions, in *Methods and Achievements in Experimental Pathology,* Vol. 12, Jasmin, G. and Simard, R., Eds., S. Karger, Basel, 1986, 141.

96. **Allen, S. L., Berezney, R., and Coffey, D. S.,** Phosphorylation of nuclear matrix proteins in isolated regenerating rat liver nuclei, *Biochem. Biophys. Res. Commun.,* 75, 112, 1977.

97. **Sevaljevic, L., Petrovic, M., Konstantinovic, M., and Krtolica, K.,** Comparative studies of rat and sea urchin embryo nuclear matrices: partial fractionation and protein kinase activity distribution, *J. Cell Sci.,* 55, 189, 1982.

98. **Henry, S. M. and Lodge, L. D.,** Nuclear matrix: a cell cycle dependent site of increased intranuclear protein phosphorylation, *Eur. J. Biochem.,* 133, 23, 1983.

99. **Ohmura, Y., Teraoka, H., and Tsukada,** A protein-tyrosine kinase in the nuclear matrix from rat liver, *FEBS Lett.,* 208, 451, 1987.

100. **Kasper, C. B.,** Nuclear envelope proteins: selected biochemical aspects, in *Chromosomal Nonhistone Proteins,* Vol. IV, Hnilica, L. S., Ed., CRC Press, Boca Raton, FL, 1984, 1.

101. **Krohne, G. and Benavente, R.,** The nuclear lamins. A multigene family of proteins in evolution and differentiation, *Exp. Cell Res.,* 162, 1, 1986.

102. **Kasper, C. B. and Lam, K. S.,** Selective phosphorylation of a nuclear envelope polypeptide by an endogenous protein kinase, *Biochemistry,* 18, 307, 1979.

103. **Agutter, P. S., Cockrill, J. B., Lavine, J. E., McCaldin, B., and Sim, R. B.,** Properties of mammalian nuclear-envelope nucleoside triphosphatase, *Biochem. J.,* 181, 647, 1979.

104. **Steer, R. C., Wilson, M. J., and Ahmed, K.,** Protein phosphokinase activity of rat liver nuclear membrane, *Exp. Cell Res.,* 119, 403, 1979.

105. **LeStourgeon, W. M., Lothstein, L., Walker, B. W., and Beyer, A. L.,** The composition and general topology of RNA and protein in monomer 40S ribonucleoprotein particles, in *The Cell Nucleus,* Vol. 9, Busch, H., Ed., Academic Press, New York, 1981, 49.

106. **Holoubek, V.,** Nuclear ribonucleoproteins containing heterogeneous RNA, in *Chromosomal Nonhistone Proteins,* Vol. IV, Hnilica, L. S., Ed., CRC Press, Boca Raton, FL, 1984, 21.

107. **Beyer, A. L., Christensen, M. E., Walker, B. W., and LeStourgeon, W. M. N.,** Identification and characterization of the packaging proteins of core 40S hnRNP particles, *Cell,* 11, 127, 1977.
108. **Beyer, A. L., Miller, O. L., Jr., and McKnight, S. L.,** Ribonucleoprotein structure in nascent hnRNA is nonrandom and sequence-dependent, *Cell,* 20, 75, 1980.
109. **Setyono, B. and Greenberg, J. R.,** Proteins associated with poly(A) and other regions of mRNA and hnRNA molecules as investigated by crosslinking, *Cell,* 24, 775, 1981.
110. **Tomcsanyi, T., Molnar, J., and Tigyi, A.,** Structural characterization of nuclear poly(A)-protein particles in rat liver, *Eur. J. Biochem.,* 131, 283, 1983.
111. **Gallinaro-Matringe, H., Stevenin, J., and Jacob, M.,** Salt dissociation of nuclear particles containing DNA-like RNA. Distribution of phosphorylated and nonphosphorylated species, *Biochemistry,* 14, 2547, 1975.
112. **Blanchard, J. M., Brunel, C., and Jeanteur, P.,** Phosphorylation in vivo of proteins associated with heterogeneous nuclear RNA in HeLa cell nuclei, *Eur. J. Biochem.,* 86, 301, 1978.
113. **Schweiger, A. and Schmidt, D.,** Isolation of RNA-binding proteins from rat liver 30S particles, *FEBS Lett.,* 41, 17, 1974.
114. **Alonso, A., Fischer, J., Konig, N., and Kinzel, V.,** Structural analysis of hnRNP particles approached by in vitro phosphorylation using exogenous protein kinase and (^{32}P)ATP, *Eur. J. Cell Biol.,* 26, 208, 1981.
115. **Dahmus, M. E.,** Purification and properties of calf thymus casein kinases I and II, *J. Biol. Chem.,* 256, 3319, 1981.
116. **Hirsch, J. and Martelo, O. J.,** Phosphorylation of rat liver ribonucleic acid polymerase I by nuclear protein kinase, *J. Biol. Chem.,* 251, 5408, 1976.
117. **Kranias, E. G., Schweppe, J., and Jungman, R. A.,** Phosphorylative and functional modifications of nucleoplasmic RNA polymerase II by homologous adenosine 3',5'-monophosphate-dependent protein kinase from calf thymus and by heterologous phosphatase, *J. Biol. Chem.,* 252, 6750, 1977.
118. **Dahmus, M. E.,** Stimulation of ascites tumor RNA polymerase II by protein kinase, *Biochemistry,* 15, 1821, 1976.
119. **Mazus, B., Szurmak, B., and Buchowicz, J.,** Phosphorylation in vitro and in vivo of the wheat embryo RNA polymerase II, *Acta. Biochim. Pol.,* 27, 9, 1980.
120. **Rose, K. M., Stetler, D. A., and Jacob, S. T.,** Protein kinase activity of RNA polymerase I purified from rat hepatoma: probable function of M_r: 42,000 and 24,600 polypeptides, *Proc. Natl. Acad. Sci. U.S.A.,* 78, 2833, 1981.
121. **Dahmus, M. E.,** Phosphorylation of eukaryotic DNA-dependent RNA polymerase, *J. Biol. Chem.,* 256, 3332, 1981.
122. **Tsiapalis, C. M.,** Chemical modification of DNA polymerase phosphoprotein from Avian Myeloblastosis virus, *Nature (London),* 266, 27, 1977.
123. **Wong, S. W., Paborsky, L. R., Fisher, P. A., Wang, T. S.-F., and Korn, D.,** Structural and enxymological characterization of immunoaffinity-purified DNA polymerase DNA primase complex from KB cells, *J. Biol. Chem.,* 261, 7958, 1986.
124. **Rose, K. M. and Jacob, S. T.,** Phosphorylation of nuclear poly(A) polymerase. Comparison of liver and hepatoma enzymes, *J. Biol. Chem.,* 254, 10256, 1979.
125. **Vidali, G., Boffa, L. C., and Allfrey, V. G.,** Properties of an acidic histone-binding fraction from cell nuclei. Selective precipitation and deacetylation of histones F2A1 and F3, *J. Biol. Chem.,* 247, 7365, 1972.
126. **Cozzarelli, N. R.,** DNA gyrase and the supercoiling of DNA, *Science,* 207, 953, 1980.
127. **Durban, E., Mills, J. S., Roll, D., and Busch, H.,** Phosphorylation of purified Novikoff hepatoma topoisomerase I, *Biochem. Biophys. Res. Commun.,* 111, 897, 1983.
128. **Gellert, M.,** DNA topoisomerases, *Annu. Rev. Biochem.,* 50, 879, 1981.
129. **Tse-Dinh, Y., Wong, T. W., and Godberg, A. R.,** Virus and cell encoded tyrosine protein kinases inactivate DNA topoisomerases in vitro, *Nature (London),* 312, 785, 1984.
130. **Sander, M., Nolan, J. M., and Hsieh, T.,** A protein kinase activity tightly associated with Drosophila type II DNA topoisomerase, *Proc. Natl. Acad. Sci. U.S.A.,* 81, 6938, 1984.
131. **Ackerman, P., Glover, C. V. C., and Osheroff, N.,** Phosphorylation of DNA topoisomerase II by casein kinase II: modulation of eukaryotic topoisomerase II activity in vitro, *Proc. Natl. Acad. Sci. U.S.A.,* 82, 3164, 1985.
132. **Cuatrecasas, P., Sahyoun, N., Wolf, M., Besterman, J., Hsieh, T.-s., Sander, M., and LeVine, H.,** Protein kinase C phosphorylated topoisomerase II: topoisomerase activation and its possible role in phorbol ester-induced differentiation of HL-60 cells, *Proc. Natl. Acad. Sci. U.S.A.,* 83, 1603, 1986.
133. **Atmar, V. J. and Kuehn, G. D.,** Phosphorylation of ornithine decarboxylase by a polyamine-dependent protein kinase, *Proc. Natl. Acad. Sci. U.S.A.,* 78, 5518, 1981.
134. **Daniels, G. R., Atmar, V. J., and Kuen, G. D.,** Polyamine-activated protein kinase reaction from nuclei and nucleoli of Physarum Polycephalum which phosphorylates a unique M_r: 70,000 nonhistone protein, *Biochemistry,* 20, 2525, 1981.

135. **Atmar, V. J., Daniels, G. R., Kuehn, G. D., and Braun, R.,** Opposing kinetic effects of an acidic nucleolar phosphoprotein from *Physarum polycephalum* on homologous and heterologous transcription systems, *FEBS Lett.,* 114, 205, 1980.
136. **Desjardins, P. R., Lue, P. F., Liew, C. C., and Gornall, A. G.,** Rat liver nuclear protein kinases, *Can. J. Biochem.,* 50, 1249, 1975.
137. **Walter, U., Uno, I., Liu, A. Y.-C., and Greengard, P.,** Study of auto-phosphorylation of isoenzymes of cyclic AMP-dependent protein kinases, *J. Biol. Chem.,* 252, 6588, 1977.
138. **Quarless, S. A.,** Identification of an ionic strength sensitive nuclear protein kinase activity from the cervical carcinoma HeLa, *Biochem. Biophys. Res. Commun.,* 133, 981, 1986.
139. **Rose, K. M., Bell, L. E., Siefkeir, D. A., and Jacob, S. T.,** A heparin-sensitive nuclear protein kinase. Purification, properties and increased activity in rat hepatoma relative to liver, *J. Biol. Chem.,* 256, 7468, 1981.
140. **Kruh, J., Moisand, F., Levy-Favatier, F., and Delpech, M.,** Rat liver nuclear protein kinases NI and NII. Purification, subunit composition, substrate specificity, possible levels of regulation, *Eur. J. Biochem.,* 160, 333, 1986.
141. **Marshall, C. J.,** Oncogenes, *J. Cell Sci. Suppl.,* 4, 417, 1986.
142. **Eisenman, R. N. and Hann, S. R.,** Myc-encoded proteins of chickens and men, in *Current Topics in Microbiology and Immunology,* Vol. 113, Springer-Verlag, Berlin, 1984, 192.
143. **Gilman, M.,** Nuclear oncogenes — a meeting review, *Genes Dev.,* 1, 137, 1987.
144. **Eisenman, R. N. and Hann, S. R.,** Proteins expressed by the c-myc oncogene in lymphomas of human and avian origin, *Proc. R. Soc. London,* 226, 73, 1985.
145. **Persson, H., Gray, H. E., Godeau, F., Braunhut, S., and Bellve, A. R.,** Multiple growth-associated nuclear proteins immunoprecipitated by antisera raised against human c-myc peptide antigens, *Mol. Cell. Biol.,* 6, 942.
146. **Hancock, D. C. and Evan, G. I.,** Studies on the interaction of the human c-myc protein with cell nuclei: p62^{myc} as a member of a discrete subset of nuclear proteins, *Cell,* 43, 253, 1986.
147. **Eisenman, R. N., Tachibana, C. Y., Abrams, H. D., and Hann, S. R.,** V-myc and c-myc encoded proteins are associated with the nuclear matrix, *Mol. Cell. Biol.,* 5, 114, 1985.
148. **Spector, D. L.,** C-myc implicated in RNA processing, *Science,* 233, 159, 1986.
149. **Sullivan, N. F., Watt, R. A., Delannoy, M. R., Green, C. L., and Spector, D. L.,** Colocalization of the myc oncogene protein and small nuclear ribonucleoprotein particles, *Cold Spring Harbor Symp.,* 51, 943, 1987.
150. **Sippel, A. E. and Klempnauer, K.-H.,** Subnuclear localization of proteins encoded by the oncogene v-myb and its cellular homolog c-myb, *Mol. Cell. Biol.,* 6, 62, 1986.
151. **Corral, M., Tichonicky, L., Guguen-Guillouzo, C., Corcos, D., Raymondjean, M., Paris, B., Kruh, J., and Defer, N.,** Expression of c-fos oncogene during hepatocarcinogenesis, liver regeneration and synchronized HTC cells, *Exp. Cell Res.,* 160, 427, 1985.
152. **Curran, T., Miller, A. D., Zokas, L., and Verma, I. M.,** Viral and cellular fos proteins: a comparative analysis, *Cell,* 36, 259, 1984.
153. **Renz, M., Verrier, B., Kurz, C., and Muller, R.,** Chromatin association and DNA binding properties of the c-fos proto-oncogene product, *Nucleic Acids Res.,* 15, 277, 1987.
154. **McCormick, P. and Harlow, E.,** Association of a murine 58,000-dalton phosphoprotein with simian virus 40 large-T-antigen in transformed cells, *J. Virol.,* 34, 213, 1980.
155. **Sarnow, P., Ho, Y. S., Williams, J., and Levine, A. J.,** Adenovirus E1b-58kd tumor antigen and SV-40 large tumor antigen are physically associated with the same 54kd cellular protein in transformed cells, *Cell,* 28, 387, 1982.
156. **Lane, D. P., Yewdell, J. W., and Gannon, J. V.,** Monoclonal antibody analysis of p53 expression in normal and transformed cells, *J. Virol.,* 59, 444, 1986.
157. **Levine, A. J., Oren, M., and Maltzman, W.,** Post-translational regulation of the 54K cellular tumor antigen in normal and transformed cells, *Mol. Cell. Biol.,* 1, 101, 1981.
158. **Rotter, V.,** p53, A transformation-related cellular-encoded protein, can be used as a biochemical marker for the detection of primary mouse tumour cells, *Proc. Natl. Acad. Sci. U.S.A.,* 80, 2613, 1983.
159. **Oren, M., Eliyahu, D. D., and Michalovitz, D.,** Overproduction of p53 antigen makes established cells highly tumorigenic, *Nature (London),* 316, 158, 1985.
160. **Wang, T. Y.,** Solubilization and characterization of the residual proteins of the cell nucleus, *J. Biol. Chem.,* 241, 2913, 1966.
161. **Wang, T. Y.,** The isolation, properties and possible functions of chromatin acidic proteins, *J. Biol. Chem.,* 242, 1220, 1967.

INDEX

A

B

C

E

F

G

Q

R

Ring-shaped bodies, 41—42
RNA polymerase, 80, 84, 88, 91, 120, 139, 177
   nucleosome, 80
   nucleosome subpopulations, 94
RNA polymerase I, 173—174, 177, 179
RNA polymerase II, 32, 60, 63, 78, 136, 177
RNA polymerase III, 32, 177
RNA polymerase reaction, control by active peptides,
     126—127
RNA polymerase reaction *in vitro,* inhibition of, 125
RNA synthesis
   inhibition, 32, 129—130
     protein B23, location of, 34
     regulatory activity in cellular systems, 127, 129—
       131
     suppression of, 29
RNAase, 55, 63, 65
RNP assembly, 168
RNP network, 41—42
RNP particles, aggregation of proteins derived from, 61
rRNA processing, role of protein B23 in, 32
rRNA synthesis actinomycin D, inhibition by, 32

S

32S RNA, spacer region, 23
[35S]-methionine labeled protein, 163
45S rRNA, spacer region, 28
Satellite DNA, 70
Scaffold, 69—70
SDS polyacrylamide gel electrophoresis, 28, 101,
     142—143
Sedimentation coefficient, protein B23, 16, 27
Selectivity of transcription initiation, 139
Sequence-specific DNA binding proteins, 141—142
Sequence specific nuclear DNA binding proteins, pat-
     tern of, 142
Serine, 48, 120
Serum starvation, protein B23, 29—32
Silver staining, 26
Single-stranded DNA, 61, 66, 108—110, 147—148
Single-stranded DNA binding proteins, 150
Skeletal structures, 44, 60—64
Skin proteins, 101
Snake venom phosphodiesterase, 49—51, 106
Soft laser densitometry, 158—164
Soft laser scanning densitometer, 158—159
Somatic cell line, 54
Somatic cells, 51
Somatic chromatin, 54
Somatic nuclear matrix, 43
South-Western blot, 142
South-Western blot mapping, 151, 153
Southern blots, 113
Specific epitope, 7
Specific β globin transcription, 139
Specific nuclear enzymes, 176—179
Spectrin, 129
Sperm, 41, 52, 56
Sperm chromatins, 52
Sperm DNA, covalent bond with, formation of, 48

Sperm matrix, 43
Sperm nuclear skeleton, tightly bound proteins, 43—46
Sperm nucleus, 54, 56
   basic proteins of, 42
   nonhistone proteins of, 42—50
   skeleton of, 43
Sperm-specific basic proteins, 42
Spermatids, 43
Spermatocytes, 43
Spermidine, 179
Spermine, 179
Spermiogenesis, 42, 56
Spleen phosphodiesterase, 49
ssDNA subunits, 108—109
Staphylococcal nuclease, 78, 84, 88
Staphylococcal V8 protease, 14
Staphylococcus aureus, 43
Steroid receptors, 171
Stoke's radius, protein B23, 27
Structural proteins, 40
Subnuclear structures, anchorage or attachment of DNA
     at, 100
Sucrose density gradient, 2, 5
Sucrose density gradient centrifugation, 27, 91
Sucrose gradient, 79, 84, 88
Supercoiled DNA loops, 109
Superphosphorylation, 4
SV 40, 4, 170, 180—181

T

T antigen, see SV 40
tbNHCp-DNA complexes, 81—84
tbNHCp proteins, 91
Template DNA, chromatin peptides, 127, 129
Terminal differentiation, 154
Thin-layer chromatography, 105—106
Threonine, 48
Tight complexes between DNA and polypeptides, 116
Tight DNA/peptide complexes, 100, 109
Tightly bound nonhistone chromatin proteins, 88
Tightly bound nonhistone proteins, 84, 92—95
   chromatin organization, 91
   nucleosome, 80—84, 89
Tightly bound polypeptides, 108
Tightly bound proteins, 40, 42
   DNA complex, 46
   electrophoretic fractionation of, 43—46
   localization, 44
   sperm nuclear skeleton, 43—46
Tightly bound sperm proteins, 54
Tissue-specific DNA-binding proteins, 154
Tissue-specific nuclear proteins, MEL cell terminal
     differentiation, 139—147
Topoisomerase I, 114—115, 178
Topoisomerase II, 114—115, 178—179
Topoisomerases, 52, 113—116, 178—179
Topologically independent domains, 60
Toyocamycin, 13, 32, 35
Transcription, 3, 7, 60—61, 78, 168
   DNA/peptide bond, 127